Newnes
Dictionary of Electronics

Newnes
Dictionary of Electronics
Fourth Edition

S. W. Amos, CEng, BSc, MIEE
R. S. Amos, BD, BSc, MISTC
Appendix by
G. W. A. Dummer, MBE, CEng, FIEE

Newnes

OXFORD   AMSTERDAM   BOSTON   LONDON   NEW YORK   PARIS
SAN DIEGO   SAN FRANCISCO   SINGAPORE   SYDNEY   TOKYO

Newnes
An imprint of Elsevier Science
Linacre House, Jordan Hill, Oxford OX2 8DP
225 Wildwood Avenue, Woburn, MA 01801-2041

First published 1981
Reprinted 1982, 1986
Second edition 1987
Third edition 1996
Reprinted 1997, 1998
Fourth edition 1999
Reprinted 2000
Paperback edition 2002

British Library Cataloguing in Publication Data
A catalogue record for this book is available from the British Library

Library of Congress Cataloguing in Publication Data
A catalogue record for this book is available from the Library of Congress

ISBN 0 7506 4331 5

Printed and bound in Great Britain by
MPG Books Ltd, Bodmin, Cornwall

For more information on all Newnes publications
please visit our website at www.newnespress.com

Preface to First Edition

The British Standards Institution and the International Electrotechnical Commission have issued a number of publications containing definitions of the terms used in electronics.

The definitions in these published standards are, however, written by experts and intended, chiefly, for use by experts: they are therefore authoritative and brief. Such definitions are not ideally suited to the needs of engineers and technicians who are working in electronics, students who are studying the subject or amateurs interested in it: this dictionary is intended primarily for their benefit.

In this dictionary definitions are supplemented by explanatory material whenever this would seem helpful. Some entries are therefore short essays including diagrams: typical examples are the entries on 'camera tube', 'digital computer' and 'thyristor'. There are also short definitions where a single sentence gives all the information likely to be required: the definition of 'B battery' is an example.

The standard circuit arrangements used in electronic equipment, such as the Hartley oscillator, are defined in words but are more readily understood from circuit diagrams. A large number of such diagrams are therefore included in this dictionary and the graphical symbols used in them conform with BS 3939.

In the form of presentation adopted definitions are arranged in alphabetical order of the initial letter of the first word. Thus 'characteristic impedance' appears under C and not under I. Cross references are provided to draw the reader's attention to related terms and definitions which should assist understanding and attention is drawn to the most helpful of these by printing them in bold italics. Technical terms used in a definition are normally defined elsewhere in the book. An appendix giving many of the abbreviations and acronyms used in the literature has been added and should serve to guide a reader to the definition of a term known only by its abbreviation.

This dictionary is associated with the 'Dictionary of Telecommunications' by S. J. Aries and the 'Dictionary of Audio, Radio and Video' by R. S. Roberts. These subjects are closely related with electronics and share many technical terms. For this reason the authors of the dictionaries collaborated during their preparation to prepare a list of technical terms to be defined, to decide in which dictionary each term should appear and what depth of treatment each should receive. North American terminology was fully represented in this list to ensure that the dictionaries have the widest-possible English-speaking readership. It was agreed that some terms for common concepts should appear in all the dictionaries to make them largely self-contained but that detailed explanations of fundamental terms should be confined to the *Dictionary of Electronics*.

Thanks are due to ITT Semiconductors and Motorola Semi-conductor Products for providing information on short-channel MOS transistors.

S.W. Amos
1981

Preface to Second Edition

In this second edition a number of new definitions have been included and some which appeared in the first edition have been revised or expanded. Most of these changes were adopted to keep the dictionary *au fait* with recent developments in semiconductor devices, digital techniques, computers and microprocessors.

I gratefully acknowledge the willing help given by my son Roger and my neighbour Pat Thwaites in the preparation of the new edition.

S.W. Amos
Broadway, 1986

Preface to Fourth Edition

In an effort to keep up-to-date with the continuous flow of developments in electronics, the publishers decided to produce this new edition of the dictionary to incorporate the new terminology introduced since the third edition was published in 1996.

Because this new material is confined chiefly to the field of computing and data-processing equipment and consists largely of acronyms, the Appendix which lists abbreviations and acronyms has been completely rewritten. My son, Roger, and I record our thanks to G. W. A. Dummer, MBE, CEng, FIEE, who supplied most of the text of the new Appendix. As for the previous edition of the dictionary, Roger prepared the camera-ready copy of the new edition for the publishers.

S. W. Amos
Broadway, 1998

A

A battery (US) A battery used to provide the low-tension supply for *electron-tube* heaters or filaments.

abend Short for *abnormal end*. In computing *fatal error* conditions that require immediate termination of the user's session.

aberration In *cathode ray tubes*, distortion of the image caused by failure of the electron beam to focus all points in the image accurately on the screen.

absolute permeability Of a material, the ratio of the magnetic flux density to the magnetising force which gives rise to it.

absolute permittivity Of an insulator, the ratio of the electric flux density to the electric field strength which gives rise to it.

absolute power level The magnitude of the power level of a signal at a point in a transmission system, usually expressed in *decibels*, relative to a power of 1 mW, known as zero power level. For example, a signal power level of 1 μW is known as −30 dB.

absolute value device A device of which the output signal has an amplitude equal to that of the input signal but always has the same polarity irrespective of the polarity of the input signal.

absolute voltage level The magnitude of the voltage level of a signal at a point in a transmission system, usually expressed in *decibels*, relative to a voltage of 0.775 V, known as zero voltage level. 0.775 V RMS is the voltage corresponding to a power of 1 mW in 600 Ω. A reference voltage level of 1 V is sometimes used in expressing the voltage level of the output of a microphone.

absorber circuit In a transmitter, a circuit, usually employing an *electron tube*, which absorbs power from the transmitter when a break occurs in an *oscillatory circuit* so preventing the formation of an arc across the break which could cause damage.

absorption frequency meter (also known as *wavemeter*) An instrument for measuring the frequency (or wavelength) of an RF signal, in which a calibrated resonant circuit is coupled to the RF source, maximum energy being absorbed from it when the circuit is resonant at the frequency of the signal.

absorption modulation A method of producing *amplitude modulation* in which the amplitude of the carrier is varied according to the power absorbed from it in a resistance the value of which is made to vary in accordance with the instantaneous amplitude of the modulating wave.

A-B test A subjective comparison of the quality of two audio or video signals by switching alternately between the two. Two items of equipment, e.g. hi-fi amplifiers, may be compared in this way. A three-head tape recorder may be tested by switching alternately between the source and playback output while recording a suitable high-quality source.

accelerated life test A form of life test so designed that the time taken to determine the probable life of a component or device is appreciably shorter than the probable life. For example, in an accelerated life test on a switch it may be operated far more frequently than would occur under normal working conditions.

accelerating voltage In an *electron tube* the voltage applied between the *accelerator* and the *cathode*.

accelerator or **accelerating electrode** An *electrode* in an *electron tube* used to increase the velocity of electrons emitted from the cathode. It is usually in the form of a cylinder or a plate with an aperture so as not to intercept electrons but to direct them towards the anode or target electrode.

accentuation In audio-frequency engineering the emphasis of a particular band of frequencies, for example emphasis of the lower end of the audio spectrum, alternatively known as *bass boost*.

acceptance test A test carried out on a device or equipment to demonstrate that it meets all the requirements of the buyer.

acceptor circuit A *resonant circuit* designed to present a low *impedance* at a particular frequency. When connected across a high-impedance circuit it acts as a *notch filter* which attenuates signals at the chosen frequency. The commonest form of acceptor circuit is a series *LC* combination. See *rejector circuit, series resonance*.

acceptor impurity In semiconductor technology, a trivalent element such as boron, atoms of which can replace the tetravalent atoms in the lattice of a silicon or germanium crystal so making *holes* available as *charge carriers*. See *p-type semiconductor*.

acceptor level In the *energy-level diagram* of a *p-type semiconductor* an intermediate level slightly above the *valence band* which is empty at absolute zero temperature and to which electrons can be thermally transferred at other temperatures.

access time Of a *store* the time taken to obtain desired information. More specifically the interval between the instant at which the data is called for and the instant at which it is available at a specified location.

accumulator (1) A battery composed of *secondary cells*. The most widely used forms of accumulator are the lead-acid battery employed in cars, and the nickel-iron battery. (2) In a computer or electronic calculator, a small-capacity *store* (or *memory*) in which the results of an arithmetical calculation are kept until required for a further stage of the calculation.

acoustical impedance The complex ratio of the alternating sound pressure applied to an acoustic system to the resulting alternating volume velocity imparted to it. By analogy with electrical impedance, acoustical impedance is made up of acoustical resistance and acoustical reactance, and the latter can arise from *inertance* (acoustical analogue of inductance) and *compliance* (acoustical analogue of capacitance). Volume velocity is the product of linear velocity and the cross-sectional area of the system.

acoustic coupler A device enabling a normal telephone handset to be used for *data transmission*. The handset is placed in a cradle forming part of the terminal equipment and is thus acoustically coupled to a microphone and loudspeaker within the terminal. The *digital signals* are converted into audio signals for transmission over the telephone circuit and at the receiving end are reconverted to digital form. The system is suitable only for low-speed transmission.

acoustic delay line A *delay line* which makes use of the time of propagation of sound waves in a solid or liquid medium.

acoustic feedback In general, *feedback* of sound from a loudspeaker to an early stage in the preceding amplifier. In the most familiar example, the sound is picked up by a microphone in a public address system but the effect can also arise when the sound is picked up by a microphonic electron tube or even a transformer. If the feedback exceeds a certain degree, sustained oscillation known as 'howlback' or 'howlround' occurs.

acoustic filter A *network* of acoustic elements designed to pass sound waves within certain frequency bands with little attenuation but greatly attenuating sound waves at other frequencies. Acoustic reactances can take the form of pipes, slots and boxes and these may be connected in series or in shunt so making possible a wide variety of types of filter as in electrical technology.

acoustics In general, the study of the production, propagation and effects of sound waves in solid, liquid and gaseous media. The term is also used to describe the properties of an environment which determine the distribution and absorption of sounds generated within it, i.e. a concert hall may be said to have good acoustics if the sound quality is good within it.

acoustic store A store utilising the properties of an *acoustic delay line*.

activation The treatment applied to *electron-tube* cathodes during manufacture to maximise electron emission. See *reactivation*.

active area Of a rectifier that area of the rectifying junction which effectively conducts current in the forward direction.

active circuit element Same as *active device*.

active current That component of the current in an alternating-current circuit which is in phase with the applied alternating voltage. The power dissipated in the circuit is the product of the RMS values of this component and of the applied voltage. See *reactive current*.

active device or **active element** A component in an electrical or electronic circuit which is capable of amplifying signals. The chief forms of active device are the *electron tube* and the *transistor*. Both require a source of power for their operation. See *amplification*.

active network See *network*.

active power In an alternating-current circuit the arithmetical mean of the product of the instantaneous voltage and instantaneous current over one period.

active transducer A *transducer* which requires a source of power for its operation other than that provided by the input signal. An example is an electrostatic microphone which requires a high-voltage source to charge the capacitor to enable the microphone to operate.

adaptive control Control which varies automatically so as to give optimum results from the controlling process. *Automatic gain control* is an example of adaptive control in which the gain of an amplifier or receiver varies automatically to give a constant level of output signal.

adder In general any device or circuit of which the output represents the sum of the inputs. In particular a combination of *logic elements* which accepts as inputs the signals to be added and the *carry* from any previous stage of addition and gives two outputs, one representing the sum of the inputs and the other the carry for a subsequent stage of addition. An adder can be composed of two *half adders*.

address In a *computer* or *data-processing equipment*, information, usually in the form of a *binary word*, which identifies a particular location in a *store*.

address register In a *digital computer* a *register* used to store *addresses*.

admittance (*Y*) A measure of the ease with which an alternating current flows through a circuit. More specifically, admittance is the complex ratio of the current in an alternating-current circuit to the EMF which gives rise to it. It is the reciprocal of *impedance* and therefore has real and imaginary components as shown by the expression

$$Y = G + jB$$

where *Y* is the admittance, *G* is the conductance and *B* the susceptance. The numerical value of the admittance is given by

$$Y = \sqrt{(G^2 + B^2)}$$

The admittance concept is useful in solving problems where components are connected in parallel because conductances are added, and susceptances added or subtracted according to their sign to give the net admittance. The unit of admittance is the mho. See *conductance*, *susceptance*.

aerial Same as *antenna*.

afterglow Same as *persistence*.

after-image In a *cathode ray tube* a visual display which lasts after the stimulus causing it has ceased. See *persistence*.

air cooling The cooling of components and active devices dissipating considerable power by transfer of heat to the ambient air by radiation and/or convection. See *conduction cooling, convection cooling, forced-air cooling*.

air gap Of an *inductor* or *transformer*, a small gap in the *magnetic circuit* deliberately introduced to increase the *reluctance* of the circuit and so prevent *magnetic saturation* of the core by the direct component of the current in the winding.

airtime The actual time spent by a subscriber using a *cellular radio system* such as a mobile telephone. The airtime is normally the basis of the charges levied by the service provider.

Algol See *computer language*.

algorithm A commonly used sequence of arithmetical or logical operations that is applied to data.

alias In computing, an assumed name for a file or an application or a user.

aliasing The chunky appearance of an oblique or curved edge as displayed on a *VDU* or in a *bitmap* (*Figure A.1*). See *anti-aliasing*.

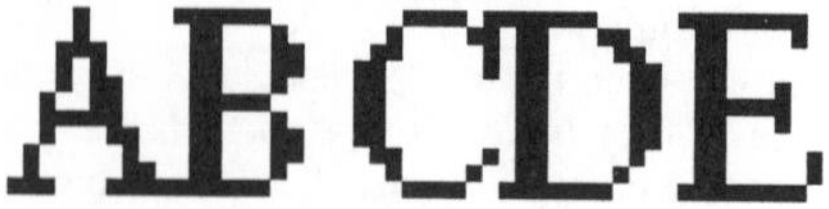

Figure A.1 Bitmapped text from a VDU display, magnified by 4, showing aliasing

aligned-grid tube A *tetrode* or *pentode* tube in which the wires forming the screen grid are arranged to be in the electron shadow of the wires forming the control grid. Such construction ensures that the screen grid intercepts very few electrons (in spite of its positive potential) and screen-grid current is thus minimised.

alignment In general the adjustment of all the *preset controls* of an equipment to give the required performance. In particular, the adjustment of the *tuning circuits* of a radio receiver to ensure that they resonate at the required frequencies as the tuning is varied over the received *waveband(s)*. In a *superheterodyne receiver* alignment is necessary to ensure the maintenance of the correct difference in frequency between the oscillator and signal-frequency tuned circuits as the tuning is varied over the waveband(s). Ideally the difference frequency should equal the *intermediate frequency* at all tuning settings, but in practice this ideal can be achieved at only two or three tuning points in each waveband.

allowed band In an *energy-level diagram* the energy range which electrons may possess. The *conduction band* and the *valence band* are examples of allowed bands. See *forbidden band*.

alloy diode A pn *diode* manufactured by fusing a pellet of a trivalent or pentavalent element to the face of a wafer of *semiconductor material*. The wafer is heated until the element dissolves some of the semiconductor. After cooling the dissolved semiconductor crystallises out but sufficient of the element remains to give the required pn structure.

alloy transistor A *transistor* manufactured by fusing pellets of trivalent or pentavalent elements to the opposite faces of a thin wafer of semiconductor material. The combination is heated until the elements dissolve some of the *semiconductor*. After cooling the dissoved semiconductor crystallises out but sufficient of the elements remains to give the pnp or npn structure required, illustrated in *Figure A.2*.

all-pass network A four-terminal *network* of which the loss is independent of frequency. A symmetrical lattice network such as that shown in *Figure A.3* is an example of an all-pass network. The *phase shift* of the network varies considerably with frequency and such networks are often used as *phase equalisers* because they can reduce *phase distortion* without affecting frequency response.

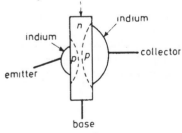

Figure A.2 Structure of a pnp alloy-junction bipolar transistor

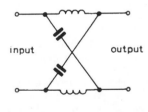

Figure A.3 A symmetrical lattice network is an example of an all-pass network

alpha cut-off frequency Of a *bipolar transistor*, the frequency at which the *current amplification factor* of the *common-base circuit* has fallen to 0.71 ($1/\sqrt{2}$) of its low-frequency value, i.e. is 3 dB down.

alphanumeric code A code using combinations of letters, numerals and other symbols to represent data.

alphanumeric display The display of information by means of letters and numbers. Such displays, usually on the screen of a *cathode ray tube*, are employed in computers and in information systems such as *teletext*.

alpha particle A positively-charged particle emitted from an atomic nucleus and consisting of two protons and two neutrons. An alpha particle is identical with a helium nucleus.

alternate display In an *oscilloscope* a means of displaying two or more signals by selecting the signals in sequence.

alternating current (AC or **a.c.**) or **alternating voltage** A current (or voltage) which periodically reverses its direction, varying sinusoidally with time about a mean value of zero. Domestic electricity supplies are usually in the form of alternating current, the frequency being 50 or 60 Hz. Alternating supplies have the advantage that they can be transformed to high voltage for long-distance distribution: high voltage implies low current and therefore low losses.

alternative denial gate Same as *NAND gate*.

aluminised screen Of a *cathode ray tube*, a form of construction in which the light-emitting phosphor of the screen is backed by a very thin coating of aluminium. This has the following advantages: (*a*) it protects the screen from bombardment by heavy negative ions so minimising ion burn, (*b*) it stabilises the potential of the screen so avoiding accumulations of charge which could produce shading effects on displays, (*c*) it increases the brightness of the display by reflecting forwards any light transmitted backwards from the phosphor.

ambient noise The noise present in the absence of a signal.

American Standard Code for Information Interchange (ASCII) A standard widely used in computers and data processing equipment to represent alphanumeric characters, punctuation marks and a selection of machine operations (such as line feed) by numbers. Originally ASCII was a 7-bit code, allowing a total of 128 characters and control codes, an eighth bit being used sometimes for *parity checking*. Many computers now use the eighth bit as part of the code; the extended character set made possible by the additional 128 characters often includes, e.g. the accented letters for most European languages and several currency symbols.

ammeter An instrument for indicating or recording the magnitude of an electric current in ampères (A). For large currents the measurement may be more conveniently expressed in kiloampères (kA) and for small currents in milliampères (mA) or microampères (μA). Current meters must be of low resistance so that, when inserted into a circuit, they have negligible effect on the current being measured.

Ampere's law Same as *Biot-Savart law*.

ampere-turn The practical unit of *magnetomotive force*. For a magnetising coil it is equal to the product of the number of turns of wire on the coil and the current (in ampères) flowing in it.

amplification The process by which an electronic device or equipment increases the amplitude of a signal. In linear amplification, the output signal is a magnified but undistorted copy of the input signal but there are circumstances, e.g. in *class-C operation*, where the output signal, although related to the input, does not have a similar waveform. Some amplifiers are designed to magnify the signal current, others the signal voltage and power amplifiers are required to give substantial current and voltage output. A wide variety of active devices and circuit arrangements are used for amplification in electronic equipment and information on these is given under appropriate headings in this dictionary.

amplification factor (μ) Of an *electron tube* the ratio of the change in anode voltage to the change in *control-grid* voltage required to maintain the anode current constant. It is usually stated for particular values of electrode voltages.

amplifier A device or assembly of devices through which the current flow is controlled by an input signal so as to give at the output a magnified signal bearing some relationship to the input.

In some applications there is in fact little relationship between input and output, the amplifier operating as a switch controlled by the input signal. In other applications strict proportionality is required between input and output to give linear amplification.

See *current amplifier, power amplifier, voltage amplifier.*

ampliphase system A modification of the *Chireix system* in which the efficiency is improved by *amplitude modulation* of the drive signals.

amplitron A microwave amplifying *electron tube* used for high-power pulsed operation, e.g. in radar transmitters. The basic form of construction of the tube is shown in *Figure A.4*. The input and output ports are connected to a slow-wave structure surrounding a cylindrical cathode. There is a strong magnetic field parallel to the axis of the cathode.

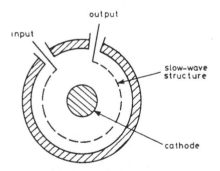

Figure A.4 Simplified sectional view through an amplitron

amplitude In general the magnitude of a signal. In particular, of a sinusoidally-varying quantity, the peak value, i.e. the maximum departure from the average value. For example, if a voltage is given by $V = V_0 \sin \omega t$, the amplitude is V_0. The value of a sinusoidally-varying quantity at any instant is termed the instantaneous amplitude.

7

amplitude discriminator In general, a circuit the output of which is a function of the amplitude of two input signals. In particular, the term is used to define a device which gives an output when the two input signals have equal amplitude.

amplitude distortion In a circuit, component or amplifier, distortion arising from variation of gain or attenuation with the amplitude of the input signal. It arises from non-linearity of the input-output characteristic and is produced by hard limiters when the input signal amplitude exceeds the extent of the linear characteristic. See *amplitude limiter, characteristic curve.*

amplitude gate Device or circuit capable of *slicing.*

amplitude limiter A four-terminal device which maintains a constant amplitude of output signal for amplitudes of input signal exceeding a certain value. There are two basic types of limiter. In one, which may be termed a hard limiter, a constant-amplitude output is ensured by chopping the peaks of large-amplitude signals, the characteristic having the form shown in *Figure A.5* (a). The resulting waveform mutilation is unimportant, for example, in the hard driven limiter often included before the detector in an FM receiver. This ensures that the signal delivered to the detector is free of amplitude variations (which could cause noise) and contains only frequency variations. A second type of limiter is required in broadcasting and recording to ensure that transmitters or the recording medium are not overloaded by signals of too great an amplitude. Limiters for such applications must maintain linearity of the input-output characteristic and the limiting action is achieved by automatic reduction of the gain of the limiter for large-amplitude input signals so as to maintain the output signal at the limiting level. The characteristics for this type of limiter are shown at *Figure A.5* (b): no distortion occurs except at the instants when the gain is changed. See *characteristic curve.*

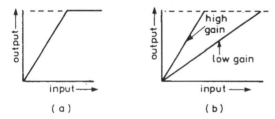

Figure A.5 (a) Non-linear characteristic of a hard limiter and (b) characteristics of a non-distorting limiter for two values of gain

amplitude modulation (AM) A method of *modulation* in which the amplitude of the *carrier wave* is made to vary in accordance with the instantaneous value of the modulating signal. In radio transmission, considerable use is made of amplitude modulation of a carrier wave and *Figure A.6* illustrates the effect of modulating a radio-frequency wave by a low-frequency sinusoidal signal. Such a modulation system is used, for example, in long-, medium-, and short-wave broadcasting and for the vision

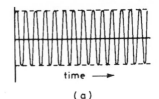
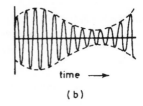

| (a) | (b) |

Figure A.6 (a) An unmodulated carrier wave and (b) an amplitude-modulated carrier wave

signal in television broadcasting. In these practical applications the difference between the carrier-wave frequency and the modulating-wave frequency is much greater than suggested in this diagram. It is also possible to amplitude modulate a pulse carrier. See *double-sideband transmission, pulse amplitude modulation, single-sideband transmission, suppressed-carrier transmission, vestigial-sideband transmission.*

analogue comparator A device used to check the output of an *analogue-to-digital converter.*

analogue computer A computer that operates on *analogue signals,* i.e. signals of which the significant property can have any value. These values are represented by corresponding values of voltage or current within the computer circuits which then carry out the required mathematical operations on the signals.

Analogue computers employ *operational amplifiers* with *negative feedback* for such operations as multiplication, addition, integration and differentiation. Potential dividers are used for division. Function generators are also needed, e.g. to give trigonometrical ratios. The final result from the computer is displayed on a cathode ray tube or on a chart.

Because of the large degree of negative feedback used with operational amplifiers, their outputs are reasonably accurate but there is inevitably some error (e.g. due to drift) and hence there is a limit to the accuracy of the results from the computer. This is normally of little significance because the inputs fed into the computer are normally from measurements which are themselves subject to error. Where precise accuracy is essential a *digital computer* is used.

analogue device A device which operates on *analogue signals*, e.g. an *analogue computer* or an audio amplifier.

analogue signal A signal, the significant property of which can have any value. The significant property may be the *amplitude, phase* or *frequency* of an electrical signal, the angular position of a shaft or the pressure of a fluid. An audio signal is an example of an analogue signal. Signals are often termed analogue where it is necessary to contrast them with digital signals. See *analogue computer, digital signal.*

analogue switch A switch with characteristics which make it suitable for use with *analogue signals.* Such signals can have a wide range of voltage and current, and switches for use with them must have values of resistance in the 'on' and 'off' states which are independent of the applied voltage and current. For solid-state analogue switches *field-effect transistors* are often used.

9

analogue-to-digital conversion (ADC) The process of converting an *analogue signal* to digital form. See *digital signal*.

analogue-to-frequency converter A device which accepts an *analogue signal* as input and gives an output with a frequency proportional to the input signal.

ancillary store Same as *backing store*.

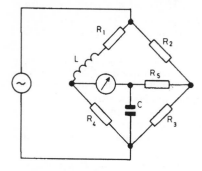

Figure A.7 Anderson bridge

Anderson bridge A modified form of *Maxwell bridge* used for measuring inductance in terms of capacitance and resistance. As shown in *Figure A.7* the bridge has an additional resistor R_5. The balance conditions (which are independent of frequency) are:

$$L = R_4C \left[R_5 \left(1 + \frac{R_2}{R_3} \right) + R_2 \right]$$

$$R_2R_4 = R_1R_3$$

and the Anderson bridge has the advantage that the two balance conditions are independent.

AND gate A *logic gate* which gives a logic-1 output when, and only when, all the input signals are at logic 1. The graphical symbol for an AND gate is shown in *Figure A.8*. See *logic level*.

Figure A.8 Graphical symbol for an AND gate

AND-NOT gate Same as *NAND gate*.

anechoic chamber A room used for testing loudspeakers and microphones in which all sound reflections have been eliminated by lining all the surfaces, including floor and ceiling, with sound-absorbing material.

angle modulation General term for *modulation* in which the *phase angle* of one signal is varied in accordance with the instantaneous value of another. *Frequency modulation* and *phase modulation* are two examples of angle modulation.

angle of current flow or **angle of flow** For an amplifier with a sinusoidal

10

input signal the fraction of each cycle, expressed as an angle, during which current flows in the amplifier. For a *class-A amplifier* the angle of flow is 360° (2π radians), for a *class-B amplifier* it is 180° (π radians) and for a *class-C amplifier* it is less than 180°.

angle of lag   See *phase angle*.

angle of lead   See *phase angle*.

angular frequency (ω)   The frequency of an alternating quantity expressed in radians per second. Because there are 2π radians in each cycle the angular frequency is equal to 2π*f*, where *f* is the frequency in cycles per second (Hertz). Thus

$$\omega = 2\pi f$$

anion   Negatively-charged *ion* formed in a gas by *ionisation* or in an electrolyte by *dissociation* and which moves towards the positively-charged electrode (*anode*) under the influence of the potential gradient. See *cation*.

anisotropic material   A material which has different physical properties in different directions. Certain crystals are anisotropic in that their optical properties depend on the direction of incident light relative to the crystal axis. Certain magnetic materials, too, are anisotropic. See *isotropic*.

annunciator   Any device, such as a lamp or a buzzer, used by equipment to attract the attention of its user.

anode   Of an *electron tube* the positively-charged electrode at which the principal electron stream from the *cathode* leaves the tube as an external current. The anode is therefore the output electrode of the tube.

anode AC resistance   See *electrode AC resistance*.

anode-bend detector   An AM detector which relies for its action on the curvature of the I_a–V_g *characteristic* of an *electron tube*.

This curvature causes the tube to respond unequally to increase and decrease in the amplitude of the modulated input signal and thus there is a modulation-frequency component in the anode current. This is illustrated in *Figure A.9*. A *pentode* is generally used as an anode-bend detector and it is

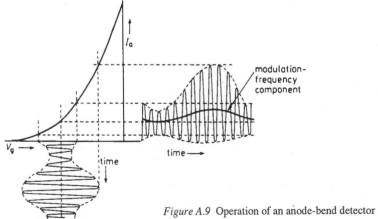

Figure A.9 Operation of an anode-bend detector

11

biased near anode-current cut off where characteristic curvature is most marked and detection, therefore, most efficient.

anode characteristic The graphical relationship between the anode voltage and anode current of an *electron tube* usually expressed for given values of *control-grid* and *screen-grid voltage*. The shapes of the characteristics for a triode and a pentode are given in *Figure A.10*.

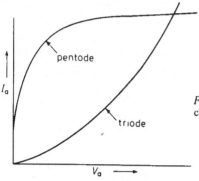

Figure A.10 Typical anode characteristics for a triode and a pentode

anode (collector or drain) dissipation That part of the power supplied to the anode (collector or drain) circuit of an electron tube or transistor which is converted into heat at that electrode. The power thus dissipated is the difference between the DC power input to the electrode and the AC power delivered to the output load.

anode (collector or drain) efficiency The ratio, usually expressed as a percentage, of the AC power delivered to the output load of an active device to the DC power input to the anode (collector or drain) circuit. For a theoretically perfect active device with straight, parallel and equidistant characteristics the maximum efficiency for a sinusoidal signal is 50%. *Pentodes* and *transistors* can give efficiencies almost equal to this but the efficiency of triodes is much less because of the limited anode voltage swing available before the onset of grid current.

anode (collector or drain) load The impedance of the external circuit between the anode (collector or drain) and the cathode (emitter or source) of an active device. The current variations in the load, the voltage variations across it or the power dissipated in it constitute the output of the active device.

anode modulation Same as *Heising modulation*.

anode stopper See *parasitic stopper*.

antenna The system of conductors used at a radio installation to radiate or receive *electromagnetic waves*. To be effective as a radiator the antenna must have dimensions comparable with the *wavelength* and thus medium- and long-wave transmitters must have very large antennas, hundreds of feet in height or length. VHF and UHF transmitting antennas can be very much smaller but must usually be mounted at the top of a hill, a tower or a high building to give adequate service area. Receiving antennas can be small because the low pickup can be compensated by high gain in the receiver circuits. There are many different types of antennas and these are defined in the *Dictionary of Audio, Radio and Video* (Butterworths, 1980).

Figure A.11 Bitmapped text that has been anti-aliased, magnified by 4

anti-aliasing The use of *pixels* in intermediate colours to smooth the jagged appearance (*aliasing*) of oblique or curved edges in *bitmaps* and *VDU* displays (see *Figure A.11*). Anti-aliased text is more readable than text displayed without anti-aliasing. The term is also applied to an analogous process used in converting digitised sound to analogue.

anti-cathode Target of *electron tube*, particularly an *X-ray* tube, on which the electron beam is focused and at which the X-rays are generated.

anti-hunting circuit A circuit incorporated in a feedback system to prevent self-oscillation. The circuit often consists of an *RC* combination designed to absorb energy at frequencies at which the system is likely to ring or oscillate.

antinode Of a *standing wave*, a point or plane at which a particular variable has a maximum value. Thus antinodes on a transmission line are points at which current or voltage is at a maximum. For a bowed string, antinodes are points at which the amplitude of vibration is at maximum.

antiresonance (US) Same as *parallel resonance*.

aperiodic circuit A circuit without natural *resonance*. The term is commonly applied to an *LC circuit* which is so heavily damped that the *impedance* is substantially constant over a wide band of frequencies centred on the resonance value. It is also applied to an *LC* circuit for which the resonance frequency is so remote from the frequency band in use that there is little change in the impedance over the working frequency range.

aperture (1) Of an optical lens the opening in the diaphragm (entrance pupil) which controls the amount of light passing through the lens. The size of the aperture is usually quoted as a fraction of the focal length of the lens, e.g. an aperture of *f/8* signifies that the diameter of the aperture is equal to one eighth of the focal length. The aperture control is used, for example, to adjust the amount of light entering a television camera to ensure that the tonal range of the scene to be televised is correctly located on the characteristic(s) of the camera tube(s). (2) In television the term is used to mean the size of the spot where the scanning beam meets the target in a camera tube, flying-spot telecine or picture tube. See *aperture distortion*. (3) The term has a third meaning when applied to antennas. See the *Dictionary of Audio, Radio and Video* (Butterworths, 1980).

aperture corrector An *equaliser* designed specifically to offset *aperture distortion*.

aperture distortion In television, distortion arising from the finite size of the *scanning spot*. Such distortion can arise at the transmitting end where the spot is effectively the cross-sectional area of the electron beam where it meets the target in camera tubes or in flying-spot telecines, or at the receiving end where the spot is the cross-sectional area of the scanning beam at the picture tube screen. Because of the finite spot size an instantaneous change in tonal value occurring along a scanning line is reproduced as a pulse with a finite rise or fall time. Thus the spot size determines the degree of detail which can be transmitted and reproduced. It is possible to compensate to some extent for the loss of definition arising from aperture distortion. See *aperture, aperture corrector, rise time*.

Applegate diagram A diagram used to illustrate the principle of electron bunching in *velocity-modulated* tubes. *Figure A.12* is a typical diagram. It is a graph in which the distance travelled by electrons from the buncher gap is plotted vertically and time is plotted horizontally. OA represents the path of a typical electron and the slope of OA measures its velocity. In a velocity-modulated tube the electrons leaving the buncher gap are accelerated by a positive voltage on which is superposed the RF signal it is required to amplify. This imparts a sinusoidal variation to the electron velocities. Suppose OA represents the velocity when the RF voltage is passing through zero and that the RF voltage executes its negative half-cycle immediately afterwards as suggested in *Figure A.12*. Then electrons leaving the gap after those represented by OA have lower velocities as suggested by BC and DE until point P is reached after which the velocities increase. The diagram shows that at a particular distance *d* from the gap the lines meet at a point showing that the electrons gather in bunches at such a distance. By repeating the construction for further cycles of RF the diagram shows that bunches occur at this same spacing from the gap at intervals of 1/*f*, where *f* is the frequency of the RF signal. See *bunching, klystron, travelling-wave tube, velocity modulation*.

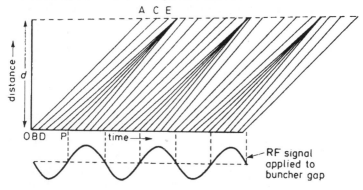

Figure A.12 The bunching of an electron beam velocity-modulated by an RF signal

application A task performed on a computer, such as desktop publishing or databasing, or a suite of software that forms an integrated package to perform that task. For example, several programs may be included together with various kinds of data file used by the programs. Often the various files that make up an application are stored together in one *directory*.

application(s) software Software that provides an *application*, in contrast with *operating systems*, *utilities* and *data*.

Application Specific Integrated Circuit (ASIC) Sometimes known as a gate array, this is a generic *integrated circuit* that has been tailored for a specific application. A single ASIC might replace several conventional *logic* integrated circuits, allowing a reduction in both circuit board size and power consumption. The term is also sometimes applied to a *microprocessor* dedicated to a single purpose.

arc Conduction of an electric current through an ionised gas usually at high current density and accompanied by emission of light. See *ionisation*, *spark*.

arcback In a *gas-filled tube*, conduction between *anode* and *cathode* resulting in current flow in the reverse direction to normal.

arc conduction See *arc*.

arc discharge In a *gas-filled tube*, conduction between *anode* and *cathode* as a result of electronic emission from the cathode. See *glow discharge*.

arc drop The voltage between the *anode* and *cathode* of a *gas-filled tube* when it is conducting normally.

architecture The overall design of a computer or data-processing system with special reference to the performance and interaction of its components, its connection to *peripherals* and other systems and facilities for future extension.

archive Any *file* in a non-volatile storage medium. More specifically, one or more files that have been subjected to a *compression* process so that they occupy less space in the storage medium.

arc-over An *arc* between two conductors or between a conductor and earth.

arc through In a *gas-filled tube*, conduction between *anode* and *cathode* in the normal direction but at a time when the anode should be non-conductive.

arc voltage Same as *arc drop*.

argument In computing additional data which is entered immediately following a *command* to specify the way in which the command is to be executed. For example, in the MS-DOS command

```
COPY A:*.*
```

the character string 'A:*.*' forms the argument, specifying that all files on drive A are to be copied to the currently selected drive and directory.

arithmetic logic unit (ALU) That part of a *central processing unit* that performs arithmetical and logical operations.

array (1) A series of similar devices arranged in a meaningful pattern. In computers *gates* and *memory cells* are usually arranged in a rectangular pattern of columns and rows, known as an array or *matrix*. Arrays of *antennas* are used at short-wave transmitters. (2) In *computing* a cluster of similar *variables* sharing the same name but distinguished by numeric suffixes enclosed in parentheses.

artificial antenna Same as *dummy antenna*.

artificial earth A system of conductors mounted a few feet above the ground under an **antenna** and used in place of an earth connection. Such a system is likely to be used where the earth conductivity is poor.

artificial intelligence Computers and computer-controlled machines so organised that they behave, to some extent, like human beings, but without the propensity for mistakes. Perhaps the closest approach has been in so-called *expert systems.*

artificial language A language based on a set of rules established prior to the use of the language.

artificial line A network of inductors and capacitors designed to simulate the electrical properties of a transmission line over a desired frequency range. See *capacitance, inductance.*

artificial load A substitute for the normal load of an equipment which has the correct *impedance* and can safely dissipate the power generated in it. The artificial load for a transmitter simulates the impedance of the transmission line and antenna but does not radiate.

aspect ratio The ratio of the width to the height of a television picture. This is now standardised at 4:3 to agree with cinema-film practice, but see *high definition television.*

assembler, assembly language See computer language

astable circuit A circuit with two possible states, both unstable, so that it alternates between the two without need of external triggering signals. Astable circuits can, however, readily be synchronised at the frequency of any regular signal applied to them. An example of an astable circuit is a multivibrator in which both inter-transistor couplings are via capacitors as shown in *Figure A.13*. The two possible states of this circuit are: TR1 on and TR2 off; TR1 off and TR2 on.

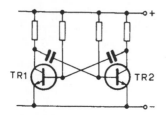

Figure A.13 An astable multivibrator circuit

astigmatism A defect in an optical or electron lens which causes focusing in different axial planes to occur at different points along the lens axis. As a result of astigmatism, a point object gives rise to an image in the form of a vertical line at one point on the axis and in the form of a horizontal line at another point as shown in *Figure A.14*. Normally the best compromise is

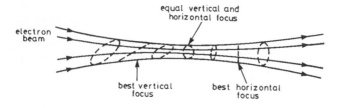

Figure A.14 Astigmatism in an electron beam

16

between these two points where the image has the form of a circle, known as the circle of least confusion, which represents equal vertical and horizontal resolution.

asymmetrical circuit A circuit the two sides of which have different electrical properties with respect to a reference potential, usually earth. The term is often applied to circuits in which one side is earthed. See *symmetrical operation*.

asymmetrical deflection Electrostatic deflection in a *cathode ray tube* where the deflecting potentials applied to the plates are not symmetrical about a reference potential (usually the final-anode potential). In particular the term is applied to electrostatic deflection where one plate of a pair is effectively earthed and the deflecting potential is applied to the other.

asynchronous logic In a *computer* or *data-processing equipment*, *logic* in which operation is started by a signal generated at the completion of the previous operation. The term is used to distinguish this form of operation from that in which operations are controlled by an internal *clock*. See *synchronous logic*.

asynchronous transmission A method of *data transmission* in which characters are sent at irregular intervals, each being preceded by a start bit and followed by a stop bit. It is the most widely used method of data transmission between microcomputers.

atomic number Of an element, the total number of unit positive charges carried by the nucleus of an atom. For a neutral atom this is the number of protons in the nucleus and also the number of orbiting electrons. See *atomic structure*.

atomic structure The atom is regarded as a dense nucleus containing *protons* and *neutrons* and with a net positive charge, around which revolve a number of electrons neutralising the positive charge of the nucleus. The total number of protons is equal to the atomic number of the element. The innermost electron orbit can contain at most two electrons, the second 8 and the third 18: the number needed for completion is given by $2n^2$, where n is the number of the orbit. The number of electrons in the outermost orbit determines the properties of the element. For example, it is difficult to remove an electron from or to insert an electron into an outermost orbit which has a complete complement of electrons. Thus atoms with completed outermost orbits are chemically inert: such atoms are those of the gases helium, argon, krypton etc. If the outermost shell has only one or two electrons these can be removed with little effort, making elements with such atoms good electrical conductors. Typical of such elements are copper and silver. If the outermost orbit has four electrons, *covalent bonds* can be formed between neighbouring atoms in a crystal lattice and the element has *semiconductor properties*. The obvious examples are *germanium* and *silicon*.

attack time The time that elapses between the instant that a signal at the input of a component or circuit exceeds the *threshold* needed to activate the component or circuit and the instant that the component or circuit begins to respond to the input.

attenuation coefficient (or **constant**) See *propagation coefficient*.

attenuation/frequency distortion or **attenuation distortion** In a system, equipment or component, variation of the gain or attenuation with the frequency of the input signal. Attenuation distortion is illustrated by the *frequency response curve* and the most commonly-encountered forms are shown in *Figure A.15*. An important feature of attenuation distortion is that it can be corrected: for example, the effects of an equipment suffering from 'top cut' can be offset by use of an equaliser with a complementary 'top lift' response.

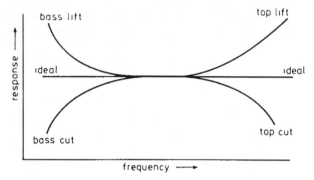

Figure A.15 Common forms of attenuation distortion

attenuator A *network* used to reduce the *amplitude* of a signal without distorting it. The reduction factor (attenuation) may be fixed or variable. A fixed attenuator is often called a pad. Attenuators may be inserted in lines or between equipments and are then designed not to introduce any impedance change, i.e. they are designed to have a particular value of *iterative impedance*. Alternatively attenuators can be designed on an *image-impedance* basis to match unequal impedances. Networks often used in attenuators are *T-networks* and π*-networks* and their balanced equivalents *H-networks* and *O-networks*. *Figure A.16* gives an example of an O network with 10-dB attenuation and with an iterative impedance of 300 Ω.

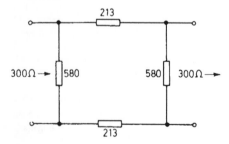

Figure A.16 An O-network 10-dB attenuator with an iterative impedance of 300 Ω

audio frequency (AF) Any frequency of sound wave which is normally audible to the human ear. The lower limit is usually quoted as 15 Hz but it is difficult at such low frequencies to differentiate between the sensations of hearing and feeling. The upper limit is often given as 20 kHz but this limit is dependent on age and falls below 10 kHz for very old people. In electronics, audio signals which are intended for reproduction by loudspeakers are regarded as high quality if the frequency range from 30 Hz to 15 kHz is covered with minimal distortion.

auto-coincidence gate Same as *exclusive-OR gate.*

autodyne or auto-heterodyne A system of *heterodyne* reception in which the same *active* and *passive* components are used for oscillation and mixing (as in a self-oscillating mixer). See *mixer, self-oscillating mixer.*

automatic brightness control In a television receiver, a circuit which automatically adjusts the brightness of the display in accordance with the level of ambient light near the receiver. A *photocell* may be used to measure the ambient light, its output, after amplification, being used to control the grid bias of the picture tube.

automatic cathode bias Use of a *resistor* in the *cathode* circuit of an *electron tube* to provide *grid* bias. A typical circuit is given in *Figure A.17.* The cathode current in R_k sets up a voltage across this resistor which biases the cathode positively with respect to the negative HT supply. The control grid is connected to HT negative by the grid resistor R_g. Thus the grid is biased negatively with respect to the cathode as required.

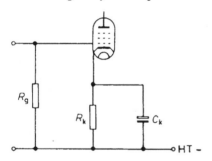

Figure A.17 Automatic cathode bias

The circuit can be used only when the mean cathode current is constant as in *class-A operation* and the value of cathode resistance required is, from Ohm's law, given by V_g/I_k where V_g is the bias voltage and I_k the mean cathode current. Thus if the grid bias is -8 V and the mean cathode current 40 mA, the cathode resistance required is 200 Ω.

For a *triode* tube the mean cathode current is equal to the mean anode current but for more complex tubes such as *pentodes*, the cathode current includes also the screen-grid current.

In addition to providing grid bias a resistor in the cathode circuit also gives *negative feedback* which reduces the gain of the tube. If this reduction is undesirable the feedback can be minimised by by-passing the

19

resistor by a *capacitor* as shown in *Figure A.17*. The capacitance should be such that the *reactance* at the lowest signal frequency is small compared with R_k.

automatic check In *computers* and *data-processing equipment* the use of *hardware* or *software* to verify the accuracy of processes carried out by the equipment.

automatic chrominance control (ACC) In a colour television receiver, an (AGC) circuit incorporated in the *chrominance amplifier* to ensure that the relative amplitudes of the *chrominance* and *luminance* signals are maintained. This is necessary to ensure correct colour rendering in spite of variations in signal level at the receiver input.

automatic contrast control See *automatic gain control*.

automatic frequency control (AFC) A circuit in a receiver which ensures that errors in the frequency of the oscillator or the tuning of a circuit are kept within certain limits. Before 1939 some amplitude modulation receivers had such a control (known as automatic tuning) and a *discriminator* was incorporated in the intermediate frequency amplifier to derive an error voltage which was applied to a reactance tube connected across the oscillator tuned circuit. In frequency modulated receivers the detector can be designed to provide the required error signal in addition to the audio frequency output. AFC is desirable in any receiver with preset (e.g. push-button) tuning to compensate for the inevitable slight drift in oscillator frequency.

automatic gain control (AGC) The maintenance of a substantially-constant input signal amplitude at the detector of a radio or television receiver (in spite of variations in amplitude of the input signal to the receiver) by automatic adjustment of the gain of the intermediate frequency and/or radio frequency stages. This adjustment is carried out by a control signal derived from the detector or from a post-detector stage. The purpose of AGC is to minimise the effects of signal fading on audio frequency output or picture contrast and to ensure that all received signals, no matter what their amplitude, give substantially the same output. When applied to radio receivers, AGC was originally known as automatic volume control and when applied to television receivers is sometimes termed automatic contrast control or automatic picture control. See *forward automatic gain control, reverse automatic gain control, variable-mu tube*.

automatic grid bias Use of a resistor in the *grid* (or *cathode*) circuit of an *electron tube* to produce grid-bias voltage as a result of grid (or cathode) current flow. Grid-current bias is commonly used in electron-tube oscillators and cathode-current bias in electron-tube amplifiers. See *cathode bias*.

automatic picture control Same as *automatic gain control*.

automatic tuning Same as *automatic frequency control*.

automatic volume control (AVC) Same as *automatic gain control*.

automation (1) The theory or technique of making a process automatic. (2) The control of processes by automatic means.

auto-transformer See *transformer*.

avalanche breakdown In a reverse-biased *pn junction*, a rapid increase in current which occurs at a particular reverse voltage as a result of cumulative multiplication of charge carriers. The effect is caused by the strong electric field across the junction which gives the few charge carriers so much energy that they liberate other *hole-electron* pairs by impact ionisation. The new carriers so created liberate further hole-electron pairs and thus the process becomes regenerative, leading to a very rapid increase in reverse current. See *charge carriers, ionisation.*

avalanche photo-diode (APD) A *photo-diode* in which the electrical output is considerably increased by *avalanche breakdown*. Such diodes are operated with a reverse bias close to the breakdown voltage so that liberated photo-electrons have sufficient energy to create *hole-electron* pairs by impact ionisation.

average detector A detector of which the output approximates to the mean value of the envelope of the input signal. See *detection.*

azimuth The alignment of a directional antenna or of a recording or playback head in a tape recorder.

B

background noise The sound, typically a faint hiss, heard from an audio system or radio receiver in the absence of a signal.

background processing or task In a *multitasking* computer system, the execution of a low-priority task in the intervals when the processor can be spared from current high-priority tasks. Tasks such as printing and disk *formatting* are often executed as background tasks.

backing store Same as *bulk storage*.

backplane A flat panel usually incorporating an earthed plate which is located at the rear of a multiboard equipment and supports the wiring and the receptacles which locate with the *edge connectors* on the *printed-circuit* or *printed-wiring boards*.

back porch In a television signal the period of *blanking level* immediately following the *line sync signal*. In colour television this period is largely occupied by the *colour burst*. See *Figure L.8*.

back-up Of equipment, to duplicate vital hardware so that, in the event of a failure, the normal service can be resumed with minimal delay. Of data, to make a copy of one or more files or a whole disk so that, if the original should be destroyed or corrupted, valuable data or applications remain available on the back-up copy. Hard disk drives are commonly backed up using *tape streamers* and sometimes using *optical disks*.

backward diode A *Zener diode* in which the doping is arranged during manufacture to give a *breakdown voltage* of zero. By contrast with normal diodes, such a diode has a reverse resistance lower than the forward resistance as shown in *Figure B.1*. Backward diodes of low capacitance are used as microwave detectors. The graphical symbol for a backward diode is given in *Figure B.2*.

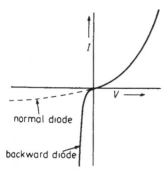

Figure B.1 Characteristics of a backward diode (solid) and a normal diode (dashed)

Figure B.2 Graphical symbol for a backward diode

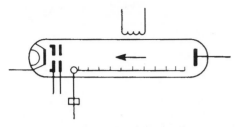

Figure B.3 Graphical symbol for a backward-wave oscillator: this example is an O-carcinotron

backward-wave oscillator or **carcinotron** A *travelling-wave tube* in which the wave propagation is in the opposite direction to the movement of the electron stream. The travelling-wave tube has a *slow-wave structure* with a matching termination at the collector end to absorb energy at the output. A backward wave set up in the tube gains energy from the electron beam and the tube can sustain oscillation when the electron-beam velocity equals the wave velocity along the tube. The frequency of oscillation can be adjusted by varying the electron-gun anode voltage and this gives a wider range of control than is possible with the *klystron* or a conventional travelling-wave tube. Power outputs of 900 mW are obtainable over a frequency range of 2500 to 5000 MHz. The graphical symbol for a backward-wave oscillator is given in *Figure B.3*: this represents in fact an O-type carcinotron.

baffle (1) In a gas-filled *electron tube*, a structure with no external connection and placed in the arc path. A baffle may help to control the distribution of current in the path of the arc or may help to de-ionise the gas after the conduction period. (2) In *acoustics*, a rigid structure such as a sheet of sound-insulating material used to improve the distribution of sound waves. In particular a structure in the centre of which a loudspeaker is mounted. The baffle increases the acoustic path length between the front and rear of the loudspeaker. By bringing the waves radiated from the rear of the loudspeaker into phase with those radiated from the front, the sound at low frequencies can be reinforced, effectively extending the low-frequency response of the loudspeaker.

balance In general the state in which the correct relationship between two quantities has been achieved. For example the signal levels in the two channels of a *stereophonic system* are said to be balanced when the correct spacial distribution of sound is achieved.

balanced modulator A circuit which accepts a carrier signal and a modulating signal as inputs and gives a *suppressed-carrier amplitude-modulated output*. *Figure B.4* gives the circuit diagram of one form of balanced modulator which consists of two matched *active devices*, the inputs of which are driven in push pull by the modulating signal but in phase by the carrier signal. The carrier input may be regarded as a signal which switches the active devices between conduction and non-conduction at carrier frequency. Thus the output of the circuit consists of a succession of half-cycles of carrier, the envelope of which has the waveform of the modulating signal. This is illustrated in *Figure B.5*: it is the waveform of a

23

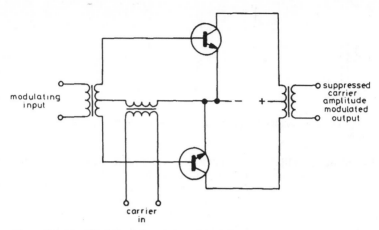

Figure B.4 Simplified circuit for a balanced modulator using two transistors in push-pull

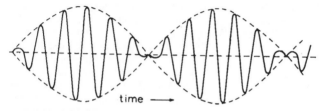

Figure B.5 Waveform of a suppressed-carrier amplitude-modulated signal for sinusoidal modulation

suppressed-carrier amplitude-modulated signal. See *amplitude modulation suppressed-carrier modulation*.

balanced operation Same as *push-pull operation*.

balanced-wire circuit A circuit comprising two conductors, electrically alike, and symmetrically arranged with respect to a common reference point, usually ground.

ballast resistor A resistor with a pronounced positive temperature coefficient so designed that when connected in series with a given load it ensures constancy in current for limited variations in supply voltage. Such resistors were used in AC/DC radio receivers where they were connected in series with the electron-tube heaters to ensure constancy in heater current. The graphical symbol for a ballast resistor is given in *Figure B.6*.

Figure B.6 Graphical symbol for a ballast resistor

ballast tube A ballast resistor enclosed in a gas-filled envelope. See *ballast resistor*.

banana tube A picture tube in which the electron beam scans a single

24

Jo Coleman

Information Update Service

Butterworth-Heinemann

FREEPOST SCE 5435

Oxford

Oxon

OX2 8BR

UK

Keep up-to-date with the latest books in your field.

Visit our website and register now for our **FREE** e-mail update service, or join our mailing list and enter our monthly prize draw to win £100 worth of books. Just complete the form below and return it to us now! (FREEPOST if you are based in the UK)

www.bh.com

Please Complete In Block Capitals

Title of book you have purchased:...

....

Subject area of interest:..

Name:...

Job title:...

Business sector (if relevant):..

Street:..

Town:... County:...

Country:.. Postcode:..

Email:...

Telephone:...

How would you prefer to be contacted: Post ☐   e-mail ☐   Both ☐

Signature:... Date:..

☐ Please arrange for me to be kept informed of other books and information services on this and related subjects (✔ box if not required). This information is being collected on behalf of Reed Elsevier plc group and may be used to supply information about products by companies within the group.

FOR OFFICE USE ONLY

Butterworth-Heinemann,
a division of Reed Educational
& Professional Publishing Limited.
Registered office: 25 Victoria Street,
London SW1H 0EX.
Registered in England 3099304.
VAT number GB: 663 3472 30.

BUTTERWORTH HEINEMANN

ℜ A member of the Reed Elsevier plc group

phosphor strip in the horizontal direction, vertical scanning being achieved mechanically by a system of cylindrical lenses mounted inside a drum which rotates around the tube. In the colour version of the tube the single phosphor strip is replaced by three parallel strips of red, green and blue phosphors and the beam is deflected vertically to scan the colour strips as necessary. The tube is not at present in commercial use.

band (1) A range of frequencies between defined limits, e.g. the audio-frequency band which may be taken as the range between 30 Hz and 20 kHz. (2) In an *energy-level diagram* a closely-spaced range of electron energy levels, e.g. the *conduction band*. (3) In *computers* or *data-processing equipment* a group of recording tracks on a *magnetic storage* device such as a drum or disk.

band gap Same as *energy gap*.

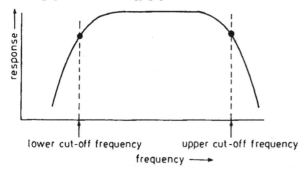

Figure B.7 General form of the response curve for a bandpass filter

bandpass filter A filter designed to pass signals with frequencies between two specified cut-off frequencies. The *frequency response* has the general form illustrated in *Figure B.7* and such a response may be obtained at radio frequencies by coupling two identical *tuned circuits*, both resonant at the same (mid-band) frequency. The block symbol for a bandpass filter is given in *Figure B.8*.

Figure B.8 Block symbol for a bandpass filter

bandstop filter A filter designed to pass signals with all frequencies except those between two specified *cut-off frequencies*. The frequency response has the general form shown in *Figure B.9* and the block symbol for a bandstop filter is given in *Figure B.10*.

bandwidth (1) The range of frequencies over which the *gain* of a *network*, amplifier or receiver falls within prescribed limits. For example the bandwidth of an amplifier is often taken as the range between the frequencies at which the response is 3 dB down compared with the response at midband. (2) The range of frequencies between the outermost side frequencies of a modulated wave. Thus if the modulating frequencies extend to 10 kHz the bandwidth of an amplitude-modulated signal is 20 kHz.

25

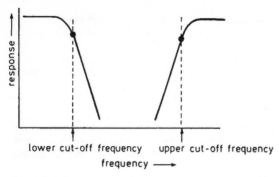

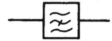

Figure B.9 General form of the response curve for a bandstop filter

Figure B.10 Block symbol for a bandstop filter

bar code An array of vertical parallel bars of varying thickness printed on packaging, publications and other products. At the checkout in a supermarket, the bar code is read by an *optoelectronic* reading device connected to the *electronic point of sale* system in order to establish the identity of the item and, from this, its price. The *host* computer can in this way make a continuous audit of how each type of merchandise is selling; this is valuable management information. See *electronic point of sale*.

Barkhausen-Kurz oscillator An early form of triode oscillator in which the grid is biased positively and the anode negatively, oscillation (usually at hundreds of MHz) depending on the *transit time* of electrons between cathode and anode. Unwanted oscillations of this type can occur in electron-tube circuits under particular biasing conditions.

barrel distortion Distortion of a television picture in which the sides of a reproduced square bulge outwards. The distortion, illustrated in *Figure B.11*, is caused by *non-linearity* of the field produced by the scanning coils. It is the opposite of *pin-cushion distortion*.

Figure B.11 Reproduction of a square by a picture tube suffering from barrel distortion

barretter Same as *ballast tube*.
barrier capacitance See *depletion layer*.
barrier diode Same as *Schottky diode*.
barrier grid (US) Same as *stabilising mesh*.
barrier layer Same as *depletion layer*.
base (region) In a *bipolar transistor*, that region between the emitter region and the collector region into which minority carriers are injected from the emitter region. It is the controlling electrode of a *bipolar transistor* and can

be compared with the control grid of an electron tube and the gate of a *field-effect transistor*. See *collector*, *emitter*, *minority carriers*, *semiconductor*.

base bias The steady current applied to the base of a *bipolar transistor* to ensure that signal excursions operate over the desired region of the transistor characteristic. See *dc stabilisation*, *operating point*.

base resistance See *electrode dc resistance* and *electrode ac resistance*.

base station A transmitting and receiving station used for radio communication with mobile stations within a restricted area. For example, in *cellular* telephone systems each cell contains a base station. A domestic cordless telephone has radio links to a dedicated base station which is connected to the telephone system.

BASIC See *computer language*.

bass The lower frequencies of the audible range extending from say 200 Hz to 20 Hz.

bass boost or **lift** *Accentuation* in which the lower frequencies of the audio range are emphasised.

bass cut *Attenuation distortion* in which the lower frequencies of the audio range are attenuated.

batch file A *text* file containing a list of commands or files for processing by a computer.

batch processing A form of data processing in which the computer is given a list of tasks and left to complete them without operator intervention. It is most often used for time consuming operations that involve, for instance, searches through large databases. Batch processing is sometimes undertaken at night or at weekends when the computer is not needed for more urgent, interactive operations.

battery A series and/or parallel arrangement of *primary cells* or *secondary cells* for supplying power. In common speech a single cell is often called a battery. See *accumulator*, *storage battery*, *voltaic cell*.

battery eliminator A unit providing power from the mains for a radio receiver originally designed for operation from batteries.

Baud In *data transmission* a unit of signalling speed equal to the number of code elements transmitted per second. It is commonly regarded as equivalent to the *bit rate*. The term is derived from the Baudot code used for over a century for the transmission of telegraph signals.

Baxandall circuit A *negative-feedback* circuit giving adjustable degrees of *bass lift*, *bass cut*, *top lift* and *top cut* and used in high quality audio amplifiers. A number of variants of the circuit are possible and one example is shown in *Figure B.12*. See *attenuation distortion*.

B battery (US) A battery used to supply the high tension for the *anodes* and *screen grids* of electron tubes.

beacon (1) A term sometimes used for an indicator, such as the stereo-subcarrier-detected indicator in a stereo radio receiver. The beacon is often a *light emitting diode*. (2) A fixed radio transmitter providing mobile stations with guidance signals. See *radio beacon*.

bead store A magnetic *store* in which the storage elements are formed by fusing *ferrite* powder directly on to the conductors forming the *matrix*. See *magnetic storage*, *storage element*.

27

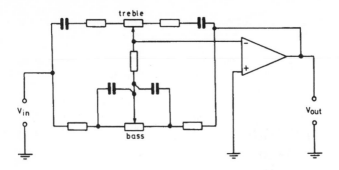

Figure B.12 An example of a Baxandall tone-control circuit

bead thermometer A thermometer using a *thermistor* as the temperature-sensing element.

beam current Of a *cathode ray tube* the current of the electron stream from the *cathode* which strikes the *screen* and is primarily responsible for the brightness of the display.

beam deflection tube A colour picture tube with a single *electron gun* and in which the *screen* is composed of horizontal stripes of red, green and blue phosphors arranged in sequence. A grid of horizontal wires is mounted close to the screen and, by applying suitable potentials to these wires the *electron beam* can be deflected so as to strike the phosphor stripe giving the required colour.

beam-indexing tube A colour picture tube with a single *electron gun* and in which the *screen* is composed of vertical stripes of red, green and blue phosphors arranged in sequence. A beam-indexing system operated, for example, by signals from vertical stripes interleaved with the red, green and blue groups ensures that at any instant the electron gun is always switched to the colour signal corresponding to the phosphor stripe on which the beam is incident.

beam parametric amplifier A *parametric amplifier* in which the variable reactance is provided by a modulated electron beam.

beam splitting A technique used in a *double-beam oscilloscope* to divide the beam from the *electron gun* into two parts which can be deflected independently to permit two different *traces* to be simultaneously displayed on the screen.

beam-switching tube Same as *beam deflection tube*.

beam tetrode A *tetrode* incorporating electrodes to direct the electron stream from the *cathode* to the *anode*. As shown in *Figure B.13* the beam-forming electrodes usually consist of two plates connected to the cathode. Beam tetrodes usually have *aligned grids* and the resulting high-density electron stream prevents *secondary emission* from the anode from reaching the screen grid so preventing the *tetrode kink* which would impair efficiency.

beanstalk A circuit consisting of a number of *transistors* connected in series across a power supply, their inter-connections being brought out as tapping points. Such a circuit is used at the input to the tetrode modulator of a

28

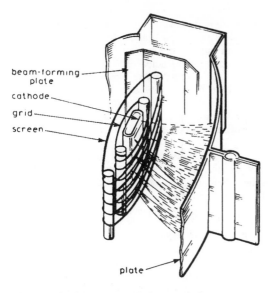

Figure B.13 Construction of a beam tetrode

sound radio transmitter, adjustable resistors connected across suitable tapping points being used to set the audio input level, the base bias and the limiter level.

beat frequency The frequency of the periodic pulsations produced by *beating*. The term is also extended to other examples where two sinusoidal inputs are combined to produce an output at the difference frequency, e.g. in *frequency changers* and *mixers*.

beat frequency oscillator (BFO) In general a variable-frequency oscillator used to interact with a fixed-frequency oscillator to produce the beat or difference frequency.

More specifically, an oscillator used in radio receivers to render audible transmissions of *continuous-wave telegraphy*. In a superheterodyne receiver the BFO beats with the signal in the *IF amplifier* to produce an audible note.

See *communications receiver, continuous wave*.

beating The combining of two sinusoidal signals to give periodic pulsations. Simple addition of the two signals, if they are of equal amplitude, gives the result shown in *Figure B.5*. The envelope consists of two overlapping sine waves with a frequency equal to half the difference between the frequencies of the original signals as shown by the identity

$$\sin \omega_1 t + \sin \omega_2 t = 2 \sin \left[\left(\frac{\omega_1 - \omega_2}{2} \right) t \right] \cos \left[\left(\frac{\omega_1 + \omega_2}{2} \right) t \right]$$

There are, however, two maxima, i.e. two beats per envelope cycle. Thus the number of beats per second is equal to the difference frequency.

If the two sinusoidal signals are combined in a non-linear device, such as an AM detector, then a multiplicity of output signals with new frequencies is generated: these are *combination frequencies* and the first two in the series are the sum and difference frequencies. The difference or beat frequency is the output from an additive *mixer* which is selected for amplification in the IF amplifier of a super-heterodyne receiver. It is significant that non-linearity is essential to generate the beat frequency in an additive mixer. This is clear from *Figure B.5*: if the wave shown here is rectified the result has a pronounced component at the difference frequency.

If the original two sinusoidal signals are combined in a device which multiplies them, e.g. a multiplicative mixer, then only two new frequencies are generated. These are the difference term—the beat frequency—and the sum frequency. This is shown by the identity

$$\sin \omega_1 t \sin \omega_2 t = \tfrac{1}{2}[\sin(\omega_1 - \omega_2)t + \sin(\omega_1 + \omega_2)t]$$

In this type of *mixer* the beat frequency is generated without need for non-linearity.

beep Onomatopoeic name for the hooting sound issued by computers and other data processing equipment encountering *error* conditions or in other circumstances where the operator's attention is required.

bel The basic unit of a logarithmic scale used for expressing ratios of powers. The bel was invented by American telephone engineers in the 1920s and was named after Sir Alexander Graham Bell, the inventor of the telephone. Two powers P_1 and P_2 are related by N bels when

$$\log_{10}(P_1/P_2) = N.$$

The number of bels is thus the common logarithm of the power ratio. It follows that an amplifier with a power gain of 100000 can be said to have a gain of 5 bels. The unit is inconveniently large and its tenth part, the decibel, is always used in practice. See *decibel*.

benchmark An industry standard test of the performance of an equipment, used for comparison purposes.

beta particles The stream of electrons emitted from certain radioactive substances.

betatron An *electron tube* capable of producing an electron beam of very high energy suitable for *X-ray* generation. A magnetic field is used to confine the beam to a circular path and its energy is continuously increased by a rapidly-changing electric field.

bias In electronics a steady current or voltage which determines the position of the *operating point* on the input-output characteristic of an *active device*. In magnetic recording a magnetic field superimposed on that due to the signal to be recorded in order to improve the linearity of the relationship between signal and remanent flux density. The bias field in an audio recorder is usually produced by an ultrasonic oscillator.

bias current Of a *differential amplifier* the current flowing into each of the two input terminals.

bidirectional thyristor A *thyristor* which can be switched to the on-state whether the anode is positive or negative with respect to the cathode. Diode

(a)   (b)

Figure B.14 Graphical symbol for bidirectional thyristors (a) diode and (b) triode (triac)

and triode types exist and the graphical symbols for both are given in *Figure B.14*. The triode type is triggered into conduction by a signal applied to the gate and this type of thyristor is alternatively known as a triac.

bidirectional transistor A *bipolar transistor* in which the connections to emitter and collector terminals can be interchanged without significant change in characteristics. Normally the emitter and collector regions are of different sizes and conductivities. Thus the terminals must be used as the manufacturer prescribes if the intended properties of the transistor are to be realised. If, however, the construction of the transistor is symmetrical, the emitter and collector terminals may be interchanged without effect on characteristics.

bifilar winding A method of winding in which two conductors are wound in close proximity on a former. This is done, for example, to give close coupling between the windings in a *ratio detector* transformer. It is also used in the construction of non-inductive resistors by arranging that the current flows in opposite directions in the two conductors so that any magnetic field set up by the current in one conductor is cancelled by that in the other. This is illustrated in *Figure B.15*. A similar technique is used in the construction of the heater of some electron tubes to minimise mains-frequency hum.

former

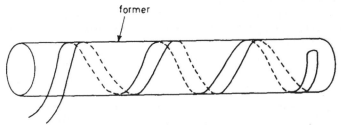

Figure B.15 Bifilar winding

bilateral impedance A mutual impedance through which power can be transmitted equally in both directions. Components such as *resistors*, *inductors*, *capacitors* and combinations of these which are used to couple circuits are examples of bilateral impedances. Any change in one circuit is reflected back to affect the current in the other circuit. See *mutual impedance*.

bimetal element or **bimetal strip** An assembly of two strips of metal with different coefficients of thermal expansion welded together so that the assembly bends into an arc when heated. Bimetal strips are used in *thermostats* and *thermal relays* to operate electrical contacts.

31

bimorph An assembly of two strips of piezo-electric material cemented together. Depending on the way in which the strips are cut from the original crystal the bimorph may bend when a voltage is applied between foil electrodes in contact with the outside surfaces, or it may twist. Bender and twister bimorphs are used in *piezo-electric loudspeakers* and disk cutter heads. If a bimorph is subjected to mechanical stress it develops EMFs between electrodes on the outside surfaces. Thus bimorphs are also used in piezo-electric microphones and disk-reproducing heads. See *piezo-electric crystal, piezo-electric effect, piezo-electric pickup*.

binary cell Same as *memory cell*.

binary-coded decimal (BCD) A method of translating decimal numbers into the *binary scale* by coding the individual figures of the number separately. For example the number 354 in binary-coded decimal is 11 101 100 whereas, if expressed in full binary form, it is 101100010. The advantage of the binary-coded decimal system is that large decimal numbers can be recognised immediately from their BCD coding once the binary equivalent of the decimal numbers from 0 to 9 are known.

binary coding A method of coding information in which the coded information consists of characters which can have only one of two possible values, e.g. in numerical coding the two possible values may be 0 and 1, the digits of the *binary scale*.

binary counter A circuit which gives one output pulse for every two input pulses. If the input pulses are irregular the circuit is usually regarded as a counter but if they are regular the circuit is generally regarded as a *frequency divider*. A *bistable circuit* such as a *multivibrator* is probably the best-known example of binary counter or divider.

binary digit One of the two digits of the *binary scale*, i.e. 0 or 1.

binary divider See *binary counter*.

binary scale A counting scale which has only two digits, 0 and 1. The digits in a binary number thus correspond to powers of 2. For example the binary number 110110 corresponds to

$$1 \times 2^5 + 1 \times 2^4 + 0 \times 2^3 + 1 \times 2^2 + 1 \times 2^1 + 0 \times 2^0$$
$$= 32 + 16 + 0 + 4 + 2 + 0$$
$$= 54$$

binary word A sequence of consecutive *binary digits* which is treated as a unit of information in the storage, transmission and processing in a *computer* or *data-processing equipment*.

binaural (1) In general, hearing by use of both ears. The significant feature of binaural listening is that the listener can locate the position of a source of sound. This can be done from the difference in the time of arrival of the sound at both ears or by detecting the difference in phase. (2) In particular a form of stereo sound transmission in which the two channels originate from microphones located at the positions of the ears in a dummy head. To obtain the full benefit of binaural transmission the two signals must be heard via *headphones* so that the left ear hears only the signal from the microphone in the dummy left ear and vice versa. The two channels must be kept separate right up to the listener's ears.

Biot-Savart law The law relating the magnetic field strength due to an electric current with the strength of the current. It can be stated thus: the summation of the magnetic force around a closed path is proportional to the total current flowing across the surface bounded by the path.

bipolar transistor A *transistor* which depends for its action on two types of *charge carrier*. It consists essentially of three layers of semiconductor material, the centre layer having conductivity complementary to that of the two outer layers. The central layer is very thin and is known as the base, the other two layers being the emitter and collector. The structure is illustrated in *Figure B.16* which also shows that there can be two basic types of bipolar transistor, namely pnp and npn.

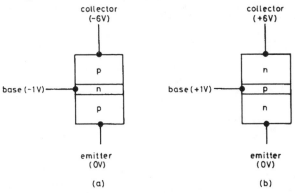

Figure B.16 Structure of (a) pnp and (b) npn bipolar transistors

Essentially the device behaves as two pn *junctions* in series and in normal usage the layers are so biased that the base-emitter junction is forward biased (and therefore of low resistance) and the collector-base junction is reverse-biased (and therefore of high resistance). Typical working voltages are shown in *Figure B.16*. In the npn transistor (b) the current crossing the base-emitter junction is carried by electrons from the emitter and holes from the base as explained under *junction*, but the impurity concentration in the emitter region is considerably greater than that of the base. Thus the electrons greatly outnumber the holes. On entering the base region a few electrons leave it via the external base connection to give a small external base current. Most of the electrons cross the very thin base region and are swept into the collector region, attracted by its positive bias, to give a considerable external collector current— despite the fact that the collector-base junction is reverse-biased. The collector (output) current can be as much as 100 times the base (input) current. This amplification is the essence of transistor action.

In the pnp transistor of *Figure B.16* (a) the holes crossing the base-emitter junction outnumber the electrons but again the external collector current is many times the external base current.

In both types of transistor electrons and holes play a part in the amplifying action, this justifying the description bipolar. Because of their low input resistance and high output resistance bipolar transistors are best regarded as current amplifiers.

33

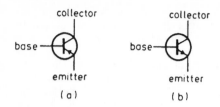

Figure B.17 Graphical symbols for (a) pnp and (b) npn bipolar transistors

The graphical symbols for pnp and npn bipolar transistors are given in *Figure B.17*.

bistable also known as **bistable circuit, flip flop** (US) or **toggle** (US) A circuit with only two possible states, both stable. When the circuit is placed in one of the states, it remains in it indefinitely unless compelled to leave it by the application of an external signal. When thus triggered the circuit enters the other possible state and here too it remains unless a further triggering signal is received and this returns the circuit to its original state. Two external signals are thus necessary to restore the circuit to its original state and this illustrates one of the applications of bistable circuits—as *pulse counters* or frequency dividers.

Because of their ability to remain in a given state, bistable circuits are extensively employed in digital switching applications, e.g. as the basic elements of *stores* and *registers* in computers. Basically a bistable circuit may consist of a *multivibrator* with two direct inter-transistor couplings but there are many possible types. See *D bistable*, *JK bistable*, *RS bistable*, *T bistable*.

bit Abbreviated form of *binary digit*.

bitmap A two-dimensional array of *pixels* forming an image or part of an image. See *graphics*, *painting software*.

bi-threshold device Same as *Schmitt trigger*.

bit rate The speed at which binary digits are transmitted. It is usually expressed in bits (kilobits or megabits) per second.

black body Same as *full radiator*.

black compression or **black crushing** In television, reduction of the gain at signal levels corresponding to black compared with the gain at white and mid-grey levels. Black compression reduces the visibility of detail in the low-light regions of reproduced pictures.

black level In television the minimum permissible level of the *picture signal* (*Figure B.18* (a)). In a properly-adjusted system the black-level voltage at the input to a picture tube gives cut-off of the electron beam as indicated in *Figure B.18* (b).

blanking (1) In general the suppression of the signals in a channel or device for a desired period. (2) In television the suppression of the electron beam in a picture tube during line and field flyback periods. (3) In television transmission the suppression of the *picture signal* component of a *video signal*, leaving the *blanking level* which may or may not contain sync signals.

34

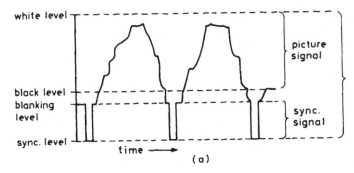

Figure B.18 (a) Waveform of a black-and-white video signal

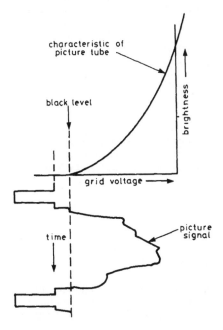

Figure B.18 (b) Application of a video signal to the characteristic of a picture tube

blanking level In a *video signal* the boundary level separating the *picture signal* from the *sync signals*, see *Figure B.18* (a).

bleeder or **bleeder resistor** A fixed resistor connected permanently across the output of a power source. It may be included to limit the rise in voltage when the load is disconnected. It may be used to discharge *capacitors* when the power supply is switched off or it may be included to improve the regulation of the supply.

blocking In electron tubes the suppression of anode current by applying a

35

large negative voltage to the grid. This may occur either fortuitously, for example, when the grid resistor is open-circuited or it may occur deliberately, e.g. when the tube is used in a *gate circuit*.

blocking capacitor A *capacitor* used between stages in an equipment to prevent the transfer of direct voltages but to provide a low-reactance path to alternating signals. The capacitor of an RC-coupled amplifier (see *Figure C.26* (a)) is an example of a blocking capacitor.

blocking oscillator An oscillator in which *positive feedback* produces so large an oscillation amplitude that the active device is cut off, usually in the first half-cycle, the circuit then relaxing until conditions again permit oscillation. The circuit thus produces regular bursts of current and is used as a pulse generator in, for example, the sawtooth generators in TV receivers.

The circuit diagram of a transistor blocking oscillator is given in *Figure B.19*. When the tuned circuit oscillates C_1 is charged by base current to a high negative voltage which cuts the transistor off. When C_1 has discharged sufficiently through R_1 oscillation commences again and C_1 is again charged. The transistor thus takes regular bursts of collector current and generates negative-going pulses across R_2 at a frequency determined by the *time constant R_1C_1*.

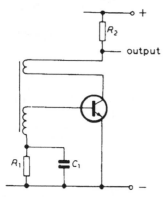

Figure B.19 A transistor blocking oscillator

Bode diagram A diagram in which the gain (and/or phase angle) of a device is plotted against frequency. Such diagrams can be used to determine the stability of the device and to show what degree of negative feedback can safely be applied.

Bode equaliser An *equaliser* used to correct the *frequency response* of TV sound circuits in which the amount of equalisation can be adjusted, without change in the shape of the equaliser characteristic, by operation of a single control.

bolometer In general a device for measuring electric current or power. The detecting device is a small resistive element, the resistance of which is strongly dependent on temperature. This is included in a bridge circuit which measures the power by the change in resistance resulting from the rise in temperature caused by the power absorbed. The temperature-sensitive element is usually a *thermistor* or a filament in an evacuated bulb.

Such instruments are useful for measuring small amounts of RF power, e.g. in a waveguide.

bombardment-induced conductivity Conductivity in an insulator or semi-conductor caused by *charge carriers* arising from bombardment by ionising particles.

bonding A low-resistance connection between conductors which ensures that they have the same voltage.

Boolean algebra An algebra in which logic relationships and operations are indicated by operators such as AND, OR, NOT etc, which are analogous to mathematical signs. It is named after its inventor George Boole (1815–64) and is used in the design of digital circuits.

booster Same as *repeater*.

booster diode A diode in the line output stage of a TV receiver which recovers much of the energy stored in the line deflection coils and makes it available as an additional source of supply. This can be added to the normal supply to provide a high-voltage (boost) supply for the line output stage and other stages in the receiver.

booting The process of starting a microprocessor or a computer. Typically, when a microprocessor is started 'from cold' it executes the set of instructions starting at a certain memory location. Clearly this location must be held in *ROM*. In a computer having a floppy-disk-based *operating system*, the ROM-based instructions may display a message to the user to insert the floppy disk containing the operating system and then to execute a *batch file* on that disk which configures the system.

bootstrap circuit A circuit in which the input signal is connected in series with the output signal and is thus 'pulled' to an extent equal to the output-signal amplitude. A simple example is shown in *Figure B.20*. The output of the amplifier is developed across R_e in the emitter circuit and the input is applied between base and emitter so that the potential of terminal b of the input moves in sympathy with the output signal. The circuit is described as 'pulling itself up by hauling on its own bootstrap'.

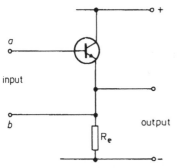

Figure B.20 Simple example of a bootstrap circuit

bottom bend Non-linear region of the I_a–V_g characteristic of an *electron tube* near anode-current cut off.

bottoming Condition in an *active device* driven heavily into conduction in which there is very little voltage drop across the output terminals of the

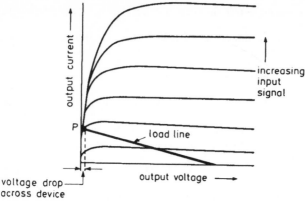

Figure B.21 Conditions in a bottomed active device

device. The device is in fact operated below the knee of the output-current/output-voltage characteristic as illustrated by the operating point P in *Figure B.21*. The output voltage is independent of the magnitude of the input signal provided this is sufficient to cause saturation of the device. When an active device is used as a switch the bottomed condition is known as the on state. See *operating point*.

bounce In television the damped oscillatory variations in signal level resulting from an abrupt change in the DC component of a *video signal*. These cause changes in brightness or vertical positioning in reproduced pictures lasting longer than a field period. See *ringing*.

boxcar circuit (US) or **boxcar lengthener** (US) A circuit for the production of *pulses* of predetermined duration, the leading edges of which are triggered by the leading or trailing edges of input pulses.

branch A single path along which current can flow between two *nodes* of a network. For example ac, ab and bd are three branches of the network of *Figure B.22*. See *node*.

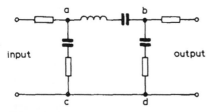

Figure B.22 Part of a network: ac, ab and bd are branches

breadboard A collection of components and/or devices interconnected as required in a new design of an equipment. The connections are temporary to facilitate changes of components and/or layout. A breadboard enables the performance of a new equipment to be checked.

breakdown voltage (1) Of an insulator the voltage at which the insulation

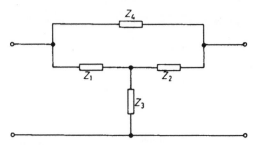

Figure B.23 A bridged-T network

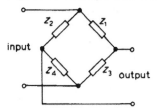

Figure B.24 A bridge network

fails and current flows through or across the surface of the insulator as a discharge. (2) Of a gas, same as *ionisation voltage*. (3) Of a semiconductor device, same as *avalanche breakdown* or *Zener effect*.

bridged-T network A *T-network* in which a fourth element is connected in parallel with the two series elements as shown in *Figure B.23*.

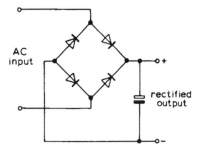

Figure B.25 A bridge rectifier with reservoir capacitor

bridge network A network of four elements connected to form a square, the input being applied to two opposite corners, the output being taken from the remaining two corners. The network is shown in *Figure B.24* and is also known as a lattice network.

bridge rectifier Four rectifiers connected to form a bridge network. The combination is extensively used for *full-wave rectification*. The rectified output has an open-circuit voltage across the reservoir capacitor approximately equal to the peak alternating input to the *bridge circuit*, *Figure B.25*. For some applications, e.g. battery charging, the reservoir capacitor is not necessary.

brightness The subjective assessment of the amount of light received from a source. The objective measured brightness is more properly called *luminance*. The brightness control of an oscilloscope or TV receiver determines the grid bias applied to the *cathode ray tube* or picture tube and hence the light output from the screen.

broadcasting The transmission of sound or TV programmes by radio or by a wire distribution system for reception by the general public.

bubble See *magnetic bubble*.

bubble-jet printer See *printer*.

bucket-brigade device (BBD) A semiconductor device consisting essentially of a series of capacitors linked by switches. In practice the switches are *bipolar* or *field-effect transistors* and by closing them appropriately by clocked or timed pulses, charge can be transferred from one capacitor to another and so move along the line of capacitors.

buffer memory In a *digital computer* an intermediate store used to compensate for the different speeds at which units can handle data.

buffer stage An *active device* included between two stages which permits signal transfer from one to the other but isolates the stages so that, for example, changes in impedance in one stage have no effect on the performance of the other.

It is common practice to use *emitter followers* as buffer stages.

bug (1) A hidden microphone or other device used for surveillance purposes. (2) A fault in hardware or software that results in a computer malfunction.

building-out network A network inserted in a circuit to modify the input impedance to a desired value. Such networks are used in *telephony* to permit accurate matching to a line which is incorrectly terminated at its receiving end.

build-up time or **building-up time** Same as *rise time*.

bulk storage A facility for storing very large quantities of data. At one time this would have involved banks of tape readers, but today is more likely to be an array of *hard disk drives*.

bulletin board system (BBS) A form of *electronic publishing* in which data and applications software is accessed over telephone lines by subscribers having a computer and suitable modem. Subscribers may also leave messages on the bulletin board for other subscribers to read. It is indeed the electronic equivalent of an office notice board.

bunching In a velocity-modulated tube the process whereby the density of the electron stream is modulated by the applied signal so as to gather the electrons into clusters at particular points along the drift space. See *Applegate diagram, drift space, velocity modulation*.

burn-in A test in which prototype equipment is operated at full load, usually in a high-temperature environment, continuously for an extended period such as 30 days. The theory is that, if the equipment survives the burn-in test, it will survive the less arduous conditions it is likely to encounter in normal service.

burst signal Same as *colour burst*.

bus In electronic equipment such as a computer, a conductor or set of conductors used as a path for conveying information or power from one or more sources to one or more destinations. A bus is required to carry a number of different types of information, e.g. for data transference, for control and for addressing. Some success has been achieved in designing standard buses for inter-board connections within a computer system, and for interconnections between various manufacturers' equipment.

bus-bar A conductor used to distribute power to or receive power from several branches.

bypass capacitor A capacitor used to provide a low-reactance path to signals within a particular frequency range.

An example is the capacitor which is connected across the resistor in a cathode bias circuit which is included to provide a low-reactance path for all signal-frequency components in the cathode current. The resistor then carries only the DC component and a steady voltage is therefore set up across the cathode resistor. See *automatic cathode bias*.

byte A sequence of consecutive *binary digits*, shorter than a word, which is treated as a unit in *computers* and *data-processing equipment*.

It is generally used for a specific purpose such as synchronising or addressing. It commonly contains 8 bits and is the smallest unit of data that can be put into or removed from an 8-bit store. There are 2^8, i.e. 256, different combinations of binary digits in an 8-bit byte and they can be used, as in the *ASCII* code, to represent the upper and lower-case letters of the alphabet, numerals, punctuation marks, arithmetical and other symbols. See *binary word*, *address*.

C

C and C++ See *computer language*.

cabled distribution system Same as *wire broadcasting*.

cache An area of high-speed *memory* used to speed up computer processing. Caches are used in disk read and write operations, in storing character bitmaps and, above all, in central processing unit operations.

calibrate To adjust test equipment against a standard to ensure that it is working accurately.

camcorder A television camera and video cassette recorder combined in one portable unit. Camcorders are widely used by broadcasting professionals for newsgathering and location work and by hobbyists for creating 'home movies'. It replaces the cine camera formerly used for these purposes and has the advantages over them that no film processing is required so that instant playback is available and that the medium is reusable.

camera tube An *electron tube* used in a TV system to produce picture signals by scanning a charge image derived from an optical image of the scene to be televised. The various types of tube that are used in black-and-white or colour TV can be classified according to the type of target used (i.e. whether photo-emissive or photo-conductive) and according to the velocity of the scanning beam (i.e. whether high or low).

The earliest tubes such as the *iconoscope* and *image iconoscope* had photo-emissive targets and high-velocity scanning beams. The optical image and the scanning beam were projected on the same face of the target and the beam was arranged to scan the target obliquely so that the electron gun did not impede the light path. The chief disadvantages of these tubes (low sensitivity and the presence of strong shading signals in the tube output) stemmed from secondary electrons released from the target as a result of bombardment by the high-velocity scanning beam.

To avoid these difficulties the beam velocity was reduced to a value at which no *secondary emission* occurred and in this way the *orthicon* and *image-orthicon* tubes were developed. The low-velocity scanning beam lands on the target and drives its potential negative until it reaches electron-gun cathode potential. When the target potential has been thus stabilised, the scanning beam is slowed to zero velocity as it approaches the target. One of the chief problems with low-velocity tubes is that of focusing and deflecting such a beam. This was solved by use of *orthogonal scanning* and a long magnetic lens: these ensured that the beam always approached the target at normal incidence so that the optical image could not be focused on the scanned face. A combined transparent signal plate and target electrode was developed which enabled the optical image to be projected on one face whilst the rear face was scanned by the beam. Orthicon tubes were first used in TV in 1946. They were free of *shading signals* but lacked sensitivity and the target stabilisation could be lost on receipt of a large light input.

Both disadvantages were overcome in the *image-orthicon* tube which included an *image section*, a *stabilising mesh* to prevent loss of target stabilisation and an *electron multiplier* into which the return scanning beam was directed. This is a highly-sensitive tube and was extensively used in black-and-white TV services for many years. It is, however, too bulky for use in colour cameras where three or four tubes are necessary. Instead, therefore, attention was directed towards photoconductive tubes (*vidicons*) which could be made very simple and compact.

In vidicon tubes, the target consists of a transparent signal plate on which is deposited a layer of photo-conductive material. The scanned face of the target is stabilised at electron-gun cathode potential as in any low-velocity tube and the signal plate is given a small positive bias. Current thus flows longitudinally through the target thickness and the magnitude of this current at any point on the target depends on the target resistance which in turn depends on the illumination of the target at that point. Early vidicon tubes suffered from lag, i.e. there was an appreciable delay between a change of light input and the corresponding change in target resistance but intensive work on photoconductive materials has resulted in satisfactory sensitivity and lag. Most modern colour television cameras incorporate three or four vidicon tubes.

cannibalise To strip serviceable components from one or more (usually non-functioning) equipments in order to build one that works.

capacitance *(C)* A property of two conductors, electrically insulated from each other, which enables them to store charge when a potential difference exists between them. The capacitance is defined as the quotient of the charge and the potential difference thus

$$\text{capacitance } (C) = \frac{\text{charge stored } (q)}{\text{potential difference } (V)}$$

The unit of capacitance is the Farad (F), this being the capacitance when the charge stored is one coulomb and the potential difference is one volt. For many purposes this unit is too large and the submultiples the microfarad (µF) equal to 10^{-6} F and the picofarad (pF) equal to 10^{-12} F are used instead.

capacitance diode Same as *varactor*.

capacitance potentiometer A number of capacitors connected in series across a source of alternating voltage, the interconnections acting as

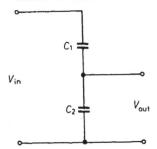

Figure C.1 A capacitance potentiometer using two capacitors

43

tapping points from which a desired value of voltage can be taken. In the simplest example there are only two capacitors as shown in *Figure C.1* and the step-down ratio is given by

$$\frac{v_{out}}{v_{in}} = \frac{1/j\omega C_2}{1/j\omega C_1 + 1/j\omega C_2} = \frac{C_1}{C_1 + C_2}$$

capacitive coupling Coupling between two circuits by virtue of mutual capacitance. The coupling capacitor may be a component connected in series with the circuits as in *Figure C.2* (a) or it may be a shunt component as in *Figure C.2* (b). In *Figure C.2* (a) the *coupling coefficient* is given approximately by C/C_1 where C is assumed small compared with C_1 and C_2. In *Figure C.2* (b) the coupling coefficient is given approximately by C_1/C, where C is assumed large compared with C_1 and C_2. In both circuits C_1 is assumed equal to C_2.

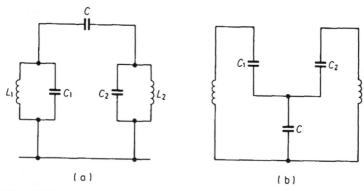

(a) (b)

Figure C.2 Capacitive coupling by (a) series capacitance and (b) shunt capacitance

capacitor A component consisting essentially of two plates or *electrodes* separated by a *dielectric* and used because of its *capacitance*. The capacitance is proportional to the area of the plates and to the *dielectric constant* of the insulator and is inversely proportional to the thickness of the dielectric. There are many different types of capacitor varying in capacitance between a few pF and thousands of µF and varying in voltage rating from a few volts to several kilovolts. The plates are often of aluminium and the most commonly-used dielectrics are impregnated paper, mica, plastics, ceramics, air and vacuum. The graphical symbol for a capacitor is given in *Figure C.3* and the component reference is C. See *electrolytic capacitor*.

Figure C.3 Graphical symbol for a capacitor

capacitor loudspeaker Same as *electrostatic loudspeaker*.

capacitor microphone Same as *electrostatic microphone.*

capacitor store A store in which each bit of data is held as a charge on a *capacitor. Dynamic RAMs* are of this type.

capture ratio The ability of a radio receiver to select the stronger of two signals at the same frequency. It is measured in *decibels*, the figure representing the ratio of the closest signal strengths that can be resolved.

carcinotron Same as *backward-wave oscillator.*

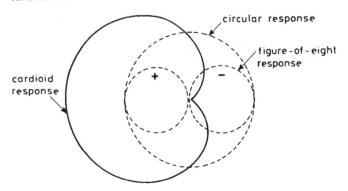

Figure C.4 Derivation of cardioid from circular and figure-of-eight diagrams

cardioid response Type of response represented by a heart-shaped *polar diagram.* Such a diagram is obtained by combining a circular diagram with a figure-of-eight diagram as shown in *Figure C.4.* A cardioid polar diagram is useful in radio direction finding and is obtained by combining the omnidirectional response of a vertical antenna with the figure-of-eight reponse of a loop antenna.

A cardioid response is also useful in a *microphone* because it gives a 180° live area, i.e. it makes the microphone effectively single-sided—ideal for positioning in stage footlights with the dead side towards the audience. This type of response can be obtained from a microphone which combines the omnidirectional (pressure) mode of operation with the figure-of-eight (pressure-gradient) mode of operation. See *pressure-gradient microphone, pressure microphone.*

card reader In early data processing equipment a device which reads the data on a punched card. Reading was by electromagnetic feelers or photo-electric scanning.

caret An I-shaped *cursor* used especially in computer displays that have proportionally spaced characters. The caret is positioned between characters, unlike a conventional rectangular cursor which is positioned directly over fixed-pitch characters.

carrier Same as *carrier wave.*

carrier storage An effect in pn *junctions* which have been driven hard into conduction whereby the current continues to flow for a brief period after the applied voltage has been removed. More *majority carriers* move towards the junction than are needed to supply the external current: these

45

are stored and are able to continue the current briefly when the applied voltage is reduced to zero.

carrier suppression A system of AM radio transmission in which the *carrier wave* is not transmitted or is transmitted at very low level. This is permissible because the information to be transmitted is carried in the *sidebands*. At the receiving end the carrier must be re-introduced to make detection possible and a suitable signal is generated locally for this purpose.

carrier wave A continuous wave, usually sinusoidal, of constant amplitude and frequency and intended for angle or amplitude modulation. The term is also used to describe the carrier-frequency component of a modulated wave. See *amplitude modulation, angle modulation*.

carry In *computers* and *data-processing equipment* the digit added to the next highest digit position when the sum of the digits in the lower position exceeds the base number.

cascade A circuit arrangement in which the output of one stage of amplification or section of a *filter*, etc, forms the input of the next stage.

cascode An amplifier consisting of a common-emitter stage feeding into a common-base stage. The arrangement avoids instability by feedback via the collector-base capacitance in the common-emitter stage and can thus be used for stable VHF and UHF amplification. The two transistors are often connected in series across the supply as shown in *Figure C.5*. The circuit arrangement can be used with *electron tubes* and with *field-effect transistors*: in fact a tetrode FET is equivalent to two triode FETs in series.

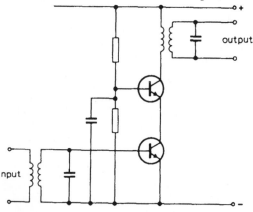

Figure C.5 A cascode circuit of two bipolar transistors

cassette A container holding sensitive storage media such as tape or film. The container protects the media against light, grease and dust and also makes it easier to handle. Audio compact cassettes and video cassettes in various formats are probably the best known examples.

catalogue A display of the contents of a *directory*.

catcher The electrodes and associated *cavity resonator* in a velocity-modulated tube which extracts energy from the bunched beam. See *velocity modulation*.

46

catcher diode A diode used to prevent the voltage at a particular point in a circuit from rising above or falling below a predetermined value.

cathode In general the electrode in any electron tube which is connected to the negative terminal of the supply. In *electron tubes* the source of electrons for the principal electron stream. In a glow-discharge tube the cathode may be cold and electrons are then liberated by bombardment of the cathode by heavy positive ions. In *photocells* electrons are liberated from the cathode under the stimulus of incident light. In most *electron tubes*, however, the cathode is heated to liberate electrons and it may take the form of a filament directly heated by the passage of an electric current through it or the cathode may be indirectly heated. See *cold cathode, glow discharge, indirectly-heated cathode*.

cathode AC resistance See *electrode AC resistance*.

cathode bias Same as *automatic cathode bias*.

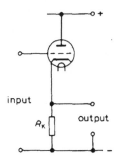

Figure C.6 A cathode follower circuit

cathode follower An *electron tube* in which the load is connected in the cathode circuit, see *Figure C.6*. This gives 100% negative feedback which reduces the voltage gain to just less than unity. Signals applied to the grid thus emerge with little loss from the cathode; in fact, the cathode voltage 'follows' that of the grid. The advantage of the circuit lies in its high input resistance and low output resistance which make it suitable as a *buffer stage*. There are corresponding *emitter follower* and *source follower* circuits.

cathode ray oscillograph An instrument for producing permanent records of the waveform of varying electrical quantities. The records are generally obtained by photographing the screen of a *cathode ray oscilloscope* and the provision for attaching a camera to the *cathode ray tube* is the only distinction between an oscillograph and an oscilloscope.

cathode ray oscilloscope (CRO) Instrument containing a *cathode ray tube* used for examining variable electrical quantities, periodic or transient, by displaying the waveform on the screen. For examining periodic waveforms the electron beam is deflected horizontally (i.e. in the X direction) by a sawtooth generator acting as a time base. The signal to be examined is applied to the vertical deflection system (Y direction) usually after amplification. By adjusting the time-base frequency to an exact sub-multiple of the fundamental frequency of the *waveform* a stationary pattern can be produced on the screen and detailed observation of the shape of the waveform is possible. To help in obtaining a stationary pattern there is provision for synchronising the time-base generator to the *waveform*.

The oscilloscope is one of the most widely used tools in electronics. In addition to its obvious application in examining waveforms to detect harmonics or the presence of spurious signals it can be used in conjunction with a *wobbulator* for the alignment of receivers. Oscilloscopes are also widely used as indicators in *radars*. They are also extensively employed in the examination of any physical quantity (e.g. mechanical stress) which can be converted into a corresponding *electrical signal*. See *double-beam oscilloscope, oscillograph*.

cathode ray tube (CRT) An *electron tube* in which an electron beam is focused on a phosphor screen and is deflected to give a visible display. Such tubes are extensively used in oscilloscopes, in radars, and as picture tubes in TV receivers. Essentially the tubes comprise an *electron gun* for generating the beam, a system (either electrostatic or magnetic) for focusing the beam on the screen, a system (also either electrostatic or magnetic) for deflecting the beam, and finally a screen coated with a phosphor which emits light when struck by the beam. *Figure C.7* illustrates the construction of a cathode ray tube with *electrostatic focusing* and *electrostatic deflection*, the type commonly used in oscilloscopes.

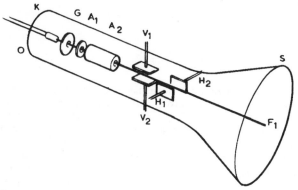

Figure C.7 Construction of cathode ray tube with electrostatic focusing and deflection: K, thermionic cathode; G, control grid; A_1, focusing anode; A_2, accelerating or final anode; V_1, V_2, plates for vertical deflection of the beam; H_1, H_2, plates for horizontal deflection of the beam; S, luminescent screen

cathode ray tuning indicator (also known as *electron ray tube* (US)). A small *electron tube* incorporating a triode amplifier and a simple cathode ray tube in which the area of *luminescence* on the screen can be controlled by the signal applied to the triode control grid. It was chiefly used as a tuning indicator in radio receivers the area of the display being controlled from the AGC line. The graphical symbol for the device is given in *Figure C.8*.

cation Positively-charged *ion* formed in a gas by ionisation or in an electrolyte by *dissociation* and which moves towards the negatively-

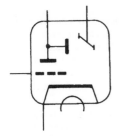

Figure C.8 Graphical symbol for a cathode
ray tuning indicator

charged electrode (cathode) under the influence of the potential gradient. See *anion*.

cat's whisker A pointed length of wire, usually copper, which was lightly pressed against a sensitive part of a crystal, e.g. galena, to form a rudimentary form of *point-contact diode*. This was used for AM detection in the early days of broadcasting in the 1920s.

cavity magnetron A *magnetron* in which the anode contains a number of *cavity resonators* facing inwards towards the cathode. The electron stream can be made to circulate around the cathode within the space between cathode and anode, and can thus become velocity modulated as a result of passing close to the mouths of the resonators. The resultant bunching of the beam can, by suitable choice of electric and magnetic field strengths give rise to *oscillation* at the resonance frequency of the cavities. The resultant energy can be directed into a *waveguide* by means of a coupling loop in one of the resonators. In this way microwave oscillations of considerable power can be generated. See *cavity resonator, velocity modulation*.

cavity resonator A cavity contained by a conducting surface the size and shape being chosen to give resonance at a particular frequency of electromagnetic waves. At microwave frequencies such resonators replace the lumped *inductance* and *capacitance* used in tuned circuits at lower frequencies.

C battery (US) A battery used to supply the grid bias voltage for *electron tubes*.

C core A *magnetic core* for a *choke* or *transformer* consisting of two C-shaped sections butted together to form a closed magnetic circuit. The C sections are made by winding magnetic alloy tape into a spiral and then cutting the resultant ring into two. C cores have better magnetic properties than laminated cores because the magnetic grain is parallel to the length of the tape. C cores are also easier to assemble with choke or transformer windings.

Ceefax The *teletext* service of the BBC.

cell (1) A device containing two electrodes and capable of generating an EMF between them. The electrodes may be immersed in a electrolyte and generate the EMF as a result of chemical action (see *voltaic cell*). Alternatively one of the electrodes may be light-sensitive and produce the EMF as a result of illumination by light (see *photocell*). (2) An elementary unit of a store, e.g. a *memory cell* (see *store*). (3) The area served by a *base station* in a *cellular radio* system.

cellular radio A radio-based mobile communications system in which

49

numerous low-power *base stations* each cover a small area or *cell*, all under the control of a central computer. As a mobile user leaves the area served by one base station, the computer determines which of the surrounding base stations now offers the best transmission path and switches the user over to that base station in a *transparent* process.

central processing unit (CPU) or **central processor** In a *digital computer* or *data-processing equipment* the unit containing the arithmetic, logic and control circuits which direct and co-ordinate operation of the computer and the *peripheral* devices.

centre frequency In frequency modulation, the carrier frequency.

Centronics printer standard The most widely used method of transferring data from a computer to a printer. The system uses 36-pin plugs and connectors with eight parallel conductors carrying data, making it faster than alternative serial transmission systems.

ceramic A class of material of very high resistivity and sometimes of high *permittivity* or *permeability* or with *piezo-electric* properties. Thus ceramic materials are extensively used in electronics in magnetic cores (as ferrite), as *dielectrics* and as piezo-electric elements. Ceramics cannot usually be machined and are cast into shapes suitable for their application.

chain code A *binary code* in which each word is derived from the previous word by shifting the bits one place to the left or the right, dropping the leading bit and inserting a bit at the end. This cyclic sequence ensures that a word cannot recur until the cycle is repeated.

channel (1) In telecommunications generally, a unidirectional signal path. (2) In particular a limited band of frequencies allocated to the transmission of a signal. The channel must be wide enough to accommodate all the significant components of the signal, e.g. *side frequencies* of a modulated signal. When channels are closely packed in a *waveband*, the spacing is made greater than the channel width to give a guard band between adjacent channels. Television channels are identified by numbers in the UK, e.g. channels 21 to 68 are in Bands IV and V. (3) Of a *field-effect transistor* the semiconductor layer along which majority carriers flow from the *source* to the *drain* terminals. The effective width of the channel and hence its electrical resistance is controlled by the voltage applied to the *gate*.

channel multiplier A type of electron multiplier of tubular construction in which the secondary-electron-emitting material is in the form of a continuous internal surface.

characteristic curve In general a curve representing the relationship between two quantities which illustrates the behaviour of a device or equipment. The term is, however, usually reserved for the curves which show how the current at an electron-tube electrode depends on its voltage or the voltage at another electrode, e.g. the I_a-V_a or the I_a-V_g curves of an *electron tube*.

characteristic impedance The value approached by the input impedance of an open-circuited *transmission line* as its length is increased. This is also the value of *impedance* which, if used to terminate the line, gives an input impedance of the same value. Two types of transmission line commonly used for RF transmission are the twin-wire and coaxial cable. At RF the *reactance* of the distributed *inductance* and *capacitance* of the conductors has more effect than the resistance and the characteristic impedance is given by

$$Z_0 = \sqrt{(L/C)}$$

where L and C are the inductance and capacitance per unit length of line. For twin-wire and coaxial cables the characteristic impedance can be calculated from the dimensions of the cross-section as indicated in *Figure C.9*.

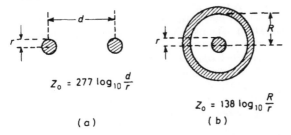

$$Z_0 = 277 \log_{10} \frac{d}{r}$$

(a)

$$Z_0 = 138 \log_{10} \frac{R}{r}$$

(b)

Figure C.9 Characteristic impedance of (a) twin-wire and (b) coaxial transmission lines

A significant point about the characteristic impedance is that if a finite length of line is terminated by an impedance equal to the characteristic impedance then the line behaves as if it were of infinite length and no standing waves are set up on it. RF energy can thus be transmitted along it without reflection.

character recognition See *optical character recognition*.

charge carrier A charged particle which, in moving under the stimulus of an electric field in a conductor, constitutes an electric current. Examples of charge carriers are the positively- or negatively-charged ions in an *electrolyte* and the *electrons* and *holes* in a *semiconductor*.

charge-coupled device (CCD) A device in which charges on a *semiconductor* surface and under *gate* electrodes can be transferred as a packet to an adjacent area of the surface by voltage pulses applied to the gates. Such a device can be used for data storage. If the gate electrodes are thin, parallel and equidistant conducting strips then, by applying suitably-phased clock pulses to the strips, any charges placed on the semiconductor surface under the first gate are transferred along the surface of the device as in a *shift register*.

The method of manufacture of such devices has much in common with that of IGFETs but is simpler in that no contacts or diffusions are required. The devices can be manufactured with very small dimensions so that great packing density is possible. They can thus be used as serial stores and can compete with magnetic disks and tape: indeed they are smaller than disk or tape for a given storage capacity but CCD *stores* are, of course, volatile. See *volatile store*.

charge image In TV the pattern of electric charges on the target surface of a *camera tube* which results from the optical-image input and which, when scanned, gives rise to the picture-signal output from the tube.

charge-storage tube An electron tube in which information is retained on its active surface in the form of a *charge image*. In most television *camera tubes* the image is stored on the target electrode in this form and is 'read off' by the scanning beam. *Storage tubes* used in VDUs also use similar principles.

charge transfer device (CTD) A *semiconductor device* which enables discrete packets of charge to be moved from one position to another. There are two main types: *bucket-brigade devices* and *charge-coupled devices*.

Chebyshev response A *response curve* in which the deviations from the ideal curve are made as small as possible and approximately equal to each other over the frequency band of interest. The parameters of *filters* are often chosen to give a Chebyshev response.

chip Popular name for an *integrated circuit*.

chip carrier A plastic or ceramic container for a *monolithic integrated circuit*, the connections to which are brought out to contacts on all four sides of the package and are designed for *surface mounting* on a *printed wiring board*. The chip carrier is an alternative to the DIP for a large number of contact pins and gives considerable saving in space because of its smaller contact spacing. It also gives a better thermal performance and higher speed of operation because the contacts are shorter than the pins of a DIP.

Chireix system An amplifying system used in AM transmitters in which two constant-voltage phase-modulated signals drive two saturated amplifiers feeding a common load. The power output is a function of the *phase difference* between the two signals and this is controlled by the modulating signal. The system was developed in France by H Chireix in the 1930s. See *ampliphase system*.

choke An *inductor*, the *reactance* of which over a particular band of frequencies is large compared with a given *impedance*. For example an RF choke may be used at the input of an audio-frequency amplifier to prevent RF signals entering the amplifier. In such an application the choke is likely to be followed by a shunt *capacitance* and for successful results the reactance of the choke must be large compared with that of the capacitance so as to give great attenuation at RF.

chopping Periodic interruption of an electrical signal or a beam of light. This can be achieved by electrical or mechanical means and is used to transform a direct-current signal into an alternating signal. *DC amplification* is often achieved in this manner, the chopped signal being amplified in conventional capacitance-coupled amplifiers and then reconverted into DC form.

chromatic aberration Defect in an optical lens due to variation of the refractive index of the material with the wavelength and causing coloured fringes in images. The defect is usually eliminated by constructing lenses of two different materials, e.g. crown and flint glass, which give refraction without dispersion. Such combinations are known as achromatic pairs.

chrominance Those attributes of a colour which can be defined by stating the *hue* and *saturation*.

chrominance amplifier The circuits which amplify the *chrominance signal*.

chrominance signal In colour television the signal which is added to the *luminance signal* to give colour information. In the *NTSC* and the *PAL systems* the chrominance signal is generated by so modulating the chrominance subcarrier that its instantaneous phase relative to that of the *colour burst* represents *hue* and its amplitude represents *saturation*.

circle diagram A set of graphs for the solution of *transmission-line* problems. Normalised *resistance* (R/Z_0) and normalised *reactance* (X/Z_0) are plotted on Cartesian or polar co-ordinates and the diagram can be used to calculate, for example, the impedance of a given length of line when terminated in any value of impedance. Z_0 is the *characteristic impedance* of the *transmission line*.

circuit (1) A path consisting of a conductor or a system of conductors through which an electric current can flow. (2) A *network* which may contain *active* and *passive devices* and may have one or more closed paths and which is intended to carry out a particular function. For example an amplifier may be said to have a circuit. (3) In telecommunications a means of bidirectional communication between two points.

circular polarisation *Elliptical polarisation* in which the amplitude of an electromagnetic wave remains constant whilst the plane of polarisation rotates. Vertically- or horizontally-polarised radio waves can become circularly-polarised after refraction in the ionosphere. Circular polarisation is used at some FM broadcasting stations to ensure that the signal can be received equally well on vertical and horizontal *antennas*.

circulating register (1) A *register* which retains data by inserting it into a delay device, by regenerating the data and re-inserting it into the register. (2) A *shift register* in which data removed from one end is put into the other end in a cyclic sequence.

clamp A circuit for ensuring constancy of the potential of a particular section of a recurrent *waveform*. Clamps are used in television transmission to ensure that the parts of the waveform which represent black in displayed pictures are maintained at a constant potential. To enable a clamp to operate at the desired instants it is driven by clamping pulses which are synchronised with the waveform to be clamped.

Clapp oscillator A *Colpitts oscillator* in which frequency stability is improved by using separate capacitors for tuning and for providing capacitive feedback. A circuit is shown in *Figure C.10* where C_1 is the tuning capacitor and C_2 and C_3 are high-value capacitors which provide the connections to the *active device* and swamp any variations in the input capacitance of the device.

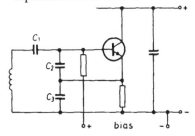

Figure C.10 Circuit diagram of one form of Clapp oscillator

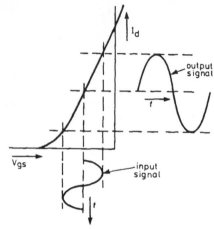

Figure C.11 I_d–V_{gs} characteristics of a junction-gate FET with input and output signals for class-A operation

class-A operation Operation of an *active device* in which the bias and signal amplitude are so chosen that output current flows throughout each cycle of input signal (assumed sinusoidal), i.e. the *angle of flow* is 360°.

This mode of operation is used in most small-signal stages of analogue equipment such as amplifiers and receivers, a linear part of the input-output characteristic being chosen to minimise waveform distortion as shown in Figure *C.11*. It is an inefficient mode of operation because the mean current of the *active device* is independent of the *amplitude* of the input signal. By utilising the maximum current and voltage swings possible the theoretical efficiency of a class-A output stage (measured by the ratio of the output power to the power taken from the supply) is 50% for a constant-amplitude signal. In AF amplification where signal amplitudes are constantly varying the practical efficiency is unlikely to exceed half this figure.

class-AB operation Operation of an *active device* which is biased to give class-A operation for small-amplitude input signals and class-B operation for large-amplitude signals, i.e. the *angle of flow* is between 180° and 360°. For *electron tube* amplifiers the suffix 1 is added after the letters AB to indicate that grid current does not flow at any period during the input-signal cycle and the suffix 2 is added if grid current is allowed to flow.

class-B operation Operation of an *active device* which is biased almost to output-current cut-off so that current flows only during one half of each cycle of sinusoidal input signal, i.e. the angle of flow is 180° This mode of operation, illustrated in *Figure C.12* is very efficient because the mean current of the active device is directly proportional to the signal amplitude and is almost zero for zero input signal. The output signal is, of course, a reproduction of only one half of each cycle of input signal but linear amplification is possible by using two class-B stages in push-pull operation and most of the output stages in transistor amplifiers and receivers are of this type.

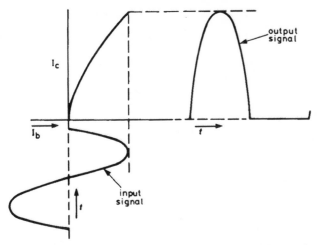

Figure C.12 I_c–I_b characteristic of a bipolar transistor with input and output signals for class-B operation

The theoretical efficiency of a class-B stage (measured by the ratio of the output power to the power taken from the supply) is 78% and efficiencies approaching this value can be obtained in practice. The chief source of inefficiency is the no-signal or quiescent current which must be great enough to minimise *crossover distortion*. For *electron tube* amplifiers the suffix 1 is added after the letter B to indicate that grid current does not flow at any time during the cycle and the suffix 2 is added if grid current is allowed to flow.

class-C operation Operation of an *active device* which is biased beyond cut-off of output current so that current flows for less than one half-cycle of sinusoidal input signal, i.e. the *angle of flow* is less than 180°. This mode of operation, illustrated in *Figure C.13*, cannot be used for linear amplification but is extensively used for RF amplification, e.g. in transmitters where a tuned output-load circuit rejects harmonics and gives a sinsoidal output signal. Efficiency (measured by the ratio of the output power to the power taken from the supply) can be very high indeed—approaching 90%. For electron-tube amplifiers the suffix 1 is added after the letter C to indicate that grid current does not flow at any period during the input-signal cycle and the suffix 2 is added if grid current is allowed to flow.

class-D operation Operation of an *active device* with a pulse-width-modulated input signal. Very high efficiency (measured by the ratio of output power to the power taken from the supply) is possible from a class-D stage because the input is of constant amplitude so permitting maximum use to be made of the current and voltage swings available in the output circuit. See *pulse-width modulation*.

clear input In *computers* and *data-processing equipment* a signal applied to a *store* to put it into a prescribed state, usually the state of zero stored information.

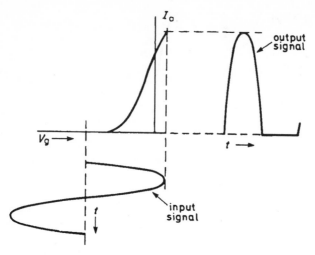

Figure C.13 I_a–V_g characteristic of an electron tube with input and output signals for class-C operation

click A *transient signal* of very short duration. The term describes the sound of the signal when reproduced by an acoustic *transducer*.

client In a computer network, a user or the user's computer or terminal or the code used at the user's end of the network.

client server In a computer network, a computer from which users (*clients*) can request stored data and applications.

clip art Computer *graphics*, usually copyright-free, supplied for the express purpose of illustrating *documents* produced using computer systems.

clipboard In computing, an area of memory reserved for the exchange of data between applications or between *documents* in a single application or within one document.

clipping The process of suppressing that part of a signal waveform which lies to one side of an amplitude boundary. In the example illustrated in *Figure C.14* the parts of the waveform which are more positive than the boundary are suppressed.

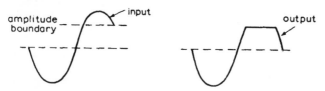

Figure C.14 Input and output waveforms illustrating clipping: positive peak clipping is illustrated

clock In *computers* and *data-processing equipment* the generator of periodic signals, known as clock pulses, which govern the timing of the *logic* and other operations carried out by the equipment.

clock speed The frequency of the *quartz crystal* that regulates the operation of a central processing unit. Although this is a widely quoted criterion when comparing the performance of computers, it is not the only factor affecting the speed at which computers execute tasks. In some applications memory access time or hard disk access time may be the limiting factor.

closed circuit A circuit which is not broken and therefore in which a current can flow. In broadcasting the term is used to describe a circuit used for rehearsal or recording of a programme which is not radiated at that time to listeners or viewers.

closed-loop control A servo system in which the controlled quantity is measured and compared with the desired value and any difference is fed back into the control system so as to minimise the difference.

clustered dither See *dithering*.

CMYK colour model See *colour model*.

C-network A network containing a single shunt element and two series elements, one in each leg as shown in *Figure C.15*. It can be regarded as a balanced form of the *L-network*.

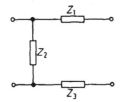

Figure C.15 A C-network

coaxial cable; coaxial line; concentric line A *transmission line* formed from two conductors, one a wire and the other a cylinder concentric with the wire, the space between them being filled with a *dielectric*. Cables of this type are extensively used for carrying RF signals at frequencies above approximately 10 MHz. Because the fields are confined within the outer conductor there is negligible loss by *radiation* from a coaxial cable. See *characteristic impedance*.

COBOL See *computer language*.

coding In general a method of representing information by means of characters suitable for the intended equipment or transmission system. For example in the Morse Code information can be expressed in normal language which can be spelled out, each letter of the alphabet being represented by a unique combination of dots and dashes. Thus A is ·— and B is —···. Such a code is suitable for signalling by flashing a light, sounding a buzzer or interrupting an RF carrier. For digital computers a binary coding system is generally used.

coercive field strength or **coercive force** The minimum reverse *magnetic field strength* which must be applied to a material to reduce the *magnetic flux density* to zero. It is represented by OC in *Figure C.16* and its value depends on the value of the magnetic flux density initially induced in the material.

coercivity The *coercive field strength* for a material which has been magnetically saturated.

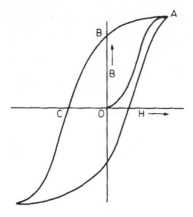

Figure C.16 A B–H hysteresis curve: CO represents the coercive field strength or coercive force

coherence The maintenance of a fixed frequency. For example a *laser* is regarded as a coherent oscillator. In interrupted *oscillations*, coherence implies the maintenance of phase continuity so that, at the beginning of each packet of oscillations the phase has precisely the value it would have had if the interruption had not taken place.

coincidence circuit or **coincidence gate** A *gate* which produces an output *pulse* when, and only when, all the inputs receive pulses simultaneously or within a prescribed time interval. See *logic gate*.

cold cathode A *cathode* which emits *electrons* without being heated. This definition strictly embraces photo-emissive cathodes but the term is usually reserved for the type of cathode used in some *gas-filled tubes*. Electron emission from the cathode arises as a result of bombardment by heavy positive *ions* released by *ionisation* of the gas and can be improved by coating the cathode with materials of low *work function*.

collector (1) In a *bipolar transistor*, that region into which *charge carriers* are swept from the base region. The carriers are *minority carriers* in the base region but *majority carriers* in the collector region. The collector-base junction is reverse-biased in normal transistor operation and hence the majority carriers are swept away from the junction and towards the collector terminal. (2) The term is also used to imply a collector of electrons, i.e. the anode in certain types of *electron tube*.

collector AC resistance See *electrode AC resistance*.

collimator A device for deflecting the paths of electrons or light rays in a non-parallel beam so as to produce a parallel beam.

colour See *hue* and *saturation*.

colour burst In colour television a *synchronising signal* consisting of a few cycles of chrominance subcarrier transmitted within the *back porch* of each line period and used to lock the frequency and phase of the *reference oscillator* in receivers.

colour cell Of a *colour picture tube* the smallest area of the screen which includes a complete set of primary colours. For example in a delta-array tube the triangle which includes one red, one green and one blue phosphor dot.

colour code A system of colours for indicating the values, tolerances and ratings of components or the connections to them. The most familiar colour code is that used on resistors. As shown in *Figure C.17* four bands of colour are used. The first (nearest the end of the resistor) indicates the first figure of the resistance value, the next band gives the second figure and the third indicates the multiplier, i.e. the number of noughts which follow the second figure. The fourth band indicates the tolerance. The code used is as shown in Tables C.1 and C.2. As an example a resistor in which the bands are coloured yellow, violet, orange and silver is of $47\,k\Omega$ resistance, tolerance ±10%. In an older system (also indicated in *Figure C.17*) the first figure of the resistance value was indicated by the body colour, the second figure by the end colour and the multiplier by a dot colour. A similar code is used for capacitors to indicate the *capacitance* in pF.

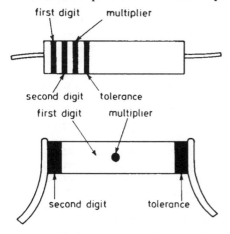

Figure C.17 Two methods of indicating the value of a resistor by means of a colour code

Table C.1
Resistance

0 black	5 green	
1 brown	6 blue	
2 red	7 purple	
3 orange	8 grey	
4 yellow	9 white	

Table C.2
Tolerance

± 1% brown
± 2% red
± 5% gold
± 10% silver
± 20% no fourth band

colour-difference signal In *compatible colour television* systems the signal obtained by subtracting the *luminance signal* from one of the three primary colour signals. Two colour-difference signals are used to modulate the colour *subcarrier, quadrature modulation* being used in the *NTSC* and *PAL* systems and *frequency modulation* in the *SECAM* system. At the receiver, after detection, the colour difference signals are added to the luminance signal to yield the red, green and blue colour signals required as inputs to the picture tube.

colour model A means of defining a colour mathematically. Colour models are used in television and in computer *graphics* and *desktop publishing*. The three colour models most regularly used are as follows.

The *RGB* (red/green/blue) model regards all colours as made up from the sum of the values of the three primary colours detected in human eyesight and used in the phosphor dots in a colour *cathode ray tube*. The values of the individual primaries are sometimes expressed as percentages and sometimes as numbers between 0 and 255 (see *true colour*). Thus white is 255,255,255 (all primaries turned fully on) and magenta is 255,0,255 (full red and full blue but no green).

The *HSV* (hue/saturation/value) model is based on the way in which the human brain interprets colour information. *Hue* represents the wavelength or perceived colour (such as green or violet). *Saturation* represents the concentration of the colour; if saturation is low the colour is pale, i.e. mixed with white. *Value* represents the overall level of the colour. All three figures are most commonly expressed as percentages, although hue is sometimes expressed as an angle (between 0 and 359°). Note that if value is 0, the colour is black irrespective of the hue or saturation figures and if value is 100 and saturation is 0, the colour is white irrespective of the hue.

The *CMYK* (cyan/magenta/yellow/key) model is based on the behaviour of the inks used in colour printing. Cyan, magenta and yellow allow a wide range of colours to be simulated by overprinting. The inks are assumed to be printed on a white background and the colours are subtracted. Thus cyan and yellow make green while magenta and yellow make the crimson sometimes called 'warm red'. Key is black and is used both to print truly black areas and to produce darker shades of other colours. Values are expressed either as pecentages or as numbers between 0 and 255.

colour television Television in which the pictures are reproduced in colours simulating those of the original scene. At the transmitting end an image of the original scene is analysed into its red, green and blue components and information about these components is contained in the *chrominance signal*. Modern colour TV systems are compatible and the transmitted signal is basically that of a black-and-white system to which the chrominance signal is added to give information about the *hue* and *saturation* of the colours in the scene. The chrominance information is conveyed by a colour subcarrier which, with its *sidebands*, is interleaved with the sidebands of the black-and-white (luminance) signal. A number of different subcarrier modulation systems are in current use. See *compatibility, chrominance, NTSC, PAL, SECAM*.

colour temperature Of a light source the absolute temperature of a *full radiator* (black body) which gives the same colour.

colour video signal or **composite colour signal** (US) In colour television the signal which gives complete colour-picture information and includes *synchronising signals* and *colour burst*.

Colpitts oscillator A sinusoidal oscillator in which the frequency-determining element is a parallel *LC* circuit connected between input and output terminals of an *active device*, *positive feedback* being obtained by connecting a tapping on the capacitive branch of the *LC* circuit to the common terminal (*cathode, emitter* or *source*) of the active device. The

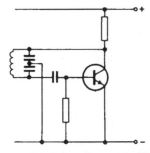

Figure C.18 A transistor Colpitts oscillator

effective tuning *capacitance* is that of the two capacitors connected in series, the common connection being used as tapping point. *Figure C.18* gives the circuit diagram of a transistor Colpitts oscillator. In VHF oscillators the inter-electrode capacitances of the active device can be used as the series-connected capacitors providing the positive feedback.

coma A defect in optical and electron lenses which causes the image of a round dot to develop a comet tail and so appear pear-shaped.

comb An assembly of discrete frequencies spaced at regular intervals within a given frequency band.

combination frequencies The frequencies generated when two sinusoidal signals at different frequencies are applied to a device with a non-linear characteristic. The combination frequencies are given by

$$mf_1 \pm nf_2$$

where f_1 and f_2 are the frequencies of the two input signals, m and n being the integers 1, 2, 3 etc.

combinational logic element Same as *logic gate*.

command An instruction given to a computer, either by typing it on the keyboard or by choosing it from a *menu*. A command normally consists of a *keyword* which may be followed by one or more *arguments* which specify how the command is to be executed. The command is interpreted and executed by the operating system or by an interactive language such as BASIC or by an application program.

command and control The process of controlling mobile personnel and equipment engaged in some project or mission, e.g. in the emergency services. Command and control equipment may integrate radio and telephone communications with computer-controlled displays to show, e.g. the present location and deployment of mobile units superimposed over maps.

command mode In *interactive computer languages* the state when no *program* is being executed and the computer will respond immediately to commands typed at the keyboard.

common-base circuit (also known as **grounded-base circuit** (US)) A *bipolar transistor* circuit in which the input signal is applied between *base* and *emitter*, and the output signal is derived from *collector* and base, the base terminal thus being common to input and output circuits. This arrangement gives a very low input resistance and a very high output

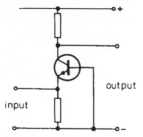

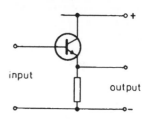

Figure C.19 Essential features of
the common-base circuit

Figure C.20 Essential features of
the common-collector circuit

resistance. Current gain is less than unity but voltage gain can be
considerable. The output signal is in phase with the input signal. The basic
form of the circuit is shown in *Figure C.19*. It is often used for VHF and
UHF amplification as part of a *cascode* circuit.

common-collector circuit A *bipolar transistor* circuit in which the input
signal is applied between *base* and *collector*, and the output signal is
derived from *emitter* and collector, the collector terminal thus being
common to input and output circuits.

This circuit arrangement has a high input resistance and a low output
resistance. Voltage gain is approximately unity but current gain can be
considerable. The output signal is in phase with the input signal. Because of
the unity voltage gain signals applied to the base emerge with negligible
loss from the emitter. For this reason the circuit is often known as an
emitter follower. The base form of the circuit is shown in *Figure C.20*. It is
used where equipment requires a high input resistance or low output
resistance and as a buffer stage.

common-emitter circuit (also known as **grounded-emitter circuit** (US)) A
bipolar transistor circuit in which the input signal is applied between base
and emitter, and the output signal is derived from the collector and emitter,
the emitter terminal being thus common to input and output circuits. This
form of connection gives a low input resistance and a high output resistance
(although the ratio is not so high as for the *common-base circuit*). Current
gain and voltage gain can be considerable. The output signal is in antiphase
with the input signal. This is the most widely used of the three basic
transistor circuits in analogue equipment probably because it is capable of
greater output power than the others. The basic form of the circuit is shown
in *Figure C.21*.

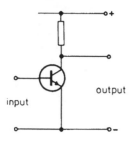

Figure C.21 Essential features of the
common-emitter circuit

common-mode rejection ratio For a differential amplifier a figure of merit which assesses its ability to reject the same signal applied to both input terminals. An ideal differential amplifier responds only to the difference between the input signals so that there should be zero output for equal input signals. The ratio is best measured by use of alternating input signals to avoid difficulties caused by *dc offset voltage*.

common-mode signal The algebraic average of the two signals applied to the inputs of a *differential amplifier*. If the two inputs are of equal amplitude and are accurately in antiphase, the common-mode signal amplitude is zero. If, on the other hand, the two inputs are in phase, then their common amplitude is also that of the common-mode signal. A perfect differential amplifier does not respond to common-mode signals but see *common-mode rejection ratio*.

common return A return *path* that is common to two or more components or circuits; in electronic circuitry one of the supply lines often provides a common return.

communications receiver A radio receiver designed to cope with difficult transmission conditions. Such receivers usually have several wavebands, facilities for controlling the bandwidth of the *IF amplifier*, a *BFO*, an efficient *a.g.c.* system and, most importantly, a good *signal-to-noise ratio*.

compact disc (CD and CD-ROM) A disk, most often 5 inches in diameter, carrying a digital recording in the form of a series of minute bumps on a spiral path. The bumps are of the order of 1 μm in diameter and the face of the disc accommodates about 20000 tracks. The disc is replayed by a low-power laser beam which is reflected back along its incident path in the absence of a bump but is scattered if it strikes a bump. Servos are used to enable the laser beam to follow the tracks accurately, to keep the beam in focus on the disc surface and to maintain constant linear speed of the disk as it passes the beam (a master oscillator being used to provide a reference standard). As a consequence of the constant track speed, the rotational speed of the disc decreases from about 500 rpm to 200 rpm as the laser beam moves from its innermost to its outermost position.

Compact discs are used to record mono and stereo sound and the single recorded side gives a maximum playing time of about 75 minutes of stereo sound. A system of PCM is used to record the two sound signals and other signals associated with the servo systems. Sound reproduction from compact discs is greatly superior to that from conventional long-playing disks in freedom from distortion, signal-to-noise ratio, dynamic range and stereo separation.

A further advantage of CDs over other recording media such as long-playing records and tape is that there is no physical contact between the rotating disk surface and the reading head. There is therefore no wear on the disk. Consequently, unless it meets with some catastrophic accident, a CD should sound the same after thousands of playings as when it was new.

Compact discs are also widely used for the mass distribution of computer *software*. Known as CD-ROMs, these discs have a capacity of over 600 megabytes, much larger than that of *floppy disks*, facilitating the distribution of very large collections of software. For major software houses it is considerably cheaper to mass produce their software on CD-

ROM than on floppy disks and this is sometimes reflected in software prices.

Other uses for compact disc include the storage of photographs (see *Photo-CD*) and video (see *video disk*).

companding The process of compressing a signal at one point in a communications system and of expanding at a later point to restore the original volume range. The process is used to improve the *signal-to-noise* ratio of the system and to be effective, the compression must take place before the point at which noise is introduced and expansion after that point. For example in a radio system the compressor is at the transmitter and the expander at the receiver, noise being introduced in the radio link. See *compression, expansion.*

comparator A circuit which compares two signals and gives an output indicating the result of the comparison.

comparator element Same as *equivalence element.*

compatibility (1) Of a colour television system that property which permits the colour signal to be reproduced by a normal black-and-white television receiver as a satisfactory black-and-white picture. (2) Of a stereophonic FM broadcasting system that property which permits the stereo signal to be reproduced by a normal monophonic receiver as a satisfactory monophonic signal.

compensation theorem If a *linear network* is modified by introducing an increment ΔZ in the impedance of one of the branches, the resulting current increment at any point in the network is equal to the current that would be produced at that point by a compensating voltage of $-I\Delta Z$ acting in series with the modified branch. I is the current in this branch before modification.

compilation See *computer language.*

complementary metal-oxide-semiconductor (CMOS or COSMOS) A form of construction of *logic monolithic integrated circuits* using complementary *insulated-gate field-effect transistors* in pairs. The logic inverter stage illustrated in *Figure C.22* is an example of a CMOS circuit.

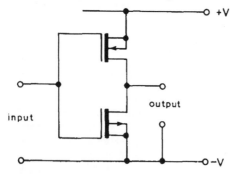

Figure C.22 A CMOS logic inverter stage

complementary transistor logic (CTL) Logic circuits which make use of pnp and npn *transistors.*

64

complementary transistors pnp and npn *transistors* with similar character-istics. A positive signal applied to the base of an npn transistor increases the collector current but the same signal applied to the base of a pnp transistor decreases the collector current. Thus if the same signal is applied to the bases of a complementary pair of transistors, the transistors operate in *push-pull* without need for a phase-splitting circuit.

complex instruction set computer (CISC) A *microprocessor* having traditional *architecture*, capable of executing a wide repertoire of commands. The processors used in IBM PC-compatible computers are of this type. The alternative is a *reduced instruction set computer* (RISC), which has a limited repertoire of commands, but is faster.

compliance In a mechanical or acoustical oscillating system a property of a spring or similar component which enables it to oppose a change in applied force and is measured by the displacement caused by unit force. It is thus the reciprocal of stiffness. Compliance is analogous to *capacitance* in electrical oscillating circuits. A spring can store mechanical energy when it is compressed just as a *capacitor* can store electrical energy when it is charged.

composite colour signal (US) See *synchronising signal*.

composite signal (US) Same as *video signal*.

compound connection Same as *Darlington pair*.

compression (1) Reduction of the volume range of a signal by varying the *gain* to which it is subjected according to the mean volume level, the gain being greater for low signal volumes than for high signal volumes. Compression permits the transmission of higher mean volume levels thus improving *signal-to-noise ratio*. (2) In computing and data processing, the reduction of the size of *files* and therefore of their storage space requirement or transmission *bandwidth*. Various compression *algorithms* are used, according to the type of data in the file. *Real-time* compression and decompression processing is sometimes built into filing systems, greatly enhancing their effective capacity.

compression ratio The ratio, usually stated in decibels, of the gain at a reference signal level to the gain at a higher stated level. See *compression*.

computer An equipment which accepts information, carries out specified mathematical and/or logical operations upon it and supplies the required answer. Computers are sometimes called data-processing equipment. The information input must be in a form which the computer can accept and there are two principal types of computer, the analogue and the digital. See *analogue computer*, *digital computer*.

computer-aided design (CAD) The use of a computer, particularly with *graphics*, to assist in the design of engineering, structural or architectural projects. As an example of the *interactive* design possible by this means, three-dimensional representations of car bodies, building frameworks or printed-wiring-board layouts can be created and modified by the use of *lightpens* and *mice*.

computer-aided engineering (CAE) Generic term embracing *computer-aided design* and *computer-aided manufacture*.

computer-aided manufacture (CAM) Production of goods with the assistance of a computer or numerically-controlled machinery.

computer-aided software engineering (CASE) A technique used in some branches of computing, especially *databasing* and business modelling, in which the *program* is generated by a computer. The problem to be solved by the new program is defined using CASE tools which may employ a *graphical user interface* to indicate relationships between *repository* members and outside events. CASE is a much faster and more powerful technique than conventional programming for certain kinds of *application*.

computer control Operation in which the variables controlling a process are applied to a computer which carries out the required mathematical and/or logical operations on the inputs, the output of the computer being used to control the process.

computer language A combination of symbols, words and statements governed by a syntax and used to construct a *computer program*. Ultimately the central processing unit only understands a limited range of instructions—*machine code* or *machine language*—which it reads in binary form. The earliest method of programming involved writing machine code directly. This was a cumbersome task since it required an intimate knowledge of the processor and provided no assistance in error finding.

Today some *applications* and *subroutines*, especially those that involve intensive arithmetic, are still programmed in machine code because of its speed of execution. Instead of writing the program directly in machine code, however, a programming language called an *assembler* is used. This uses as source code a sequence of *mnemonics* which represent the processor instructions; these are easier to understand than the binary numeric codes themselves. The source code is written in a word processor or text editor and when finished is converted to executable machine code in a process called *assembly*. Because assembler source code inevitably is closely related to the machine code to which it is converted, an assembler is regarded as a *low-level* computer language.

For many applications computer programs are written in *high-level languages* such as C and BASIC in which the source code bears no direct relationship to the machine code that is ultimately executed. High-level languages use *keywords* to represent common operations such as displaying output on the VDU or searching text for a certain character sequence. They also allow the programmer to define *subroutines* or procedures for frequently used sequences of operations and these can be called as required from any point in the program. In most high-level languages the completed source code must be converted to executable machine code in a process called *compilation* before it can be executed. Some high-level languages, however, do not require compilation. Instead the source code itself is executed in *real time* by a program called an *interpreter*. Inevitably execution is slower than in a compiled language, but interpreted languages are interactive and easy to use; moreover run-time errors usually generate informative error messages such as 'Variable not found at line 360'.

Some of the computer languages in common use are:

C and C++ A general-purpose compiled language using a rigid syntax and structure. One benefit is the comparative ease with which one set of source code can be differently compiled for use on computers having widely differing central processing units and operating systems.

BASIC (Beginner's All-purpose Symbolic Instruction Code). English keywords and simple syntax make this one of the most widely-used languages, especially in microcomputers where it is generally interpreted rather than compiled.

Algol in which the instructions are presented in the form of algorithms.

FORTRAN (FORmula TRANslation) which is mainly used for instructions designed to solve mathematical or logic problems.

COBOL (COmmon Business-Oriented Language) intended for business and commercial use.

Forth used primarily for process control and scientific applications, is unusual in being a compiling language that is interactive. Forth consists of a dictionary of keywords and a compiler; applications are implemented by repeatedly defining new keywords in terms of the original or previously-defined keywords until one keyword is defined which embraces the entire application.

Pascal uses BASIC-like keywords in a very rigid and logical syntax. Because its use fosters an orderly and systematic approach to programming it is much used in education although originally developed for scientific applications.

Occam has been specially developed for use with *transputers*.

computer terminal An item of equipment at which data can be fed into a (possibly-remote) computer or can be received from it. The data may be fed in via a keyboard and may be received in alphanumeric or graphical form.

concentric line Same as *coaxial cable*.

concertina phase-splitter Circuit in which an active device is located at the midpoint of the load resistor so that equal and opposite signal voltages are developed at the *anode* (collector or drain) and *cathode* (emitter or source). *Figure C.22* shows a pnp transistor used in such a circuit.

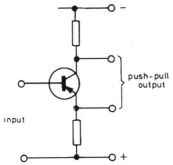

Figure C.23 Simplified concertina phase-splitter using a pnp transistor

condenser An obsolete term for *capacitor*.

condenser microphone See *electrostatic microphone*.

conditional implication gate Same as *IF-THEN gate*.

conductance (G) The reciprocal of resistance and the real component of admittance. For a purely-resistive AC circuit, the voltage and current are in phase and the conductance is the ratio of current to voltage.

67

conduction band In an *energy-level diagram* the energy range of electrons which have broken free of the parent atom and can thus act as *charge carriers*, i.e. they can move freely under the influence of an electric field.

conduction cooling Of a device or equipment dissipating considerable power a method of removing heat by use of a *heat sink* in good thermal contact with the device. The heat sink may be a finned radiator or a prepared surface on a panel.

conductivity In general a term expressing the ability of a material to carry an electric current. More specifically conductivity is the reciprocal of *resistivity*. The conductivity of metals, for example, is better, i.e. lower than that of non-metals.

conductor Any element in a component or a *circuit* providing a path for the passage of electric current, e.g. the strips of copper on a *printed circuit* board.

conjugate impedances Two *impedances* with equal values of resistive component and in which the reactive components are equal but of opposite sign.

constant-amplitude recording In disk recording a characteristic for which constant-amplitude sinusoidal signals applied to the recording head give constant recorded amplitudes. See *constant-velocity recording*.

constant-current characteristics Graphical relationship between the voltages at two electrodes of an *electron tube* necessary to maintain the current at one of them constant. The term is generally applied to the V_a-V_g characteristics of an *electron tube* for constant I_a which are used in assessing the performance of *high-power triodes* as output stages in transmitters.

constant-current modulation Same as *Heising modulation*.

constant-*k* filter A ladder network composed of *reactances*, the series reactances being of opposite sign to the shunt reactances. Because positive (i.e. inductive) reactance is directly proportional to frequency and negative (i.e. capacitive) reactance is inversely proportional to frequency, the product of the series and shunt reactances is independent of frequency. The constancy of this product has given this type of filter its name.

Constant-*k* filters may be low-pass, high-pass, band-pass, etc. types but the steepness of fall-off in response at the edges of the passbands is not good and the *characteristic impedance* is not constant over the passband so that a purely-resistive termination is not satisfactory. Both disadvantages can be overcome by the inclusion of m-derived sections or half-sections. See *m-derivation*.

constant luminance system A colour TV system in which the brightness of the reproduced picture depends solely on the transmitted *luminance signal* and is unaffected by the *chrominance signal* transmitted with it. This is an ideal which is not perfectly achieved in the *NTSC*, *PAL* and *SECAM* colour television systems currently in use.

constant-resistance network A network which, when terminated in a pure *resistance* of value R, gives an input resistance of R which is independent of frequency. Attenuators and equalisers are usually designed as constant-resistance networks so that they can be inserted in circuits without affecting resistance terminations. An example of a constant-resistance network is the

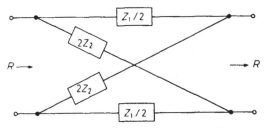

Figure C.24 A lattice network in which $Z_1Z_2 = R^2$ is an example of a constant-resistance network

lattice network of *Figure C.24* in which Z_1Z_2 must equal R^2. Z_1 and Z_2 must hence be inverse impedances to give the desired constant-resistance property.

constant-velocity recording In disk recording a characteristic for which constant-amplitude sinusoidal signals applied to the recording head give constant recorded velocities, i.e. the recorded amplitude is inversely proportional to frequency. In practice such a characteristic cannot be used over the whole of the audio spectrum because of the possibility of exceeding the groove spacing at low frequencies and giving inadequate *signal-to-noise ratio* at high frequencies. Moreover the chosen characteristic must take into account the fact that sound amplitudes are smaller at high frequencies than at low. Practical recording characteristics tend therefore to be a compromise between constant amplitude and constant velocity. See *constant-amplitude recording*.

contact The conductive areas which touch to establish *continuity* in a switch or a relay.

continuity A continuous low-resistance electrical path between two points.

continuous wave (CW) Radio waves of constant amplitude and frequency. The transmission and reception of such a wave establishes a link between transmitter and receiver but to enable information to be sent over the link the continuous wave (carrier wave) must be keyed on and off in accordance with a code (e.g. Morse) or must be modified in amplitude, phase or frequency, i.e. must he modulated. See *interrupted continuous wave*, *modulation*.

contrast (1) In sound transmission the *dynamic range*, i.e. the ratio of the loudness of the loudest to the quietest passages in a programme. This range must be compressed before a sound programme can be recorded or transmitted. Methods of expansion are available to re-establish the original dynamic range. (2) In TV, the ratio of the brightness levels of two parts of a displayed picture. It is determined by the amplitude of the *video* signal input to the picture tube and hence by the gain of the preceding amplifiers.

contrast gradient Same as *gamma*.

control code An *ASCII* character in the range 0 to 31. Such characters represent instructions to perform some operation on the output device rather than printable characters. For example code 7 ('BEL') emits a beep and code 9 ('TAB') moves the insertion point to the next tab stop. From a keyboard these codes may be issued by using the **Ctrl** key in conjunction with alphabetic and other keys.

control grid An *electrode* situated between the *cathode* and the other electrodes of an *electron tube*, the potential of which determines the magnitude of the electron current flowing from the cathode to the other electrodes. The control grid is usually situated very close to the cathode and is constructed in the form of a wire spiral the pitch of which can be adjusted during manufacture to give the required degree of control over the density of the electron stream. Thus the pitch of the grid determines the *mutual conductance* of the tube. In a *cathode ray tube* the control grid (sometimes known as the modulator) is often in the form of a disk containing a small aperture through which the electron beam passes.

control impedance Same as *unilateral impedance*.

controller In *computers* and *data-processing equipment* an item of *hardware* which controls the operations of *peripheral units* according to instructions from the central processor.

controlled-carrier modulation Same as *floating-carrier modulation*.

controlled silicon rectifier (CSR) Same as *thyristor*.

control ratio In a *thyratron* the slope of the characteristic relating the anode voltage with the critical grid voltage. *Figure C.25* shows that the characteristic has a long linear section showing that the control ratio is constant over a wide range of operating voltages.

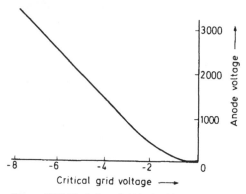

Figure C.25 Variation of critical grid voltage with anode voltage for a thyratron

control unit Of a *digital computer* that section which is responsible for interpreting and acting upon the input instructions and for applying signals to the *arithmetic unit* and other sections of the computer to enable it to carry out the allotted task.

convection cooling Of a device or equipment dissipating considerable power a method of removing heat which relies on natural convection and radiation from the device. An adequate clear space must be provided around the device for this method to be successful.

convergence In a colour picture tube the adjustments which ensure that the three primary colour images are correctly superposed.

conversion conductance (g_c) or conversion transconductance (US) The ratio of a small change in the wanted-signal component in the output current of a *frequency changer* to the small change in signal-frequency input voltage which gives rise to it under specified operating conditions.

converter General term for an equipment which changes: (1) The nature of a current, e.g. an AC to DC converter (a device changing DC to AC is usually known as an *inverter*). (2) The voltage of a power supply, e.g. one converting a 12 V DC supply into 240V DC. (3) The frequency of a signal, e.g. the *frequency changer* in a *superheterodyne receiver*. (4) The form in which information is coded, e.g. a television standards converter which converts a 525-line signal into a 625-line version of the same picture.

coprocessor An additional *processor* in a *computer* designed to carry out specific functions, such as mathematical operations or handling graphical output, relieving the burden on the main processor.

cordless telephone A telephone in which the link between the handset and the base is by *radio*, both containing *transceivers*. The advantage is that the user may roam freely within the unit's range during a telephone conversation. The handset is powered by an internal rechargeable battery which is recharged when it is plugged into the base unit; this requires a mains supply.

corona discharge Visible discharge of electricity surrounding a charged conductor when its potential exceeds the critical value at which the surrounding air is ionised. The discharge is often accompanied by a hissing noise and becomes more marked as the radius of curvature of the conductor is reduced. Sharp points, therefore, give pronounced corona discharge and to minimise the effect the surface of the conductor must be smooth.

counterpoise Same as *artificial earth.*

counting The process of deriving one output pulse for every n input pulses. If $n = 2$ the device is a *binary counter* and if $n = 10$ it is a *decade* counter. If the input pulses are at regular intervals the device is more properly known a *frequency divider*. Usually a counter is a circuit which can take up a number of different states and which changes from one state to another in a particular sequence each time an input pulse is received.

coupling Any means whereby power can be transferred from one circuit to another. The circuits may be physically separated, power being

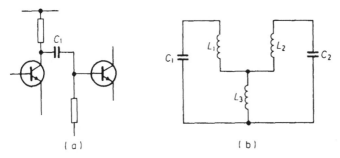

Figure C.26 Examples of coupling by (a) series capacitance and (b) shunt inductance

71

transferred between them by magnetic or capacitive means, or they may be connected, power being transferred via components common to both circuits. The common component can be in series as C_1 in *Figure C.26* (a) or in shunt as L_3 in *Figure C.26* (b).

coupling coefficient (*k*) A quantitative measure of the degree of coupling between two circuits. It is equal to the ratio of the ***mutual impedance*** (resistive, inductive or capacitive) between the circuits to the geometric mean of the two impedances of like kind in the two circuits. The maximum possible value of *k* is unity and values used in practice are often as low as 0.01.

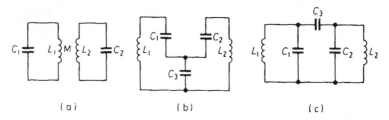

(a)　　　　　　　(b)　　　　　　　(c)

Figure C.27 Examples of coupling by (a) mutual inductance, (b) shunt capacitance and (c) series capacitance. The coupling coefficient for each circuit is given in the text

If two tuned circuits L_1C_1 and L_2C_2 are coupled by mutual inductance M as shown in *Figure C.27* (a) the coefficient of coupling is given by

$$k = \frac{M}{\sqrt{(L_1L_2)}}$$

and if the two circuits are identical $L_1 = L_2 = L$, so that

$$k = \frac{M}{L}$$

For shunt-capacitance coupling shown in *Figure C.27* (b) the coefficient of coupling is given by

$$k = \frac{\sqrt{(C_1C_2)}}{C_3}$$

in which C_1 and C_2 are assumed small compared with C_3. If $C_1 = C_2 = C$

$$k = \frac{C}{C_3}$$

For series-capacitance coupling shown in *Figure C.27* (c) provided C_1 and C_2 are large compared with C_3, the coefficient of coupling is given by

$$k = \frac{C_3}{\sqrt{(C_1C_2)}}$$

approximately and if $C_1 = C_2 = C$

$$k = \frac{C_3}{C}$$

The examples given are of intentional power transfer, but coupling is sometimes unwanted because it can give rise to *instability* or *oscillation*. To avoid such coupling it may be necessary to employ *electric* or *magnetic screening* or *decoupling*.

covalent bonds   The bonds between two similar atoms which share two *electrons*, one from each atom. This occurs in the tetravalent atoms of *germanium* and *silicon* which therefore behave in some respects as though they had eight electrons in the outermost shell. This is sufficient to fill the shell which accounts for the very low electrical conductivity of pure germanium and silicon.

CPS emitron tube   Same as *orthicon*.

crash   In computers and data processing equipment, an *error* which results in total loss of control of the system. A common cause is that the processor begins executing an endless loop of instructions to which the user has no input. To the user, the system often appears to 'freeze'. His only recourse is to reset and restart the system. Any processed data that had not been stored at the time of the crash is usually lost irretrievably.

critical coupling   The degree of coupling between two circuits tuned to the same frequency which permits maximum transfer of energy from one to the other at the *resonance frequency*. For degrees of coupling smaller than the critical value the combination has a single-peaked response curve and for degrees of coupling greater than critical the *frequency response* has two peaks equidistant from the resonance frequency, the peak separation

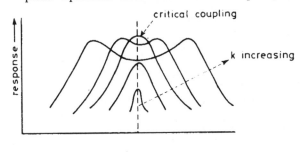

Figure C.28 Response curves for two identical tuned circuits showing the effect of varying the degree of coupling

increasing with increase in coupling as shown in *Figure C.28*. Coupled circuits giving a two-peaked response, with the peaks near to the resonance frequency, are often employed as bandpass filters in RF and IF amplifiers in a receiver. The critical value of the coupling coefficient is given by

$$k = \frac{1}{\sqrt{(Q_1 Q_2)}}$$

73

where Q_1 and Q_2 are the Q values of the two tuned circuits. If the circuits are identical and $Q_1 = Q_2 = Q$

$$k = \frac{1}{Q}$$

Thus if $Q = 100$, a common value, the critical value of k is 0.01.

critical damping The condition where the rate of loss of energy is just sufficient to prevent free oscillation of an oscillating system. For a resonant electrical circuit the value of series resistance which gives critical damping is equal to $2\sqrt{(L/C)}$. As an example of a mechanical system the friction in the moving parts of a measuring instrument is commonly adjusted to give critical damping so that the pointer takes up its final reading very quickly and without overshoot (which is the start of *oscillation*).

critical grid current In a *thyratron* the value of the grid current corresponding to the *critical grid voltage*.

critical grid voltage In a *thyratron* with a particular value of anode voltage the minimum value of grid voltage at which anode current begins to flow. The value of the critical grid voltage depends on the anode voltage, becoming less negative as the anode voltage is reduced as shown in *Figure C.25*.

critical-space tetrode A *tetrode* in which the spacing between *screen grid* and *anode* is large so that the negative charge of the electron stream in it at any instant is sufficient to act as a *suppressor grid* and so prevent secondary electrons released from the anode from reaching the screen grid. In this way the *tetrode kink* can be eliminated so improving the ability of the tube to handle large output voltage swings.

cross coupling Unwanted *coupling* between two signal paths which can result in *crosstalk*.

cross modulation Transfer of the *modulation* of one signal to another. In AM reception this can be caused by non-linearity in RF stages. As a result the carrier of a wanted signal becomes modulated by the unwanted modulation of another signal.

crossover distortion In a class-B amplifier, waveform distortion caused by too low a value of standing current. It takes the form of a discontinuity at the datum level of a sinusoidal signal as shown in *Figure C.29* caused by the slight delay which occurs whilst each half of the output stage takes over from the other. See *class-B operation*.

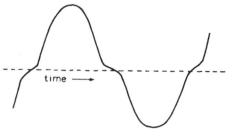

time $\longrightarrow$

Figure C.29 Crossover distortion

crossover frequency (1) In *dividing networks* the frequency at which equal power is delivered to each of the adjacent frequency channels (assumed correctly terminated). It is the frequency, in fact, at which the signal crosses over from one channel to the other. In two-unit *loudspeakers* the crossover frequency is usually between 500 Hz and 2 kHz. (2) See *turnover frequency*.

crossover network Same as *dividing network*.

crossover voltage In *cathode ray tubes* the voltage of a secondary-emitting surface at which the *secondary emission ratio* is unity.

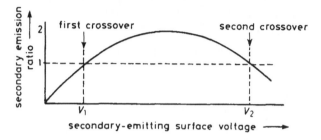

Figure C.30 Variation of secondary emission ratio with surface voltage

There are generally two such voltages as shown at V_1 and V_2 in *Figure C.30*. From *Figure C.30* it is possible to predict the voltage at which the target of a TV camera tube or the screen of a CRT will stabilise. For example if the initial target potential is below V_1 the secondary emission ratio is less than unity and the number of *electrons* striking the target exceeds those lost from it. Thus the target potential is driven negative until it stabilises at a potential near that of the *electron-gun cathode*. This is the type of target stabilisation used in all low-velocity camera tubes.

If, however, the initial target potential is between V_1 and V_2 and if the final anode potential of the electron gun is also between these two voltages then the secondary emission ratio is greater than one and the target releases more electrons than it gains from the *electron beam*. If the target potential is initially below that of the final anode all the secondary electrons released are collected by the final anode and the target potential rises until it is approximately equal to that of the final anode. If the target potential is initially above that of the final anode the retarding field between target and anode returns the secondary electrons to the target and the target potential falls until again it is approximately equal to the final anode potential. This is the type of stabilisation which occurs in high-velocity TV camera tubes and in *cathode ray tubes*.

crosstalk Unwanted signals appearing in one signal path as a result of *coupling* with other signal paths. An example is the *induction* of unwanted speech signals into one pair of conductors from neighbouring pairs in a telephone cable.

crowbar circuit Protective circuit which, in the event of a fault that could damage equipment by excessive voltage or excessive current, applies a short circuit across the supply to the equipment, causing the *fuses* to blow or a cut-out to operate thus isolating the equipment from the supply. The short-circuiting device must have very low resistance when triggered into conduction by the fault-detecting circuit and may be, for example, a *thyratron* or a *thyristor*.

cryogenics The study of the properties of devices at temperatures near absolute zero (–273°C or 0°K). See *superconductivity*.

cryotron (cryogenic store) A *store* which makes use of the ability of a magnetic field to control the transition from the normal to the superconductive state of resistive elements.

crystal See *piezo-electric crystal*.

crystal bimorph See *bimorph*.

crystal-controlled oscillator See *crystal oscillator*.

crystal diode A *semiconductor* crystal in contact with a metal probe (see *point-contact diode*) or two different types of semiconductor crystal in contact (which may be regarded as a crude form of *junction diode*). These were used for detection in the early days of broadcasting and successful results depended on finding and retaining sensitive points on the crystal surfaces. See *cat's whisker*, *crystal receiver*.

crystal filter A filter network containing one or more *piezo-electric crystals* as frequency-determining elements. The resonant properties of the crystals are used when the filter is required to accept or reject a narrow band of frequencies.

crystal growing See *crystal pulling*.

crystal loudspeaker A *loudspeaker* which relies for its action on the *piezo-electric effect*. The AF signal is applied to a crystal *bimorph* and the resulting vibrations of the bimorph are mechanically coupled to the diaphragm. Such loudspeakers generally operate over a limited frequency range at the upper end of the AF spectrum and form one unit (the tweeter) of a multi-unit loudspeaker. The graphical symbol for a crystal loudspeaker is given in *Figure C.31* (a).

(a)   (b)   *Figure C.31* Graphical symbols for (a) crystal loudspeaker and (b) crystal microphone

crystal microphone A microphone which relies for its action on the *piezo-electric effect*. In one type the sound waves strike a diaphragm and the resulting mechanical vibrations are coupled to a crystal *bimorph*. The voltages developed across the faces of the bimorph constitute the microphone output. In another type of crystal microphone there is no diaphragm and the sound waves impinge directly on a *sound cell* which consists of two crystal bimorphs mounted back-to-back. The graphical symbol for a crystal microphone is given in *Figure C.31* (b).

crystal oscillator An *oscillator* the frequency of which is controlled by a *piezo-electric crystal* usually of quartz. Such oscillators have great stability and are used as frequency standards, in signal generators and as master oscillators in radio transmitters.

crystal oven A thermostatically-controlled oven containing a *crystal-controlled oscillator*. By maintaining it at a constant temperature, a crystal oscillator can give very high *frequency stability*. Such ovens have been used in broadcasting transmitting stations to achieve the required stability of *carrier* frequency.

crystal pickup A disk-reproducing head which relies for its action on the *piezo-electric effect*. The movement of the stylus point caused by the groove modulation is coupled to a crystal *bimorph* and the voltage developed between its opposite faces constitutes the pickup output. A stereo pickup includes a mechanical device for resolving the movement of the stylus into two directions at right angles to each other, each movement driving a separate bimorph. The symbol for a crystal pickup is given in *Figure C.32*.

crystal pulling A method of producing large single crystals such as are required in the manufacture of *semiconductor* devices by slowly withdrawing a developing crystal from the molten semiconductor material.

Figure C.32 Graphical symbol for a crystal pickup

crystal receiver (1) A radio receiver comprising a *tuned circuit*, a crystal detector and a pair of *headphones*. A typical circuit diagram for such a simple receiver is given in *Figure C.33*. Receivers of this type were

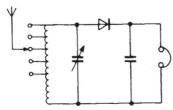

Figure C.33 Circuit diagram of crystal receiver of the 1920s

commonly used in the 1920s, before *electron tubes* were available, for the reception of broadcast programmes. They required no batteries or mains supply, the AF output depending on the magnitude of the signal picked up by the receiving antenna. A high outdoor *antenna* was therefore desirable for good results. (2) A *waveguide* incorporating a *crystal diode* and used for the detection of received radio signals.

crystal set Same as *crystal receiver* (1).

crystal triode Early name for *transistor*.

cumulative grid detector Same as *grid-leak detector*.

current A unidirectional movement of *charge carriers*. In a conductor the current is a drift of *electrons* towards the positive pole of the applied electric field. In a semiconductor, in addition to the electrons moving to the positive pole, current is also carried by *holes* moving in the opposite direction. In vacuum electron tubes current is again carried by electrons attracted to the anode: such currents are sometimes called *convection currents*. In a gas-filled tube *ionisation* causes the medium to divide into positive and negative *ions* which move towards the cathode and anode respectively. A similar process occurs in electrolytes as a result of *dissociation*. There is, however, no division into charged particles in the *displacement current*, a hypothetical concept introduced to explain the passage of *alternating current* through the dielectric of a capacitor.

current amplification factor Of a *bipolar transistor* the ratio of a small change in output current to the change in input current which gives rise to it. In a *common-base circuit* the current amplification factor relates changes in collector current with changes in emitter current: this is represented by α or h_{fb} and is necessarily less than unity. In a *common-emitter circuit* the current amplification factor relates changes in collector current with changes in base current. This is represented by β or h_{fe} and practical values may lie between 10 and 500. α and β are related thus

$$\beta = \frac{\alpha}{1 - \alpha}$$

current amplifier A circuit incorporating one or more *active devices* and designed to amplify current waveforms. The term distinguishes such amplifiers from those designed to amplify voltage waveforms or to deliver power to a load. It is the ratio of source *impedance* to input impedance and of output impedance to load impedance which determines whether a device or amplifier is best regarded as a current amplifier, *voltage amplifier* or *power amplifier*. If both ratios are large compared with unity the signal transferred from source to input and output to load is best regarded as a current waveform. The associated signal voltage waveform in a current amplifier is of minor interest and is usually of small amplitude. Because bipolar transistors have a low input impedance and a high output impedance they are usually regarded as current amplifiers.

current dumping An amplifying technique in which a *class-A stage* feeds the output load at low signal levels, a *class-B stage* taking over progressively as signal level increases, an overall negative feedback system ensuring linearity. The class-B stage is regarded as a current dumper.

current feedback A *negative feedback* system in which the feedback signal is directly proportional to the current in the load. The feedback signal may be derived from a *resistor* in series with the load as shown in *Figure C.34*, the resistor being small compared with the value of the load. A simple way of applying current feedback to an *active device* is to omit the decoupling capacitor which is normally connected across the cathode or emitter bias resistor. The effect of current feedback is to improve linearity, decrease *gain* and to increase the effective output resistance of the active device

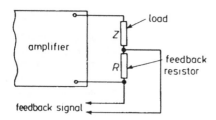

Figure C.34 Basic principle of current feedback

driving the load, i.e. it tends to make this device a constant-current source. See *voltage feedback*.

current mirror A circuit arrangement in which a current in one of a pair of matched active devices causes an equal current in the other. In the basic form of the circuit shown in *Figure C.35* the current I_{c1} in TR_1 is determined chiefly by the supply voltage and the value of R_1. This sets up a particular value of base-emitter voltage in TR_1 (dependent on the type of transistor and its temperature) which is directly communicated to the matching transistor TR_2 causing an equal collector current I_{c2} in R_2. If TR_2 is

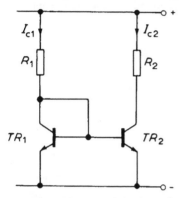

Figure C.35 Basic current-mirror circuit

replaced by a number of transistors in parallel their total collector current is equal to I_{c1}. This technique is used in integrated circuits to adjust and stabilise the currents in the individual transistors.

current mode logic (CML) A form of logic giving high switching speed and good immunity from noise, particularly from disturbances on the supply line. It employs *bipolar transistors* arranged in balanced pairs and fed with complementary input signals. The supply to each pair is via a series transistor arranged to maintain a constant current.

cursor The symbol, often a flashing character-sized block, that shows where on a VDU the next character to be typed will appear.

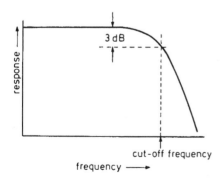

Figure C.36 Response curve for a low-pass filter

79

cut-off attenuator   A length of *waveguide* used at a frequency below its *cut-off frequency* to introduce non-dissipative attenuation which can be controlled by adjusting the length of the waveguide.

cut-off frequency   (1) Of a filter the upper or lower limit of a passband. Outside the passband the attenuation should ideally increase rapidly and the cut-off frequency is sometimes defined as the frequency at which the attenuation is 3 dB as shown for a lowpass filter in *Figure C.36*. (2) Of a *waveguide* for a given mode of transmission the frequency below which transmission of RF signals along the guide is not possible. The cut-off frequency depends on the mode and on the cross-sectional dimensions of the waveguide.

cycle   An interval of space or time in which a periodically varying quantity completes one set of changes of the variable. For a *sinusoid* a cycle contains one complete set of positive values followed by an equal set of negative values.

cyclic code   Same as *Gray code*.

cyclic store   A *store* to which access is available only at fixed intervals during a cycle, e.g. a *magnetic drum*.

cylinder   The position of one track on a *floppy* or *hard disk*. The term originates from its use on hard drives consisting of a number of disks on a common spindle. Each disk contains two tracks at that position, one on each side. Consequently that position, repeated for each disk on the spindle, could be considered as constituting a hypothetical cylinder. A cylinder-shaped symbol, as shown in *Figure C.37*, is widely used in the computer industry to represent *mass storage*.

Figure C.37   The cylinder symbol used widely in the computer industry to represent mass storage

D

D-type connector A standard for nine-, 15- and 25-pin connectors that support RS-232-C, RS-432 and RS-422 communications. The connector is so named because of its shape.

daisy chaining A method of linking *peripheral* devices in series so that only one communications *port* is used on the host computer. A *SCSI* port, for example, can support up to seven daisy-chained devices, signals being passed from device to device until they reach their destination. See *Small Computer Systems Interface (SCSI)*.

daisy wheel *See printer*.

damping That property of an electrical, mechanical or acoustical resonant circuit which causes a progressive fall in the amplitude of free *oscillations*. Damping is caused by resistance in electrical circuits and by friction and viscosity in mechanical systems. In electronic circuitry resonant circuits are often deliberately damped to increase the *bandwidth* by the addition of *resistance*.

damping coefficient (or **factor**) (1) A measure of the rate of decrease of *amplitude* in a damped *oscillation*. It is equal to the ratio of one amplitude to the next in the same direction. It is also equal to the *logarithmic decrement* divided by the periodic time of the oscillation. In an electrical resonant circuit the damping coefficient is related to the circuit constants, being given by $R/2L$ where R is the series resistance in the circuit and L is the inductance. (2) Of an AF *power amplifier* the ratio of the load resistance to the output resistance of the amplifier.

dark current The current which flows in a *photocell*, *photo-transistor* or TV *camera tube* when there is no light incident on the photo-sensitive surface. Ideally this current should be zero but in practice there is usually a finite current.

dark-trace tube A *cathode ray tube* with a screen composed of potassium chloride which darkens when struck by the electron beam. It is used in *radars*.

Darlington pair An amplifier stage using two *bipolar transistors* in which the base of the second transistor is directly connected to the emitter of the

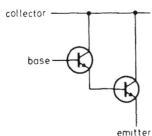

Figure D.1 A Darlington pair

first, the two collectors being commoned as shown in *Figure D.1*. The combination can be regarded as a single transistor with a gain equal to the product of the gains of the individual transistors.

data In electronics and computing, a *signal*, often in *digital* format, that relates to something in the outside world. The information technology revolution that has blossomed since the early 1980s has been based on the fact that electronic equipment can perform all manner of useful operations on data that represent entities in the real world, quickly and accurately. The word *data* is, strictly speaking, a plural noun, the singular being *datum*, from the Latin for 'a given thing'.

databank Same as *database*.

database A computer application that sorts, stores and retrieves data, sometimes in very large quantities. For example, a doctor might use a database to store the personal details and medical histories of thousands of patients. A bank might use a database to store details of the accounts of millions of customers. For retrieval data may be selected using combinations of specific factors. The doctor could request a list of all his female patients between certain age limits who are due for a certain medical check-up. The bank could retrieve details of all its customers in a certain region whose accounts exceed a certain sum. The information retrieved could be fed into a *mailmerge* application for incorporation into a special mailing to each person selected.

data compression See *compression*.

data dictionary See *repository*.

data processing equipment Automatic machines which carry out operations on items of data in order to obtain wanted information. The operations may include sorting, arranging, relating and interpreting the data, and the carrying out of calculations, and is generally achieved today by a *digital computer*.

data transmission The automatic transfer of data between two points, usually a computer and a data terminal or between two computers. The transfer is usually by a telegraph, telephone or radio link.

data warehouse A computer application that provides access to data in several different *databases* in an integrated manner.

D bistable A *bistable* circuit with an input (*D*), the logic state of which is faithfully copied at the output on the application of a clock pulse to a separate input. However the output can be forced to logic 1 or 0 by the clear and reset inputs. The graphical symbol for a D bistable is given in *Figure D.2*.

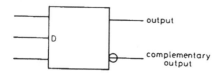

Figure D.2 Logic symbol for a D bistable

DC amplifier An amplifier using direct coupling between stages and thus capable of amplifying a direct current or a signal with a very slow rate of variation.

DC component The mean value of a *signal*. A signal which has equal areas above and below the time axis, such as one composed solely of sine waves, has a mean value of zero, i.e. has no DC component. An asymmetrical signal has a mean value and this is one of the components (the zero-frequency component) of a complex signal obtained on *Fourier analysis* of the signal.

DC coupling Inter-stage coupling via a path which permits DC (*zero-frequency*) signals to be transferred from the output of one stage to the input of the next. The coupling may simply be a physical connection between the output and input electrodes of the *active devices* or may consist of a resistive *potential divider*.

DC restoration A technique for re-inserting the *zero-frequency* or low-frequency components of a *waveform* after these have been lost or attenuated, e.g. as a result of passing the signal through capacitance-coupled circuits. A repetitive feature is necessary in the waveform such as the negative-going sync signals in television and a diode can then be used to effect restoration as indicated in *Figure D.3*.

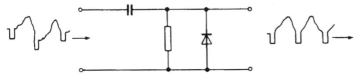

Figure D.3 DC restoration of a television waveform by a simple diode circuit

DC stabilisation Of the operating point of an active device the process of ensuring that the mean current through the device is substantially unaffected by variations in the characteristics of the device such as may occur with changes in temperature or when the device is replaced by another of the same type. Stabilisation is necessary particularly when equipment is mass-produced to ensure that the device can always give the planned performance and to prevent *thermal runaway*.

A typical circuit for DC stabilisation of a *bipolar transistor* circuit is given in *Figure D.4*. The potential divider R_1R_2 stabilises the base voltage of the transistor and hence the emitter voltage; the resistor R_3 then determines the mean current through the device.

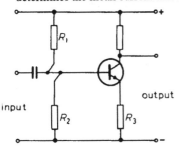

Figure D.4 Circuit often used for DC stabilisation of the operating point of a bipolar transistor

dead room Same as *anechoic chamber.*

debug (US) To detect errors or faults (bugs) in a procedure or an equipment and to correct them.

decade In electrical circuit theory the interval between two frequencies with a ratio of 10:1.

decade box An item of test equipment which enables a specific value of *resistance, capacitance* or *inductance* to be introduced into a circuit. The box has a number of sections, each with ten switched positions, each section having ten times the value of the previous section. Thus any value of resistance, capacitance or inductance within the range of the box can be set up.

decay coefficient (or **factor**) Same as *damping coefficient* (or *factor*).

decay time (1) Of a charge storage tube the time taken for the stored information to fall to a stated fraction of its original value. (2) Of a *pulse* same as *fall time.*

decelerating electrode or **decelerator** An electrode in a *cathode ray tube* biased so as to slow down the electron beam. It is usually in the form of a mesh or a short cylinder. Such an electrode is used in low-velocity TV *camera tubes* such as the *image orthicon* in which the electron beam is required to approach the target at a very low velocity. See *image orthicon, orthicon, orthogonal scanning.*

decibel One tenth of a *bel*, the unit used for comparing power levels and voltage levels in electronics and for expressing transmission gains and losses. Two powers P_1 and P_2 are said to differ by n decibels when

$$n = 10 \log_{10}\frac{P_1}{P_2}$$

n being positive when P_1 exceeds P_2 and negative when P_1 is less than P_2. Similarly two voltages V_1 and V_2 are said to differ by n decibels when

$$n = 20 \log_{10}\frac{V_1}{V_2}$$

n being positive when V_1 exceeds V_2 and negative when V_1 is less than V_2.

The decibel is a logarithmic unit and therefore gains and losses expressed in decibels can be added and subtracted arithmetically.

decineper A transmission unit equal to one tenth of a *neper.*

decoder Any device that converts data from one format to another, especially if the input data is encoded. In logic circuitry a decoder may convert binary coded decimal signals (on four lines) to binary (that is, selecting one of 10 or 16 lines). See *decoding.*

decoding In general the recovery of the original signal from a coded form of the signal. In particular in *stereophonic* radio reception the recovery of the left and right signals from the multiplex received signal and in colour TV reception the recovery of the three primary colour signals from the colour *video signal.*

decoupling The process of confining signals to the paths where they are required and thus preventing them from entering other areas of a circuit.

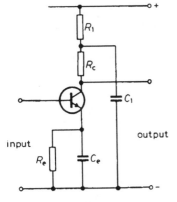

Figure D.5 R_1C_1 are collector-decoupling components and C_e is an emitter decoupling capacitor

In multi-stage electronic equipment this is necessary to minimise the *coupling* which otherwise might exist between stages by virtue of the finite *impedance* of the common power supply, and the usual method of decoupling is by the inclusion of *RC* or *LC* circuits. For example, R_1, C_1 are *collector*-decoupling components (see *Figure D.5*); the reactance of C_1 is made small compared with R_1 at all operating frequencies and thus the signal-frequency components of the collector current of the transistor are confined to the path which includes R_c, C_1 and C_e and do not enter the power-supply circuits where they could cause *feedback* problems, possibly resulting in instability, if the power supply also serves other stages.

C_e provides a low-reactance path for signal-frequency components of the emitter current and thus C_e may be described as a capacitor decoupling R_e.

dedicated Reserved for a special purpose and therefore unavailable for other uses. For example, a fax machine contains a microcomputer which has many of the features of a general-purpose microcomputer, but is dedicated to the communication operations of the fax machine.

de-emphasis The use of a network with a falling high-frequency response to reduce the subjective effect of noise introduced by a transmission system or a record/reproduce system. The upper frequencies are accentuated on recording or at the transmitting end (see *pre-emphasis*) to ensure an overall level frequency response and the two curves must be complementary. In FM broadcasting the curves are governed by a *time constant* of 75 µs.

default In hardware or software, the selection of settings made by the manufacturer. They determine how the system behaves if you choose not to over-ride the settings with others that you prefer.

definition In television in general a statement of the degree of detail in reproduced images. The ability of a TV system to reproduce abrupt changes in tonal value occurring along the scanning lines is known as the horizontal definition, i.e. horizontal definition measures the sharpness of reproduction of vertical edges or lines in the image. Similarly the vertical definition is the ability of the system to reproduce abrupt changes of tonal value occurring along a line at right angles to the scanning lines. The vertical definition measures the sharpness of reproduction of horizontal edges or lines in the image.

deflection sensitivity The linear displacement of the spot on the screen of a *cathode ray tube* for a given deflecting field. For *electrostatic deflection* it is stated in millimetres for 1 V applied between the deflector plates. Because the deflection sensitivity depends on the final-anode potential of the tube, this must be quoted in any numerical statement of deflection sensitivity.

deflector coils Coils carrying currents to generate the magnetic deflecting field for a *cathode ray tube*. Two coils are used to give horizontal deflection of the beam and are commonly of rectangular section which are then moulded around the neck of the tube in the form of a saddle. For a TV picture tube or camera tube two further coils are required to give vertical deflection and these are also moulded around the neck overlapping the horizontal deflector coils. All four coils are mounted on an assembly known as a *yoke*. To concentrate the magnetic field within the *cathode ray tube* the yoke is often constructed of magnetic material such as *ferrite*.

deflector plates In general, electrodes carrying the voltages to generate an electric deflecting field in a *cathode ray tube*. For horizontal and vertical deflection of the beam in a cathode ray tube the plates are mounted within the neck of the tube, one pair being used for horizontal deflection (x direction) of the beam and a second pair for vertical (y) deflection. The two sets of plates are arranged in tandem as shown in *Figure C.7*.

degassing Of an *electron tube* the removal of any traces of remanent or occluded gases after the normal evacuation process. These last traces are removed by heating the electrodes by eddy currents induced in them by a coil carrying RF and surrounding the tube whilst pumping proceeds. An extremely good vacuum can be obtained in this way but a *getter* is also used to further improve the vacuum.

degaussing Of a colour picture tube the process of minimising residual magnetisation of the structure of the tube and nearby metalwork. This is necessary to avoid possible deflecting effects on the scanning beams which could cause misregistration and colour fringing of reproduced images. Degaussing can be achieved by use of a coil surrounding the face of the tube and which is energised by an alternating current, the amplitude of which is reduced to zero at the beginning of each period of use of the tube.

de-ionisation The return of an ionised gas to the neutral or un-ionised state by recombination of positively- and negatively-charged *ions* after the ionising source has been removed.

de-ionisation time Of a *gas-filled tube*, the time taken for the *grid* to regain control after interruption of the anode current. De-ionisation takes a significant time and must be complete before the grid is effective again.

dekatron A cold-cathode gas-filled tube used for *counting*. It has a central anode and ten cathodes disposed at regular intervals around it with transfer electrodes between the cathodes. In operation a *glow discharge* surrounding one cathode is transferred to the next cathode each time a *pulse* is applied to the transfer electrodes. After ten pulses the discharge is back at the original cathode and every tenth pulse can be applied to another dekatron thus making a complete counter registering visibly in the decimal system.

delay In general, the time taken for a signal to pass through a system or equipment. In particular for a receiver or amplifier with a step-signal input, it is the time taken for the output signal to reach one half of its final amplitude.

delay distortion or **delay/frequency distortion** Distortion arising from the lack of constancy of the *phase delay* of a system over the frequency range required for transmission. To avoid distortion of a complex wave, phase delay should be independent of frequency, i.e. *phase shift* should be directly proportional to frequency over the frequency range occupied by the signal. Even small departures from linearity of the phase-frequency curve can cause serious waveform distortion if they occur over a limited frequency range. Thus deficiencies in phase response are better indicated by variations in the slope of the ϕ/ω curve than by variations in the value of ϕ/ω itself. See **group delay**, **phase distortion**.

delayed automatic gain control An AGC system so designed that it does not come into operation until input signals reach a predetermined (threshold) amplitude. AGC is therefore inoperative for weak signals which therefore receive the full gain of the amplifier or receiver. Such a system does not therefore limit the sensitivity of the receiver.

delay generator An active circuit with a controllable delay between input and output signals. Monostable pulse generators such as *multivibrators* can be used as delay generators, the input signal being used to trigger the leading edge of the generated pulse, the trailing edge acting as the output signal. The circuit constants are chosen to give a duration of generated pulse equal to the required value of delay.

delay line or **delay network** A length of transmission line or an *equivalent network* of lumped inductance and capacitance designed to introduce a desired value of delay in the transmission of a signal through it. The lumped-constant network is sometimes termed a delay network.

delta-array colour picture tube A shadow-mask picture tube the screen of

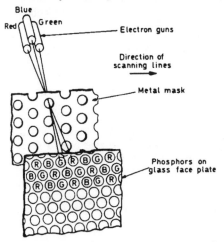

Figure D.6 Delta-array colour television tube: basic arrangement of electron gun, shadow mask and screen

which is composed of a very large number of colour cells, each consisting of a dot of red, green and blue phosphor arranged in a delta (triangular) formation as shown in *Figure D.6*. See *shadow-mask picture tube*.

delta network A network of three branches connected in series to form a closed loop or *mesh*, the three external *nodes* providing the three connections to the network as shown in *Figure D.7*. A three-phase circuit may be delta connected. A delta network is the same as a π network. See *branch*, *star connection*, *star-delta transformation*.

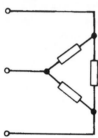

Figure D.7 Three circuit elements connected in delta formation

demagnetisation The reduction of the *magnetic flux density* to zero by taking the material through *hysteresis loops* of gradually-diminishing amplitude.

demodulation The process of abstracting from a modulated wave a replica of the original modulating signal. It is the converse of *modulation*.

de Morgan's theorem In *Boolean algebra* a principle of duality which states that the values 0 and 1 may be interchanged with AND and OR operations.

demountable Designed to be easily removed from its host equipment. For example, a demountable *hard disk* can be easily removed from one computer and installed in another having a suitable drive. This contrasts with conventional hard disks which are permanently installed.

depletion layer A region in a *semiconductor* material which is free of charge carriers. Such a region occurs in a reverse-biased pn junction, the charge carriers being attracted towards the terminals by the positive or negative charge due to the applied EMF. The depletion layer bounded by the conductive zones of the massed *charge carriers* resembles the structure of a charged capacitor and in fact this capacitance is made use of in *variable-capacitance diodes*.

depletion mode A mode of operation of a *field-effect transistor* which is designed to take a considerable drain current at zero gate bias, a reverse bias being necessary to reduce the current to a normal working value.

depolarisation Of a *primary cell* the removal of the gas which collects on the electrodes during discharge and limits the maximum current that may be drawn from it. In the *Leclanché cell*, for example, manganese dioxide is used as a depolarising agent, being packed around the carbon (positive) electrode to oxidise to water the hydrogen liberated from this electrode during discharge of the cell.

derivative equaliser An *equaliser* which operates by adding to the signal to be equalised controllable amounts of the time derivatives of the signal. Such equalisers are used in TV transmission for correcting waveform distortion in video circuits and normally only the first and second derivatives are used.

desktop In computing, the operating environment of a *GUI*. Effectively it is the layout of the *VDU* display when no applications are active and includes the built-in utilities such as a calculator, calendar/clock, text editor and file management system.

desktop computer A self-contained microcomputer which is small enough to be used on a desktop. See also *laptop computer* and *personal computer*.

Desktop Publishing (DTP) A computer application which allows the user to create and edit on screen any type of *document* from a simple letter to a full-colour illustrated magazine. Through the use of *outline fonts* DTP allows a virtually unlimited range of typefaces, type styles and type sizes in almost any preferred page layout. Documents can also include *bitmap graphics*, such as *scanned* photographs, and *vector graphics* such as line drawings, schematics and maps. DTP is strictly *WYSIWYG* ('what you see is what you get') in that the screen display accurately reflects the appearance of the printed document. A *GUI* is a prerequisite of serious DTP applications. The finished document may be printed on a desktop *printer* such as a laser printer or the output may be in the form of a *PostScript* file which can be read by professional photo-typesetting equipment. Many publications are now prepared using DTP in this way.

destructive reading A *reading* process which also erases the data.

detection The process of abstracting information from a *carrier wave*. In the AM signal detector of a radio receiver the information abstracted is that of the original modulating signal (see *demodulation*) but an AGC detector abstracts an output proportional to the unmodulated carrier amplitude. There are many different types of detector. See *anode-bend detector, grid-leak detector, infinite-impedance detector, ratio detector*.

deviation Same as *frequency deviation*.

deviation ratio In *frequency modulation* the ratio of the *rated frequency deviation* to the maximum value of the modulating frequency. In FM broadcasting the rated frequency deviation is ±75 kHz and the upper AF limit is 15 kHz giving a deviation ratio of 5.

diamagnetism A property of certain materials subjected to a magnetising field which causes induced magnetisation which opposes the applied field. Such materials are repelled by a magnetic pole and have an effective permeability of less than unity. This effect is typical of elements with an atomic structure containing completed electron shells such as bismuth. See *ferromagnetism, paramagnetism*.

diaphragm (1) In an electrolytic cell, a porous membrane which separates two electrolytes but permits the passage of *ions* between them. (2) In a *microphone* a thin resiliently-mounted metal disk or ribbon which vibrates when exposed to sound waves and communicates its motion to the *transducer* of the microphone. (3) In a *loudspeaker* a thin resiliently-mounted paper structure, usually in the shape of a shallow cone, which radiates sound waves when driven by the *transducer*. The diaphragm needs to be fairly large, particularly for radiation at low audio frequencies, but there are now ingenious designs of small cabinet loudspeakers which give good bass response.

dichroic mirror A form of optical filter which transmits part of the light spectrum and reflects the remainder. It consists essentially of a plain sheet of optically-flat glass on one face of which a structure of very thin alternate layers of high and low refractive index are deposited by vacuum evaporation. It relies on interference for its operation. Such filters are used in colour TV cameras to split the light from the optical image input into its red, green and blue components. One arrangement which can be used for this purpose is illustrated in *Figure D.8*.

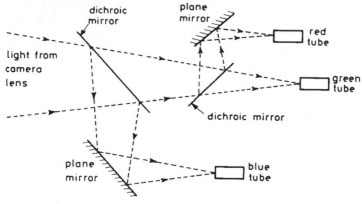

Figure D.8 Simplified diagram showing the use of dichroic mirrors in a colour television camera

dielectric An insulator in which an electric *field* persists after the inducing field has been removed. When an insulator is subjected to an electric field there is a redistribution of electric charges within it, the molecules becoming polarised and aligning themselves in the direction of the field. Dielectrics are used as the medium separating the plates of a *capacitor*.

dielectric constant Same as *permittivity*.

dielectric dispersion The variation with frequency of the dielectric constant (*permittivity*) of a material.

dielectric hysteresis That property of a *dielectric* which causes the electric flux density to be dependent on the previous state of the material in addition to the present value of the electric field strength. For example a dielectric takes a finite time to recover from the effects of an electric strain.

dielectric loss The energy dissipated as heat in an insulator subjected to a varying electric field. In an ideal insulator there is no dielectric loss but in practice changes in the polarisation of the molecules generate some heat.

dielectric strength The ability of an insulator to withstand electric stress without breakdown. Quantitatively it is the *voltage gradient* (in volts per centimetre) at which breakdown occurs under standardised conditions.

difference signal In stereo broadcasting the signal formed by subtracting the B signal (in the right channel) from the A signal (in the left channel). In the pilot-tone system of transmission this difference signal is used to modulate the subcarrier. See *pilot-tone stereo, stereophonic system.*

differential amplifier A *DC amplifier* with two independent inputs and one output, the output signal being proportional to the difference between the two input signals. There is hence no output if the two input signals are equal. *Long-tailed pairs* are often used in differential amplifiers.

differential electrode resistance Same as *electrode AC resistance.*

differential gain In a video transmission system the difference in the gain of a small high-frequency signal at two stated levels of a low-frequency signal on which it is superposed. In colour TV the colour subcarrier is used as the high-frequency test signal. Differential gain may be expressed as a percentage or in *decibels.*

differential permeability Of a material the *permeability* as measured by the slope of the curve of *flux density* plotted against magnetising force. This value of permeability applies to very small excursions in magnetising force about a steady value and it varies over a wide range depending on the steady value of magnetising force. See *magnetic field strength.*

differential phase In a video transmission system the difference in phase of the output of a small high-frequency signal at two stated levels of a low-frequency signal on which it is superposed. In colour TV the colour *subcarrier* is used as the high-frequency test signal.

differential resistance Of a *diode* the ratio of a small change in the voltage drop across the diode to the associated small change in forward current.

differentiating circuit A circuit the output of which is approximately proportional to the rate of change of the input signal. A common example is a circuit comprising a series capacitor followed by a shunt resistor, often used to select the leading or trailing edges of a pulse as shown in *Figure D.9.* The time constant RC must be small compared with the period of the pulses.

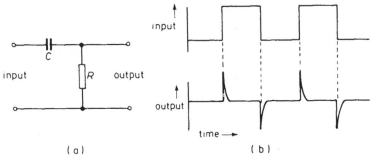

Figure D.9 (a) A simple differentiating circuit (b) Input and output waveforms

diffraction The bending of the path of an *electron beam* or an *electromagnetic wave* when it passes the edge of an obstacle.

diffusion In semiconductor-device manufacture, a technique by which a required amount of an impurity can be introduced into a *semiconductor* by exposing it to the vapour of the impurity in a furnace. The depth of penetration of the impurity can be controlled by the duration of the exposure and the graded impurity concentration which results is often helpful in transistor performance. See *graded-base transistor*.

digital audio broadcasting (DAB) A broadcasting system in which a number of digitally-coded audio signals are used to modulate a closely spaced multiplicity (a *comb*) of carriers. The system was developed by a consortium of European Broadcasting Authorities to overcome the dropouts and *multipath distortion* experienced in conventional FM reception on mobile and portable receivers. DAB transmissions can be from terrestrial or satellite stations.

The system uses approximately 1500 carriers spaced at 1-kHz intervals in Band III (217.5–230 MHz). The audio signals are digitised and the bit rate is reduced to about one sixth of that used on compact disks by a technique which makes use of the fact that a quiet sound is masked by a loud one and therefore need not be broadcast. The signals so processed are then interleaved in time and the total bit rate is spread over the comb of carriers in a precisely controlled manner. Thus the interval between successive bits on any particular carrier is greatly increased. The comparatively small time delays caused by reflected signals are of no account and are absorbed in a guard period inserted between transmitted bits. The advantage of using a large number of carriers is that heavy interference on individual carriers does not interrupt or affect reception.

digital audio tape (DAT) A standard for the digital recording of audio on cassette tape. Theoretically this results in a high-quality recording, free from background noise and distortion, as on *compact disc*. DAT is also sometimes used in computing for making *backups* of *hard disks*.

digital computer Equipment for performing calculations and/or logical operations on input data in digital form under the control of a stored program.

Computers play an important role in science and engineering where use is made of the high speed at which they can perform complex calculations. They are extensively used in industry and commerce where their ability to store and process large quantities of information is valuable. They also control and monitor industrial processes. The graphics displays which computers can give are invaluable in design work (CAD) and in providing information and entertainment in the home and in education. Finally mention must be made of the important role played by computers in manufacture (CAM).

The principles of the modern computer can be traced to the work of Babbage who started work on a mechanical machine in 1822. The development of the electronic computer is deemed to have occurred in four distinct steps termed generations as follows:

(a) First generation. The first electronic computer, constructed immediately after the Second World War, employed 18000 electronic tubes as switching elements and used delay lines for storage. It was very large, occupying several rooms and consumed vast quantities of power. Input and output were by teleprinter and punched cards.

(b) Second generation. These computers, developed during the 1950s, used discrete transistors as switching elements and storage was by magnetic cores, drums and tapes. Line printers were used for output. These machines were much smaller and more reliable than the first generation but were still very large compared with modern computers. For the first time a programming language was used.

(c) Third generation. Integrated circuits were used in these computers giving greatly reduced size, higher operating speeds and improved reliability. There were enormous increases in storage capacity. Languages such as FORTRAN, COBOL and BASIC were in common use. The individual minicomputer operated by a few or even one man gained favour over the large centrally-managed machine.

(d) Fourth generation. Large-scale integration has permitted yet further reductions in size and complete processors can now be formed on a single monolithic integrated circuit known as a *microprocessor*. Semiconductor storage has virtually replaced core stores so making possible the microcomputer.

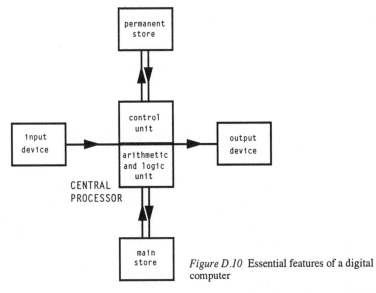

Figure D.10 Essential features of a digital computer

All computers have the five basic sections shown in *Figure D.10*. The main store consists chiefly of random-access memory (RAM) but there is also some read-only memory. The RAM section is used for holding the current program and transient data being processed by the program. The ROM section may hold only the routines used at start-up, but in some machines also holds the *operating system* which contains frequently-used routines that do not require alteration. The capacity of the main store is sometimes inadequate and the permanent store (such as a hard disk) may be used to hold additional data which can be transferred to the main store if and when required. The permanent store also stores programs and data between operating sessions.

The arithmetic and logic unit performs such operations as addition, subtraction, left and right shifts, AND, OR and NOT.

The control unit determines which command in the program is next executed. Usually commands are executed sequentially, but sometimes instructions require a jump to some other section of the program or to a routine in the operating system.

The input device is a means of feeding data into the computer in the digital form which it can accept: this could be a punched-card reader, a keyboard or a mouse.

The output device displays the final result of the program in an intelligible form: this could be a printer or a VDU.

There is a vast range of different types of computer but all belong in one of the following categories:

A microcomputer consists of a microprocessor, main and permanent storage facilities, a keyboard and a visual display unit or, occasionally, circuitry for displaying output on the screen of a domestic television receiver. Usually a BASIC interpreter is included in permanent storage, or even in ROM, allowing users to program the machine in an easily learned programming language.

The minicomputers and midicomputers at one time extensively used in offices, laboratories and industry were multi-user machines having several terminals served in rotation. They have now been largely replaced by networks of microcomputers which offer greater computing power, speed and versatility.

Mainframe computers are used by large organisations such as national banks and public utilities for processing vast quantities of data, such as millions of accounts. Usually they support a large number of time-sharing users. The central processing unit is usually situated in a room with a temperature- and humidity-controlled environment.

digital signal In general a signal of which the significant property can have only a limited number of discrete values. The term is widely used in binary transmission where there are only two discrete values. The signal can then take the form of the presence or absence of a voltage, the closing or opening of a contact, the presence or absence of a hole in a punched card, etc. See *analogue signal.*

digital-to-analogue converter (DAC) Circuitry for converting a *digital signal* to analogue form.

digitise (US) Same as *analogue-to-digital conversion.*

diode See *diode electron tube, junction diode, point-contact diode.*

diode detector An AM detector which makes use of the unidirectional conductivity of a diode. Two typical circuit diagrams for diode detectors are given in *Figure D.11.* At (a) the diode is in parallel with the load resistor R_1 and at (b) it is in series. Both circuits operate in the same way. On one half-cycle of the carrier-wave input from the tuned circuit LC the diode conducts and charges C_1 to the peak value of the carrier voltage. During the following half-cycle the diode is reverse-biased and cut off and C_1 begins to discharge through the diode load resistor R_1 (and L in (a)). However the time constant R_1C_1 is such that little of the charge is lost before the charge is restored to peak value again when the diode conducts

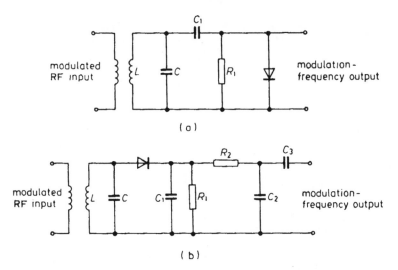

(a)

(b)

Figure D.11 Diode detectors (a) shunt-connected and (h) series-connected

during the next half-cycle. Thus the voltage across C_1 is maintained at the peak value of the carrier input.

When the carrier amplitude varies during modulation, the voltage across C_1 varies correspondingly and is thus a copy of the modulation envelope. The voltage across C_1 is therefore the detector output (in *Figure D.11* (a), L has negligible impedance at modulating frequencies).

The time constant R_1C_1 must not be too long otherwise the voltage across C_1 cannot accurately follow downward swings of carrier amplitude. This problem is most acute at high modulating frequencies and at great modulation depths; in fact these factors set an upper limit to the value of time constant which should be equal approximately to the period of the highest modulating frequency.

Figure D.11 (b) represents a sound radio detector and R_2C_2 is an RF ripple filter, C_3 being a blocking capacitor to remove the direct component of the detector output.

diode electron tube   An *electron tube* with two electrodes, a *cathode* and an *anode*. Such a tube cannot amplify and its significant property is that it can conduct only when the anode is made positive with respect to the cathode. It is therefore used as a rectifier of alternating voltages in power supply units, as a detector of radio signals and as a switching device in analogue and digital equipment. *Diodes* often form part of multiple electron tubes, e.g. a double-diode-triode includes two diodes. Typical I_a–V_a characteristics for a diode are given in *Figure D.12* (a) and the graphical symbol for an indirectly-heated diode is given in *Figure D.12* (b). See *diode detector*.

diode-transistor logic (DTL)   A system in which use is made of *logic elements* comprising diodes and transistors. See *logic, logic gate, logic level*.

direct access storage   Same as *random access memory*.

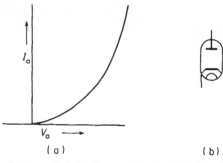

(a)   (b)

Figure D.12 I_a–V_a characteristic of a diode (a) and graphical symbol for an indirectly-heated diode (b)

direct component Same as *DC component*.

direct coupling Same as *DC coupling*.

direct current (DC or d.c.) A current which flows only in one direction, such as the current drawn from a battery. See also *alternating current*.

directly-heated cathode Same as *filament*.

directory In computers and data processing equipment, data is stored in *files*. Essential information about each file such as its name, type, length, date of creation and physical location in the storage medium is stored in a special file called a *directory*. Besides details of files a directory may contain details of other directories, allowing the organisation of stored files in a logical, hierarchical manner. Each storage medium has a *root directory* in a fixed location; this must be accessed before any data can be read from or written to the medium.

direct voltage A voltage which does not change in polarity, e.g. the voltage available from a battery. Such supplies are used in electronic equipment, e.g. for the *collectors* and *drains* of transistors. Direct supplies can be obtained from the a.c. mains by using a *transformer* to give the required voltage, by *rectification* and *smoothing* the output.

disassembler A computer program which analyses *machine code* and converts it to the mnemonics used in an *assembler*. Disassemblers are used mainly for diagnostic purposes.

discharge In general the removal of all or some of a stored electric charge. In particular: (1) Of a *primary* or *secondary cell*, the conversion of the chemical energy into electrical energy; (2) Of a *gas-filled tube* the passage of an electrical current through the device; (3) Of a charged *capacitor* the release of some or all of the stored charge through an external circuit.

discriminator In a frequency- or phase-modulated *receiver* the stage in which a voltage (derived from the received signal) is substantially proportional to the deviation of the frequency or the phase from the unmodulated value. There are a number of circuits which can be used for this purpose. See *Foster-Seeley discriminator, ratio detector, Round-Travis discriminator*.

disk See *compact disk, floppy disk, floptical disk, hard disk, magneto-optical disk, video disk*.

Disk Operating System (DOS) An *operating system* that resides on a *hard* or *floppy disk*, in contrast with one that resides in *ROM*. Probably the best known is Microsoft DOS (MS-DOS) which is the most commonly used operating system in IBM *PC*-compatible microcomputers.

disk-seal tube An *electron tube* made suitable for operation at UHF in which the closely-spaced *electrodes* are extended as annular disks which project through the glass envelope to locate with the ends of the coaxial lines used as tuning elements. The fine spacing of the electrodes is necessary to minimise *transit time* and so raise the upper frequency limit. *Figure D.13* shows the form of construction used in one type of disk-seal triode.

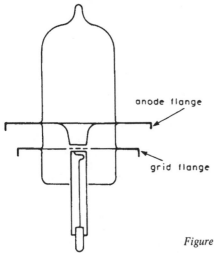

anode flange

grid flange

Figure D.13 Construction of a disk-seal triode

dismount In computing to terminate the use of a *drive*. This involves closing all open *files* and any *directory* displays that relate to that drive. There are many possible reasons for dismounting. When an application has been saving data byte by byte on a demountable medium such as *floppy disk*, the data is written first to a *buffer* which is copied to the disk later. If the disk is removed from the drive before the buffer is copied, some of the data is lost. Dismounting forces the completion of such data transfer and therefore ensures an orderly updating and shutdown of open files. See *mount*.

dispersed dithering See *dithering*.

dispersion Of any kind of radiation the spreading of a beam according to the value of some property. For example, the spectrum obtained by passing a beam of white light through a prism is caused by dispersion, the beam being deflected in the prism to an extent dependent on frequency.

dispersion gate Same as *NAND gate*.

displacement current A concept introduced to explain the flow of current through the *dielectric* of a *capacitor* subjected to a varying voltage. This hypothetical current is equal to the rate of change of the electric flux in the dielectric but there is, of course, no conduction current, i.e. no movement of charge carriers through the dielectric.

display The visual presentation of data in alphanumeric or graphical form or as a drawing, usually on the screen of a cathode ray tube.

dissipation Energy wasted in the resistive component of an electric circuit. The resistance may be that of resistors forming an essential part of the circuit, e.g. the collector load of a transistor. Alternatively the resistance may be incidental, e.g. that of the windings of a transformer. However energy can be dissipated by a circuit even though there is apparently no resistance present, intentional or incidental. A capacitor with poor-quality dielectric or a circuit radiating RF power are two examples. It is common practice to represent the losses in such circuits as due to a resistor which is shown as a discrete component in equivalent circuits. The lost energy is manifested as heat and, for a resistance R carrying a current I, is given by I^2R, V^2/R or VI, where V is the voltage across the resistance.

dissociation The division of a metallic salt into charged ions on solution. Copper sulphate, for example, divides into positively-charged copper ions and negatively-charged sulphate radical ions. If the solution contains electrodes connected to a DC supply, the ions act as *charge carriers* and a current passes through the solution. The copper ions move towards and are deposited on the cathode and the sulphate ions migrate to the anode with which they form a new salt which goes into solution. If the electrodes are of copper, the effect of passing a current through the solution is that the cathode increases in size, the anode decreases in size and the concentration of the solution remains constant. If the cathode is of iron, for example, it becomes copper plated when a current is passed and this is the basis of the electroplating process.

distortion (1) Of an electrical signal any unwanted change in the *waveform*. There are, in general, three causes of waveform distortion:
(a) Variations with frequency of the *gain* or *attenuation* of the signal path. See *aperture distortion, attenuation distortion*.
(b) Variation with signal amplitude of the gain or attenuation in the signal path. See *amplitude distortion, harmonic distortion, intermodulation distortion, non-linear distortion*.
(c) Lack of proportionality between *phase shift* and frequency in the signal path. See *delay distortion, phase distortion*.
(2) Of a displayed image any unwanted change in the tonal *value, hue* or *saturation* or the introduction of spurious picture details. This can arise, for example, from waveform distortion as detailed under (1) (see *overshoot, preshoot, ringing, streaking, undershoot*). Distortion can also arise from any errors in the relative positioning of reproduced picture elements arising from shortcomings in the scanning system. See *barrel distortion, pincushion distortion, trapezium distortion*.

distributed amplifier A wide-band amplifier in which the input *capacitances* of a number of *active devices* form a *delay line* with a number of inductors, the device inputs being excited in sequence by a signal applied to the input of the line. The output capacitances similarly form a delay line with a number of inductors, the output signals from the active devices arriving at the end of the output line in phase.

The advantage of such an amplifier is that its gain can be increased indefinitely by adding more active devices; there is no upper limit to the gain-frequency product as in conventional cascaded amplifiers.

distributed connection In logic circuitry a method of paralleling the outputs of a number of logic elements which gives the effect of an AND or OR operation so avoiding the necessity for including a gate specifically for the purpose. The output terminals of logic elements intended for this form of connection are free transistor *collectors* or *emitters* destined for connection, with a number of similar elements, to a common external load. The graphical symbols for distributed AND or OR connection are given in *Figure D.14*.

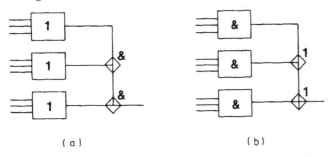

(a)   (b)

Figure D.14 (a) A distributed AND connection shown with three OR gates and (b) a distributed OR connection shown with three AND gates

distributed constants The inherent *capacitance* (*inductance* or *resistance*) of an electrical circuit which has a significant length or area, as opposed to the lumped capacitance (inductance or resistance) of a component specifically included in a circuit to provide that property.

For example, a *transmission line* has distributed capacitance and inductance and these, in fact, determine the *characteristic impedance* of the line.

disturbance See *interference*.

dithering In computer *graphics* the simulation of colours not available in the current *palette* using patterns of discrete *pixels* in the palette colours. With a careful choice of palette colours (optimising procedures are available) most *true-colour bitmaps* can be rendered satisfactorily on the screen in 256-colour video modes. An algorithm known as *Floyd and Steinberg error diffusion* allows the overall colour content of areas of an image to be matched with very high accuracy. In *VGA* and higher-resolution displays dithering with discrete pixels (known as *dispersed dithering*) is not obvious and a casual observer might imagine that he was viewing a true-colour display. Such dithered bitmaps, however, are not amenable to many *image processing* operations (such as sharpening) and they may also yield unsatisfactory printouts on some colour *printers*.

Colour printers also use dithering to simulate most colours since they can only print in eight colours (the ink colours cyan, magenta, yellow and black; the three colours obtained by overprinting pairs of inks: red, green and blue; and the option of leaving the paper, assumed to be white, unprinted). Colour printers may use *clustered* dithering which organises the

99

output into clusters of, say, 36 pixels which may be in any combination of the eight printable colours. This offers a wide range of simulated colours, but reduces the effective resolution of the printout (to 60 'lines per inch' in a 360 dots-per-inch printer). Some high-resolution colour printers, however, use dispersed dithers with Floyd and Steinberg error diffusion, resulting in sharp printouts sometimes described as 'near photographic quality'.

diversity Method of radio reception in which the effects of fading are minimised by combining or selecting from two or more signals carrying the same programme, the signals being obtained via receivers from (*a*) spaced **antennas**, (*b*) differently-polarised antennas or (*c*) channels on different frequencies. A common technique is to common the AGC lines of the receivers so that weak signals are suppressed and only the strongest is received.

dividing network In multi-unit *loudspeakers* a frequency-selective network which divides the frequency range to be radiated into two or more adjacent bands each of which is fed to the appropriate unit.

document A term often used in computing to describe the set of data being processed, whether or not it is destined to be printed and therefore to form a conventional paper document.

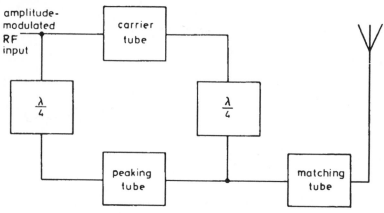

Figure D.15 Block diagram of Doherty circuit

Doherty amplifier A linear amplitude-modulated RF amplifying circuit giving high efficiency and used in sound transmitters. It consists of a carrier tube and a peaking tube with inputs and outputs coupled by quarter-wave *networks* as shown in *Figure D.15*. Matching and operating conditions are so adjusted that for inputs up to half the maximum the carrier tube delivers power to the *antenna*, the peaking tube being inoperative. For these low inputs the load presented to the carrier tube is twice the optimum. For inputs exceeding half the maximum the peaking tube comes progressively into operation, so reducing the carrier-tube load to the optimum value, at peak modulation the two tubes contributing equally to the output.

Dolby system A system for improving the *signal-to-noise ratio* of audio recordings. The system is based on the principle of compressing the signal before recording and using complementary expansion on reproduction but the audio spectrum is divided into four bands each with its own compressor and expander. Complementary operation is ensured by using the same non-linear *network* for compression and expansion, its output being added to the direct signal before recording and subtracted on reproduction.

The system is capable of excellent results and has had wide acceptance in professional sound recording. It is, however, too complex for general use with domestic tape recorders and for these simpler systems have been designed. These are single-band systems intended chiefly to minimise the effects of tape hiss.

domain In a magnetic material a microscopic region regarded as an elementary magnet. When the material is unmagnetised the directions of the *magnetisation* of the domains are randomly distributed so that their effects cancel. When a magnetising force is applied the domains rearrange themselves so that their magnetic axes align with the applied field.

dongle See *hardware key*.

donor impurity In *semiconductor* technology a pentavalent element such as arsenic, atoms of which can replace the tetravalent atoms in the lattice of a germanium or silicon crystal so making *electrons* available as *charge carriers*. See *n-type semiconductor*.

dopant The impurity used for *doping*.

doping The addition of an impurity to a *semiconductor* material to obtain a particular value of p-type or n-type conductivity. Only a very small addition is needed to produce material suitable for use in the manufacture of semiconductor devices. An impurity content of 1 part in 10^8 increases the conductivity of pure germanium and silicon sixteen times and impurity concentrations of a few parts in 10^8 are common in semiconductor-device manufacture. The initial semiconductor material must hence be very pure indeed and the purification of germanium and silicon is one of the most difficult processes in the manufacture of semiconductor devices. See *n-type semiconductor, p-type semiconductor*.

Doppler effect The apparent change in frequency of a sound or radio wave when there is relative motion between the source and the observer. The frequency increases as the source approaches the observer and decreases as it recedes from him. The effect is used in radio navigation aids, in intruder alarms and various speed measuring instruments.

dot Same as *picture element*.

dot-matrix printer See *printer*.

dot pitch A measure of the resolution of a *monitor* screen. It indicates the size of the phosphor dots that make up the display. The smaller these are, the sharper the image.

dot-sequential colour television system A system in which signals from the primary colour sources (e.g. the three tubes of a colour camera) are transmitted in sequence so rapidly that each signal persists for only the duration of scanning one picture element. Thus neighbouring elements are reproduced in different primary colours but the elements are so small and so closely spaced that at normal viewing distances the eye cannot resolve them and sees the colour formed by the addition of the primary components.

double-base diode Same as *unijunction transistor.*

double-heam oscilloscope An oscilloscope in which the *electron beam* of the *cathode ray tube* is split into two parts which can be deflected independently to permit two different traces to be simultaneously displayed on the *screen*. This facility is useful, for example, for indicating the phase relationship between two signals or for comparing the input and output waveforms of an amplifier. See *oscilloscope.*

double-diffused metal-oxide-semiconductor transistor (DMOS) An enhancement *planar insulated-gate field-effect transistor* in which successive diffusions by opposite types of impurity through the same opening in the silicon dioxide layer is used to produce a very short channel, the length of which can be closely controlled by the diffusion process. A simplified diagram showing the structure of a double-diffused transistor is given in *Figure D.16*. It has a very small p-region and the channel formed on the upper part of this region under the gate contact is much shorter than in earlier types of insulated-gate transistor.

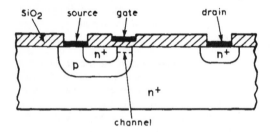

Figure D.16 Simplified structure of DMOS transistor

In *Figure 1.7*, for example, the channel occupies all the lateral distance between the two n-regions. The channel length determines the drain-source resistance in the on-condition and the speed with which the drain current can be switched on and off, and to achieve low resistance and high speed a short channel is needed. Current flow in the double-diffused transistor is lateral as in earlier field-effect transistors but in other short-channel transistors such as *TMOS* and *VMOS* current flows through the thickness of the wafer as in planar bipolar transistors.

double-sideband transmission (DSB) A system of amplitude modulation in which both *sidebands* are fully transmitted. This is the system used in sound broadcasting on the long, medium and short *wavebands.*

double superheterodyne reception A method of reception in which two intermediate frequencies are employed, a high one to minimise second channel interference and a low one to give high adjacent-channel selectivity. See *superheterodyne receiver.*

download To receive data from another computer or other data source.

downtime The period for which equipment is out of action as a result of faults. For equipment providing a public service such as broadcast transmitters downtime must be as low as possible and such equipment is designed to permit very rapid diagnosis of faults and speedy remedy.

downward modulation (US)   Same as *negative modulation*.

Dow oscillator   Same as *electron-coupled oscillator*.

drain   In a *field-effect transistor* the connection to the *channel* which *majority carriers* enter on leaving the channel. It corresponds with the collector in *bipolar transistors* and the anode in *electron tubes*.

drain AC resistance   See *electrode AC resistance*.

drawing software   Software for creating or editing *vector graphics* images.

drift space   In a velocity-modulated tube the region between the buncher and the *catcher* which is free from alternating fields and in which the velocity-modulated beam gathers into bunches. See *Applegate diagram*, *bunching*, *velocity modulation*.

drift transistor   Same as *graded-base transistor*.

drive   (1) In general the input signal for an *active device* or equipment. In particular the source of carrier signal for a transmitter. (2) In computing usually used of the mechanical receptacle for a *storage* device such as a *floppy disk drive*, a *hard disk drive*, a *tape streamer drive* or a *CD-ROM drive*. All of these are electromechanical devices in which a motor physically drives the storage media past the read/write head, but the term is also used in media which are purely electronic. RAMdisks are sometimes called RAMdrives and *PCMCIA* card interfaces are also called drives although there are no moving parts to 'drive' mechanically.

driver   In computing, software that controls a *peripheral*, such as a printer or a *scanner*, and that converts the data being sent to it or received from it into the required format. A driver usually provides control functions such as choice of resolution in a printer or scanner driver.

driver stage   Stage providing the input signal for the following stage, usually the output stage of an amplifier or transmitter.

driving-point impedance   At any pair of terminals in a *network* or point on a transmission line the ratio of the applied voltage to the resulting input current. It is usually measured with all other pairs of terminals correctly terminated. The concept is useful in the design of *feeders* to *antennas*.

drop-out   In a tape recording system momentary loss of output signal on replay caused by a temporary loss of contact between the recording head and the magnetic coating of the tape or of the replay head and tape. Drop-outs are also caused by imperfections in the magnetic coating or by excessive wear removing the magnetic coating from the tape. In radio reception, loss of signal caused when waves reaching the receiving antenna via two or more paths are out of phase causing a net cancellation. See *multipath distortion*.

dry joint   An imperfectly soldered joint in which the solder does not adhere properly to the metals to be joined. Such a joint is weak mechanically and may come apart. It can also have a high electrical resistance.

dry reed   Same as *reed relay*.

dual-in-line (DIL)   A concept for the standardisation of the arrangement and spacing of the connections to components intended for mounting on *printed-wiring boards*. It requires two rows of pins at 0.1-inch spacing. See *dual-in-line package*.

dual-in-line package (DIP) A plastic or ceramic container for a *monolithic integrated circuit*, the connections to which are brought out in a DIL format.

As the complexity of ICs has increased, the number of pins required on the DIP has risen to 64 (32 each side). At the standard spacing of 0.1 inch such an IC requires a package over 3 inches long. These considerations have led to the introduction of chip carriers which are much smaller.

dumb terminal A purpose-designed computer *terminal* having only limited data processing and storage capability and therefore not able to run any other applications or serve any other purpose. It is therefore a slave to the *host* computer which offers full data storage and processing functions. In contrast, an *intelligent terminal* offers some local data processing capability.

dummy antenna A network of *resistance* and *reactance* used to simulate the impedance of an *antenna*. For transmitter testing the dummy antenna must, in addition to simulating the antenna impedance, be capable of safely dissipating the maximum power output of the transmitter but must not radiate.

For receiver alignment a dummy antenna is connected between the *signal generator* and the antenna input terminals of the receiver. A dummy antenna for this purpose need not, of course, be capable of dissipating great power.

dummy load Same as *artifical load*.

duplex operation A method of operation in which communication between two stations can take place in both directions simultaneously. In a radio system, for example, a receiver at A can receive signals transmitted from B on a carrier frequency f_1 at the same time as a receiver at B is picking up signals radiated from A on a carrier frequency f_2. See *simplex operation*.

duty cycle Of a system in which there are regular variations in load or in switching, one complete cycle of variations.

duty factor or **duty ratio** Of a regular pulse *waveform* the ratio of the pulse duration to the pulse period. Thus for a square wave the duty ratio is 0.5. Duty ratio may also be defined as the ratio of the average pulse power to the peak pulse power.

dynamic characteristics Of an *active device*, characteristics representing the relationship between electrode current and voltage under specified operating conditions, in particular with a given load. For example a *resistor* included in the anode circuit of a *triode* causes the anode voltage to fall as anode current increases. The dynamic I_a–V_g characteristic thus crosses the static characteristics for a number of different anode voltages as shown in *Figure D.17*. A significant point is that the slope of the dynamic characteristic is less than that of the static characteristics. For a *pentode* the anode voltage has little effect on the anode current and thus the dynamic characteristics are coincident with the static characteristics.

For transistors (bipolar and field-effect) the output voltage also has little effect on the output current. This is shown by the near-horizontal nature of the output characteristic (*Figure B.21*). Thus the dynamic characteristics for transistors differ little from the static characteristics.

dynamic error In a signal varying with time, an error caused by an

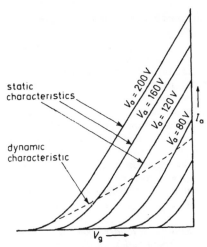

static characteristics

dynamic characteristic

$V_a = 200\,V$
$V_a = 160\,V$
$V_a = 120\,V$
$V_a = 80\,V$

I_a

$V_g \longrightarrow$

Figure D.17 Dynamic and static characteristics of a triode

inadequate high-frequency response. Such errors can arise in **analogue computers** from shortcomings in **transducers**. See **static error**.

dynamic impedance The **impedance** of a parallel LC circuit at its resonance frequency. At this frequency the impedance is purely resistive and the term dynamic resistance is often used instead. For a circuit of inductance L, capacitance C and series resistance R the dynamic impedance is given by L/CR, $QL\omega$ or $Q/\omega C$.

dynamic limiter A limiter circuit capable of adjusting its limiting level automatically to compensate for variations in input-signal level. The basic circuit for a dynamic limiter is given in *Figure D.18*. It is assumed that this is fed from an amplifying stage, e.g. the final IF stage in an FM receiver, LC being resonant at the intermediate frequency. R_1 is chosen to give heavy damping of LC and C_1 is made large so that the time constant R_1C_1 is an appreciable fraction of a second.

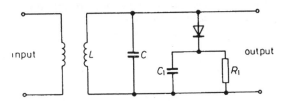

input

L C C_1 R_1 output

Figure D.18 Circuit diagram of a simple dynamic limiter

In operation, C_1 is charged to the peak value of the carrier input from LC which is effectively damped by a parallel resistance of $R_2/2$. Now suppose there is a momentary increase in the input to the preceding stage caused, for example, by a spike of noise. This is so fleeting that the voltage across C_1 does not have time to rise to equal it. As a result the diode conducts heavily during this instant and applies very heavy damping across LC so reducing the gain of the previous stage and maintaining the voltage across LC constant.

If, on the other hand, there is a momentary negative voltage spike in the input to the previous stage, again the voltage across R_1C_1 cannot change quickly enough to register it and this time the diode is cut off for the duration of the spike. LC is thus momentarily relieved of the damping due to R_1 and the gain of the previous stage rises, maintaining constant the voltage across LC. Thus instantaneous upwards and downwards changes in the input to the previous stage do not affect the voltage across LC. The circuit is an effective limiter for upward and downward changes in input amplitude.

If the input to the previous stage alters permanently due to a change in received signal strength then the voltage across R_1C_1 alters slowly to the same extent and the limiting action still applies as before: the circuit is self-adjusting. A dynamic limiter of this type is incorporated in the *ratio detector*.

dynamic loudspeaker (US)   Same as *moving-coil loudspeaker*.

dynamic microphone (US)   Same as *moving-coil microphone*.

dynamic pickup (US)   Same as *moving-coil pickup*.

dynamic random access memory (DRAM)   A *volatile store* in which the fundamental storage devices are capacitors arranged in *matrix* formation. Associated with each capacitor are *field-effect transistors* which act as switches when data is put into the store or withdrawn from it. To prevent loss of data as the capacitors discharge through the inevitable leakage paths, their charges are regularly 'topped up'—a process known as *refreshing*. Despite the need for refreshing the dynamic RAM is simpler, more compact and cheaper than a *static RAM* and, although its speed is lower, is more widely used.

dynamic range   (1) Of a sound programme the ratio, usually expressed in *decibels*, of the loudest to the quietest passage.

(2) Of a transmission or recording system, the ratio, usually expressed in decibels of the overload level of the system to the minimum acceptable signal level and the latter must exceed the *noise level* of the system sufficiently to give the required *signal-to-noise* ratio.

dynamic resistance   See *slope resistance*, *dynamic impedance*.

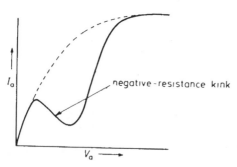

Figure D.19 A typical dynatron I_a–V_a characteristic

106

dynamic store A *store* from which data can be retrieved only at fixed time intervals, e.g. a rotating magnetic disk or drum.

dynatron A *tetrode* in which the *anode* is at a lower voltage than the screen grid so producing a region of negative slope in the I_a–V_a characteristic as shown in *Figure D.19*.

The effect arises as a result of secondary emission at the anode, the *secondary electrons* collected by the screen grid exceeding the primary electrons received from the cathode. The effect is useful in *dynatron oscillators* but is a disadvantage when the tube is used as an amplifier where large anode voltage swings are required. The negative-resistance kink can be eliminated by the inclusion of an earthed (or negatively-biased) suppressor grid between anode and screen grid (so giving rise to the *pentode*) or by using beam-forming plates and aligned grids. See *beam tetrode*.

dynatron oscillator An oscillator which utilises the region of negative slope of the I_a–V_a characteristic of a tetrode which occurs when the anode voltage is below that of the screen grid. The frequency-determining circuit is connected to the anode as shown in *Figure D.20*. Oscillation then occurs if the dynamic resistance of the tuned circuit has a suitable value, the voltage swing being limited to the extent of the negative-resistance kink. See *dynatron*.

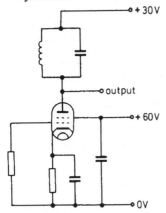

Figure D.20 Circuit diagram of a dynatron oscillator

dynode An electrode included in an *electron tube* to increase the electron stream by *secondary emission*. See *electron multiplier*.

E

earphone   A small electro-acoustic *transducer* designed to fit into or onto the ear.

earth (also known as **ground** (US))   A low-resistance connection, intentional or accidental, to the mass of the earth or to a large conducting body serving as the earth such as a ship's hull or the chassis of a vehicle or of electronic equipment. To make low-resistance contact with the earth a conductor, for example a system of copper wires, may be buried in the ground; this system is commonly used at radio transmitters. For low-power equipment such as receivers an earth connection may be provided either by a copper spike driven into the ground or a connection to a metal cold-water pipe.

The potential of the earth or chassis is taken as a reference level of zero and the potential at any point in the equipment is measured with respect to it. Thus a 180-V supply line means that the voltage is 180 V with respect to earth or chassis potential. A point such as a supply line which is at a constant potential above or below earth potential is often termed 'earthy' because it is usually decoupled to earth by a capacitor giving low reactance to earth. At signal frequencies such a point may be regarded as at earth potential.

earthed-base, -emitter, -grid, etc circuit   Literally a circuit in which the base, emitter, grid, etc is connected to earth but the term is generally used to mean a *common-base, -emitter, -grid* etc *amplifier*.

earth loop   A loop of conductors formed accidentally when the various components of a system are earthed in different places. If mains conductors pass through the loop, the loop can act as a pick-up coil, introducing undesirable levels of noise into the system. Earth loops are a common cause of mains hum in *hi-fi* systems.

earth return or **ground return** (US)   Use of the conducting surface of the earth as one leg of an electric circuit, the other leg being provided by a conductor.

Eccles-Jordan circuit   A bistable *multivibrator* comprising two *triodes*, the anode of each being coupled to the grid of the other. It was used for generating pulses, the circuit being switched between its two stable states by external triggering signals.

echo   An acoustic or electromagnetic wave which is so delayed as a result of *reflection* or *refraction* that it is received as a signal distinct from that directly transmitted.

economy diode   Same as *booster diode*.

eddy current   A circulating current flowing in a conductor as a result of a voltage induced in it by a moving or varying magnetic field. Such currents generate heat in the conductor and this is put to useful purpose in such applications as the eddy-current brake (see also *degassing*). In transformer cores, however, these currents represent a loss of power and to minimise such losses the cores of low-frequency transformers are constructed of stacked laminations, each electrically insulated from its neighbours.

At radio frequencies, eddy-current losses are so serious that magnetic core materials, if conductive, must be reduced to powder, the individual grains being insulated from each other to minimise losses. Alternatively non-conductive materials such as *ferrites* can be employed in the cores of RF inductors and transformers.

edge connector A linear multi-contact plug or socket designed to mate with the contacts at the edge of a printed-circuit or printed-wiring board.

Edison effect Electrical conduction between a heated filament and an independent cold electrode in the same envelope. In experiments on the blackening of the inside walls of lamps with carbon filaments Edison put a metal plate inside a lamp and found that a current flowed between filament and plate when the latter was made positive with respect to the filament. This was the first observation of *thermionic emission*.

effective capacitance Same as *equivalent capacitance*.

effective inductance Same as *equivalent inductance*.

effective resistance (also known as **radio-frequency resistance**) (1) Same as *equivalent resistance*. (2) The total resistance of a conductor to alternating current including the ohmic resistance and the resistance due to skin and *proximity effects*. At RF the effective resistance can be many times the ohmic resistance. Also known as *RF resistance*. See *skin effect*.

efficiency In general of an electrical device the ratio expressed as a percentage of the output power to the input power. For example the efficiency of a power transformer approaches 100%. The efficiency of an electronic device such as an amplifier is the ratio of the useful output power to the power taken from the supply. For example the efficiency of a class-A output stage reaches a maximum of 50%.

efficiency diode Same as *booster diode*.

eigentones Standing sound waves set up between opposite surfaces of a rectangular enclosure such as a room. These waves can occur at the frequency for which the wall spacing is half a *wavelength*, one wavelength, one and a half wavelengths, etc so that there is a series of frequencies at which eigentones can occur. Standing waves can occur across the length, width and height of a room so that the eigentone structure can be complex. The frequencies of the eigentones largely dictate the room acoustics and it is best if there is no simple relationship between these dimensions so that the eigentones are spaced evenly over the audio spectrum.

EITHER-OR gate Same as *OR gate*.

elastance The reciprocal of *capacitance*.

electret A dielectric solid in which the voltage applied to induce an electric field persists after the supply has been removed, leaving the dielectric with separated electric poles of opposite sign. It is the electrostatic analogue of a permanent magnet. Electrets have obvious applications in electrostatic *microphones* where they avoid the need for a polarising supply.

electrically-alterable read-only memory (EAROM) Same as *electrically-erasable programmable read-only memory*.

electrically-erasable programmable read-only memory (EEPROM) A *non-volatile read-only memory* of *MOSFET* structure in which the stored data can be repeatedly altered in whole or in part by the user.

Data can be erased and reprogramming achieved with the EEPROM in situ (c.f. EPROM) by using Fowler-Nordheim tunnelling—a principle whereby certain electrons subjected to an electric field can cross the forbidden gap of an insulator to enter the conduction band and thus flow freely for a short distance to a positively-charged area.

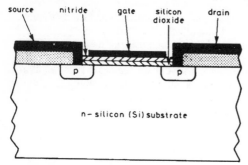

Figure E.1 Cross section of MNOS structure

The earliest EEPROMs used the metal-nitride-oxide-semiconductor (MNOS) structure shown in *Figure E.1*. By applying a voltage of around 20 V to the gate, charge can tunnel through to the silicon-dioxide-nitride interface. Here the charge remains trapped because of the high insulation of these regions and renders the transistor insensitive to normal operating voltages of say 5 V.

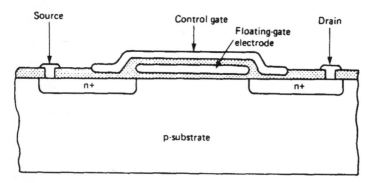

Figure E.2 Cross section of an EEPROM

As shown in *Figure E.2* more recent EEPROMs have, in addition to the normal input gate, a floating gate which can be charged and discharged by tunnelling. For an explanation of the behaviour of the floating gate see **EPROM**. A *memory cell* usually consists of such a (storage) transistor together with a second (select) transistor. The two transistors are connected in series in a *cascode* arrangement and, depending on the voltage applied to their gates and to the drain terminal of the select transistor, the memory cell can be put into one of four states: unselected, erased, written and read.

electric deflection Same as *electrostatic deflection*.

electric displacement Same as *electric flux density*.

electric field A region near a charged body in which an electric charge experiences a mechanical force caused by the charged body. In particular there is a field between two charged bodies of opposite sign, e.g. between the plates of a charged capacitor.

electric field strength At a point in an electric field a vector representing the magnitude and direction of the mechanical force on a unit positive charge at that point. It is normally expressed in terms of the potential gradient, i.e. in volts per centimetre.

electric flux The lines of force constituting an *electric field*.

electric flux density The electric flux per unit area normal to the direction of the *lines of force*.

electric force Deprecated term for *electric field strength*.

electric intensity Same as *electric field strength*.

electric screen A screen of conducting material used to reduce the penetration of electric fields into a particular region.

electric shield (US) Same as *electric screen*.

electric strength Same as *dielectric strength*.

electrode Of an *active device*: an element which emits, collects or controls the movement of electrons or holes. Electrodes have a variety of terms such as *anode*, *base*, *cathode*, *collector*, *dynode*, *emitter*, *gate*, *sink* and *source*.

electrode AC resistance Of an *active device* the ratio of a small change in electrode voltage to the resulting change in electrode current. Its value is usually quoted for a specific operating point. Thus for the electrode characteristic shown in *Figure E.3* the electrode AC resistance is given by the reciprocal of the slope of the tangent to the characteristic at the point P, i.e. it is given by RS/RQ.

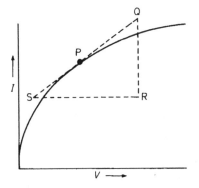

Figure E.3 The electrode AC resistance is given by the reciprocal of the slope of the tangent SO to the curve at the operating point P

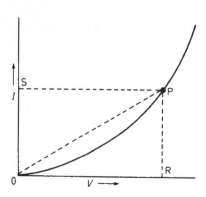

Figure E.4 The electrode DC resistance is given by the reciprocal of the slope of the chord OP from the origin to the operating point P

electrode DC resistance Of an *active device* the ratio of the direct voltage applied to an *electrode* to the resulting direct current through the electrode terminal. Its value is usually quoted for a specific operating point. Thus for the electrode characteristic shown in *Figure E.4* the electrode DC resistance is given by the reciprocal of the slope of the line OP, i.e. it is given by PS/PR.

electrode differential resistance Same as *electrode AC resistance*.

electrode impedance The ratio of the voltage applied to an *electrode* to the resultant current through the electrode terminal of an *active device*. Its value is usually quoted for a particular set of operating conditions. This is a general definition and the electrical characteristics of electrodes are generally given in more specific definitions such as *electrode AC resistance* and *electrode DC resistance*.

electrode slope resistance Same as *electrode AC resistance*.

electrodynamic instrument An instrument which relies for its operation on the interaction between two magnetic fields, one due to a current in one or more movable coils and the other due to current in one or more fixed coils. Some power factor meters operate on this principle.

electroluminescence Emission of light from certain materials when stimulated by an electric potential. One material used consists of a phosphor powder embedded in an insulating material. If a layer of such construction is sandwiched between transparent conducting sheets, light is produced when an alternating voltage is applied between the conducting sheets. Electroluminescent panels can be made of almost any size and to emit a wide variety of colours depending on the phosphor used.

electrolyte A conductor in the form of a liquid or paste in which the *charge carriers* are positively- and negatively-charged ions which migrate to the electrodes under the action of an applied EMF. Two obvious examples of electrolytes are those in electroplating baths and in *voltaic cells*. Sometimes the term is applied to the chemical which, when dissolved in a liquid (usually water), gives rise to the *ions* by *dissociation* and thus promotes electrical conductivity.

electrolytic capacitor A fixed capacitor in which the *dielectric* is a thin film of oxide formed by electrolytic action on aluminium foil which acts as the positive plate, the *electrolyte* acting as negative plate. Because of the minute thickness of the dielectric this form of construction gives high *capacitance* in a component of small dimensions and the dimensions can be further reduced by etching the foil to increase its surface area. In the so-called 'dry' electrolytic capacitors the liquid electrolyte is replaced by a layer of paper saturated with electrolyte. Electrolytic capacitors differ from other types in that they are polarised, i.e. one particular terminal must always be biased positively with respect to the other. This is not a serious disadvantage; because of their high capacitance such capacitors tend to be used in *smoothing* and *decoupling* circuits where a polarising voltage is present. The irreversible nature of the electrolytic capacitor is indicated in the graphical symbol by representing the positive plate as an open rectangle as shown in *Figure E.5*.

Figure E.5 Graphical symbol for an electrolytic capacitor. The positive sign is not always necessary

electromagnet A device comprising a ferromagnetic core within a winding which displays magnetic properties only when a current flows in the winding.

electromagnetic deflection In a *cathode ray tube*, deflection of the *electron beam* by the magnetic field set up within the tube by current-carrying coils clamped to the outside of the tube. This is the method of deflection favoured in TV camera tubes and picture tubes. Two sets of coils are used, one fed with a line-frequency *sawtooth* current to deflect the beam horizontally and the other fed with a field-frequency sawtooth current to give vertical deflection.

electromagnetic induction In a *conductor* linked with a magnetic field, the generation of a voltage in the conductor as a result of a change in the magnitude of the field or of relative movement between conductor and field.

The *transformer* is an application of electromagnetic induction: the voltage induced in the secondary winding is a result of the continually-changing magnetic field produced by the current in the primary winding. The *moving-coil microphone* or pickup is an application in which voltages are induced in a conductor which moves in a static magnetic field. Reference is sometimes made to 'induced currents' but electromagnetic induction causes voltages to be set up in conductors and these can give rise to currents only if the conductors form closed circuits.

electromagnetic lens A lens in which an *electron beam* can be brought to a focus at a desired point by means of a *magnetic field*. There are two basic types; in one type, used with TV camera tubes, the entire tube is included within a coil carrying DC which produces a magnetic field parallel to the tube axis. This field has no effect on *electrons* travelling along the tube axis but those moving at angles to the axis are rotated about the axis and return to it at regular intervals along the axis as shown in *Figure E.6*. Thus there

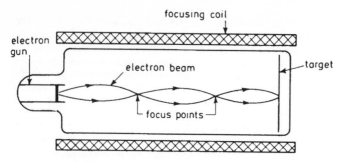

Figure E.6 A long magnetic lens

are a number of points at which the electron beam is in focus and by adjustment of the accelerating electric field or of the magnetic field one of these focus points can be made to coincide with the target.

The second type of magnetic lens employs a very short magnetic field which can be produced by a permanent ring magnet. The field from such a magnet has pronounced radial components and the interaction of these with the electron beam causes the beam to rotate about the tube axis. The axial component of the field causes the beam to be deflected towards the tube axis so that an image, usually rotated, can be produced on the target.

electromagnetic screen A screen of conducting and possibly magnetically-permeable material used to reduce the penetration of electric and magnetic fields into a particular region.

electromagnetic shield Same as *electromagnetic screen.*

electromagnetic waves Waves consisting of linked magnetic and electric fields in which each field is at right angles to the other and to the direction of propagation of the waves. Such waves travel in free space at a velocity of 3×10^{10} cm/s and are known as light, radio, gamma, etc waves according to the frequency as indicated in *Figure E.7.*

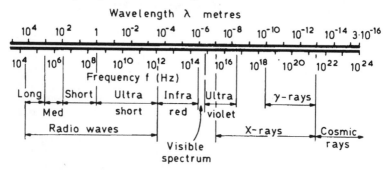

Figure E.7 Spectrum of electromagnetic waves

electrometer Originally an instrument in which the mechanical force between electrically-charged bodies was used to measure potential difference. Thus an *electrostatic voltmeter* is an example of an electrometer. The unique feature of the electrometer is that it takes no current from the source of voltage being measured, i.e. the input resistance of the device is infinite. The term is now used for instruments in which the input signal is applied between the grid and cathode of an *electron tube*. This also has a high input resistance provided that the grid is negatively biased. The amplified output from the tube can be measured by a conventional measuring instrument such as a moving-coil milliammeter.

electrometer tube An *electron tube* specifically designed to have a very low value of grid current for negative grid voltages so as to be suitable for the measurement of small direct currents or voltages in an *electrometer*.

electromotive force (EMF) The property of a physical or chemical device which enables it to drive an electric current around a circuit. Most of the devices are energy converters and the energy manifested as an EMF results from an input in another energy form. For example in an alternator the EMF is generated by *electromagnetic induction* arising from mechanical movement of the rotor. In a transformer secondary winding the EMF is generated by electromagnetic induction due to the changes in *magnetic flux* produced by the signal in the primary winding. In a *thermocouple* the EMF results from heat applied to the junction and in a photovoltaic cell it is generated by incident light. In a *voltaic cell* the energy providing the EMF arises from chemical action within the cell. The unit of EMF is the volt, symbol V.

electron One of the two basic charged particles of which the atoms of chemical elements are assumed to be composed. The other charged constituent is the *proton* and an atom is regarded as a central nucleus composed of positively-charged protons and uncharged neutrons. The net positive charge of the nucleus is offset by the negative charge of an equal number of electrons which are assumed to orbit the nucleus in a series of shells or rings as indicated in *Figure E.8*. In this diagram, for simplicity, the

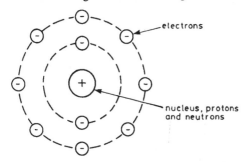

Figure E.8 Simplified diagram showing the construction of the atom

shells are shown as coplanar. Electrons in the outermost orbit of certain elements can be removed with little effort. Such electrons are known as free. By movement of these electrons it is possible to achieve a transfer of negative charge through a material: this constitutes an electric current and

115

is the mechanism of conduction in metals and other good electrical conductors. It is sometimes known as ohmic conduction. Free electrons are also available in *n-type semiconductors* and here too current can flow through the material under the stimulus of an EMF, the electrons acting as negative *charge carriers*. A somewhat different conduction process occurs in *p-type semiconductors*.

Another way in which a unidirectional flow of electrons can be obtained is by heating a material rich in free electrons such as tungsten in a vacuum and by collecting the electrons by a positively-charged electrode. In this way a stream of electrons, i.e. an electric current, can be made to cross an empty space, a phenomenon known as *thermionic emission* which is extensively exploited in *electron tubes*.

electron beam A stream of *electrons*, all moving with approximately the same velocity, and confined to a particular value of cross-sectional area. In TV camera tubes and picture tubes such a beam originates in an *electron gun* and is focused by an electron lens on the target or screen.

electron-coupled oscillator (ECO) A *tetrode* or *pentode* circuit in which the cathode, control grid and screen grid operate as an earthed-screen-grid oscillator, so eliminating any capacitive coupling between the output (anode) circuit and the oscillator circuit. A typical circuit diagram utilising a *Hartley oscillator* is given in *Figure E.9*. The only link between the

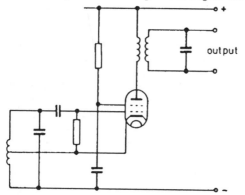

Figure E.9 Circuit diagram of an electron-coupled oscillator using a Hartley circuit

oscillator circuit and the output circuit is provided by the electron stream from the cathode. Thus the frequency of oscillation is unaffected by changes in loading on the anode circuit and can also be made substantially independent of HT supply variations by making the screen-grid voltage a particular fraction of the anode voltage.

electron device A device which operates by virtue of electrons moving through a vacuum, a gas or a *semiconductor*.

electron gun An assembly of *electrodes* which generates the *electron beam* in a *cathode ray tube*. As shown in *Figure E.10* the gun consists typically of an indirectly-heated cathode, a *control grid* and an *anode*. The negative

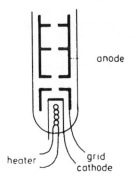

anode

heater   grid
          cathode

Figure E.10 One possible form of electron-gun structure

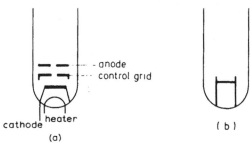

- anode
--- control grid

cathode heater

(a)

(b)

Figure E.11 Graphical symbols for an electron gun. (a) detailed; (b) simplified

voltage on the control grid determines the density of the beam leaving the gun (and hence the brightness of the display). The anode is normally a cylinder containing a number of apertures which limit the cross section of the beam leaving the gun. *Figure E.11* gives detailed and simplified graphical symbols for an electron gun.

electronic calculator   Digital equipment usually with a keypad input and a numerical display output capable of carrying out arithmetical operations.

The design of calculators has greatly benefitted from the developments in *monolithic integrated circuits* since 1960 and they use many of the techniques employed in computers. Early machines were bulky desk-top models using *seven-segment light-emitting diode* displays. These were superseded by portable personal machines, battery-driven, still with LED displays. Modern calculators can be as small as a visiting card, have a *liquid-crystal display*, and require so little power that they operate from built-in photocells in normal room lighting.

They are capable of addition, subtraction, multiplication, division, square roots, trigonometrical and other functions. They even have an internal memory to hold intermediate results in complex calculations. They deliver answers correct to ten significant figures!

electronic mail   A system whereby messages are sent from one computer terminal to another. The message is not normally routed directly to the recipient's terminal, because it may be switched off or in use for other

117

purposes. Instead the received message is stored in the *host* computer providing the recipient's mail service. The recipient may be informed that there is a message for him when he next logs on to the host or the onus may be on him to check his 'electronic mailbox' regularly for incoming messages.

Electronic Point of Sale (EPOS) An intelligent electronic cash register which offers other functions besides those normally expected of a cash register. For example, it may automatically establish the identity of an item of merchandise by reading a *bar code* on it and then retrieve its price from a *host* computer which monitors the transactions at several such EPOS terminals.

electronic publishing In general any system using electronic equipment to make quantities of data available on demand to a wide public. More specifically the term is often applied to *bulletin board systems* and to the posting of information on large computer networks such as the *Internet*.

electronics The study of the conduction of electricity in a vacuum, in gases and in *semiconductors*. Electronics is therefore concerned with methods of generating and controlling *charge carriers* such as *electrons*, *holes* and *ions* in, for example, *electron tubes* and *transistors*, and with applications of such devices.

electronic switching Use of *active devices* to perform switching operations. A *transistor* is a good approximation to a switch because it has nearly infinite *impedance* when non-conductive, and a very low impedance when conductive. Moreover it can be switched from one state to the other by a signal applied to the *base* (or *gate*). Two transistors with a common emitter or collector connection can perform a changeover operation by suitable signals applied to the bases. Such circuits are better than mechanical switches because they have no moving parts and can operate more quickly. Electronic switching is extensively employed in logic circuitry, for example in *digital computers*.

electron image Same as *charge image*.

electron lens An assembly of *electrodes* or of permanent or *electromagnets* which can be used to focus an electron beam at a given point, e.g. on the target of a cathode ray tube or a camera tube. See *electrostatic lens*, *magnetic lens*.

electron microscope An electron-optical instrument which gives a greatly-enlarged image of an object and with magnification and resolution superior to those of an optical microscope. The image may be formed by an *electron beam* which passes through a slice of the object or is reflected from it and is magnified by electron lenses for projection on to a screen or photographic film. Sometimes the image is formed from *electrons* emitted by the subject itself.

electron multiplier An *electron tube* in which the signal current is amplified by means of secondary emission. The tube contains a *cathode*, a series of *dynodes*, and an *anode* arranged as shown in *Figure E.12*. *Electrons* released from the cathode (e.g. as a result of light falling on it) are attracted to dynode 1 by its positive potential and, on striking it, release secondary electrons from it. These are, in turn, attracted to dynode 2 and release further secondary electrons from this. Thus the process continues, the

118

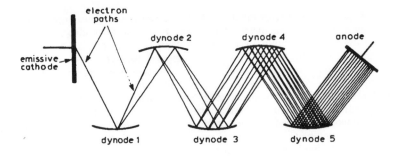

Figure E.12 The principle of the electron multiplier

number of secondary electrons growing at each stage until the anode is reached. In this way current amplification factors of several thousand can be obtained. Such multipliers are used in photocells (see *photo-multiplier*) and in the *image-orthicon tube*.

electron optics The study of the behaviour of *electron beams* subjected to electric and magnetic fields, particularly the use of such fields to deflect and focus electron beams. By applying suitable potentials to a system of electrodes it is possible to produce an electric field with *equipotential surfaces* shaped like convex or concave lenses. The behaviour of an electron beam entering such a system is similar to that of a light beam entering an optical lens system and this analogy has prompted the adoption of the term electron optics. See *electron lens*.

electron ray tube (US) Same as *cathode ray tuning indicator*.

electron tube General term for an evacuated or gas-filled envelope containing electrodes and in which conduction occurs by virtue of *electrons* or *ions* which move in a controlled manner according to the voltages or currents applied to the *electrodes*. The US term 'tube' has a wider meaning than the English 'valve' because it embraces cathode ray tubes, camera tubes, picture tubes and photocells which are not usually regarded as valves. See *gas-filled tube*, *vacuum tube*.

electrostatic coupling Same as *capacitive coupling*.

electrostatic deflection Deflection of the *electron beam* in a *cathode ray tube* by the electric field between two charged deflector plates. The beam is deflected towards the positively-charged plate. See *deflection sensitivity*.

electrostatic field Same as *electric field*.

electrostatic flux Same as *electric flux*.

electrostatic focusing Focusing of the *electron beam* in a *cathode ray tube* by the use of an *electrostatic lens*.

electrostatic induction The process by which an electrically uncharged body acquires a charge when placed near (but not touching) a charged body. The process is a manifestation of the repulsion between 'like' charges. If the charged body is negative it has a surfeit of electrons and these repel electrons on the uncharged body to its tar side, leaving the nearside deficient of electrons, i.e. positively charged—a charge arising from electrostatic induction.

119

electrostatic lens A lens in which an *electron beam* is brought to a focus at a desired point by means of an electric field. In general an electrostatic lens consists of an assembly of *electrodes* usually of cylindrical form and concentric with the beam. By suitably biasing these electrodes an electric field is produced which has an effect on the electron beam similar to that of a convex lens on a beam of light and by adjustment of the cylinder potentials the electron beam can be brought to a focus at the desired point.

electrostatic loudspeaker A *loudspeaker* in the form of a charged parallel-plate *capacitor* with a very small air gap, one plate being fixed and the other free to vibrate under the action of the AF signal applied between the plates. This simple form of loudspeaker is illustrated in *Figure E.13* and

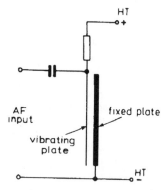

Figure E.13 A simple electrostatic loudspeaker

gives considerable *harmonic distortion* unless the amplitude of vibration is kept very small. Its use is therefore limited to the upper end of the AF spectrum where signal amplitudes are small. If, however, the vibrating plate is mounted between two perforated fixed plates, the AF signal being applied in push-pull to the outer plates, the distortion can be so reduced that successful loudspeakers operating on this principle over the whole audio range have been developed.

electrostatic memory A device in which data are stored in the form of electric charges, e.g. on the screen of a *storage tube*.

electrostatic microphone A *microphone* in the form of a small charged parallel-plate *capacitor* with a very small air gap, one plate being fixed and the other free to vibrate under the influence of any incident sound wave. When the moving plate (the diaphragm) vibrates, the capacitance of the microphone changes in sympathy with the pressure of the incident sound wave. Because the charge on the plates is maintained constant, the voltage between the plates varies similarly and this is the output of the microphone.

If such a microphone is connected to a long length of cable, the capacitance of the cable is in parallel with the microphone capacitance and effectively reduces the microphone output. To avoid this effect a common technique is to incorporate a head amplifier in the microphone casing to isolate the microphone from the cable and, of course, to provide amplification of the low microphone output.

electrostatic screen Same as *electric screen*.

electrostatic shield (US) Same as *electric screen*.

electrostatic voltmeter An instrument for measuring high direct voltages which relies for its action on the attraction between the opposite charges on the plates of a *capacitor*. One form of the instrument resembles a multiplate variable capacitor, one set of plates being fixed and the other free to move against the pull of a spring. The extent of the movement is an indication of the magnitude of the applied voltage. The instrument, being capacitive, takes no current from the source of voltage after the initial charging current and is therefore useful for measuring voltages with a very high source resistance such as the EHT supply in TV receivers.

elliptical polarisation Property of an *electromagnetic wave* of which the plane of *polarisation* is rotating and for which the *amplitude* varies according to the direction of the plane of polarisation. Vertically- and horizonally-polarised radio waves can become elliptically polarised after reflection in the ionosphere. If the amplitude of the wave is constant whilst the plane of polarisation rotates the wave is said to be circularly polarised.

emission General term for the release of electrons from a surface when suitably stimulated. See *cold cathode, field emission, photo-electric emission, secondary emission, thermionic emission*.

emitron tube Same as *iconoscope*.

emitter That region of a bipolar transistor from which *charge carriers* flow into the base region. The carriers are *majority carriers* in the emitter region but *minority carriers* in the base region.

emitter AC resistance See *electrode AC resistance*.

emitter-coupled logic (ECL) A form of logic using *bipolar transistors* in which the output signals are derived from the emitters. The low output impedance thus obtained permits rapid charging and discharging of output circuits. Moreover by using *Schottky diodes* connected between collectors and base the transistors can be prevented from becoming saturated. A saturated transistor takes appreciable time to turn off because of *carrier storage*. Thus very low rise and fall times are possible with ECL making this one of the fastest logic systems.

emitter follower Same as *common-collector circuit*.

emulation The use of one device or system to play the part of another. More specifically, the *programming* of a *computer* or *processor* to behave as though it were a computer or processor of a different type. In computing this is useful when a user wishes to access data stored in a format peculiar to a machine which is now obsolete and has been discarded. It is common practice for *printers* to offer a range of emulations of other popular printer types. This ensures that the user will have suitable *driver* software.

enabling signal A signal which prepares a circuit for some subsequent action. For example the signal may open a normally-closed gate.

Encapsulated PostScript (EPS) A *subset* of *PostScript* which is widely used for transferring *vector graphics* images between different applications and different computer types.

encapsulation The process of encasing and sealing a component or an assembly of components in moulded insulating material such as resin. The process is usually carried out in a vacuum and under pressure to prevent the formation of voids and air bubbles.

encoding Same as *coding*.

endurance test A test on a component to see if it can withstand the changes in stress it is likely to encounter during its life. Depending on the nature of the component the test consists of repeated changes in mechanical or electrical stress, the number of changes being comparable with that likely to occur during the life of the component.

energy gap In an *energy-level diagram* the difference in electron energy between the upper boundary of the *valence band* and the lower boundary of the *conduction band*. This is equal to the energy of the *forbidden band* and may be defined as the minimum amount of energy required to raise an electron from the valence band to the conduction band.

energy-level diagram A diagram in which electron energy is plotted on a vertical axis.

Electrons have different energies depending on the orbit they occupy. Thus the energies occupy vertical bands such as the conduction band and the valence band as shown in *Figure E.14*. See *conduction band, forbidden band, valence band*.

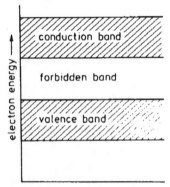

Figure E.14 Typical energy-level diagram

enhancement mode A mode of operation of a *field-effect transistor* which is designed to take zero drain current at zero gate bias, a forward bias being necessary to increase the current to a normal working value.

entity In computing an item of data stored in a *database* or a *repository* member or a relationship between repository members.

envelope delay The time taken for a particular point in a *waveform* to pass though a system or equipment. It is equal to the first derivative of the phase shift ϕ with respect to the angular frequency ω. Thus

$$\text{group delay} = \text{envelope delay} = \frac{\delta\phi}{\delta\omega}$$

See *delay, delay distortion, phase distortion*.

envelope detector A detector of which the output waveform is substantially that of the *modulation envelope* of the input signal.

envelope velocity The velocity of propagation of the envelope of a wave occupying a frequency band over which the *envelope delay* is approximately constant.

122

epitaxy A method of depositing a thin layer of *semiconductor* material on a crystal of the same material so that the layer takes up the same crystal orientation as the crystal. The method is used in transistor manufacture so that the bulk of the emitter and collector regions can be of low *resistivity* (and hence low-loss) material, the epitaxial layer being of resistivity suitable to form the pn junctions with the base layer.

equaliser A *network* designed to offset the effects of attenuation distortion or phase distortion occurring in *lines*, equipment or *transducers*. Thus an equaliser intended to compensate for a loss in high-frequency response of a line requires a complementary 'top-lift' frequency characteristic. Equalisers are often made variable so that they can be adjusted to suit individual line requirements.

equality gate Same as *equivalence element*.

equipotential surface In an electric field an imaginary surface all points on which have the same electric potential. Lines of electric force always cross equipotential surfaces at right angles. In the electric field surrounding a point charge the equipotential surfaces are spherical and concentric with the point charge. In a charged parallel-plate *capacitor* the equipotential surfaces are planes parallel to the plates.

equivalence element In digital circuitry an element with two inputs which gives a logic-1 output when and only when both inputs are at logic 1 or at logic 0. The output of an equivalence element is therefore the complement of that of an *exclusive-OR gate*. See *logic gate, logic level*.

equivalent capacitance That value of *capacitance* which can replace a given *two-terminal network* without affecting the voltage or current at the terminals. For a simple network of capacitances C_1, C_2, C_3, etc in parallel the equivalent capacitance C_{eq} is equal to the sum of the individual capacitances thus

$$C_{eq} = C_1 + C_2 + C_3 + \text{etc}$$

For a simple network of capacitances C_1, C_2, C_3, etc in series the reciprocal of the equivalent capacitance C_{eq} is equal to the sum of the reciprocals of the individual capacitances thus

$$\frac{1}{C_{eq}} = \frac{1}{C_1} + \frac{1}{C_2} + \frac{1}{C_3} + \text{etc.}$$

equivalent circuit A simple theoretical circuit of *resistance, inductance, capacitance*, voltage or current sources which has approximately the same electrical characteristics as a practical circuit or *active device* over a particular amplitude or frequency range. Such equivalent circuits are convenient for analysis of circuit behaviour or circuit design. *Figure E.15* is an example of one form of equivalent circuit for a *bipolar transistor*.

equivalent inductance That value of *inductance* which can replace a given *two-terminal network* without affecting the voltage or current at the terminals. For a simple network of inductances L_1, L_2, L_3, etc in series (and without mutual coupling) the equivalent inductance L_{eq} is equal to the sum of the individual inductances thus

$$L_{eq} = L_1 + L_2 + L_3 + \text{etc}$$

123

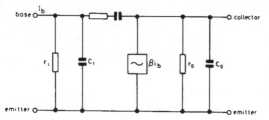

Figure E.15 Example of an equivalent circuit for a bipolar transistor which may be used at radio frequencies

For a simple network of inductances L_1, L_2, L_3, etc in parallel (and without mutual coupling) the reciprocal of the equivalent inductance L_{eq} is equal to the sum of the reciprocals of the individual inductances thus

$$\frac{1}{L_{eq}} = \frac{1}{L_1} + \frac{1}{L_2} + \frac{1}{L_3} + \text{etc.}$$

equivalent network A *network* which, under certain conditions, can replace another network without effect on the electrical performance of the system of which it forms part. For example a *pi-network* can replace a *T-network* at a particular frequency provided the elements have suitable values. See *star-delta transformation*.

equivalent noise resistance Of a circuit or equipment the value of a fictitious resistor included at the input of the circuit (assumed noise-free) which at 290°K gives the same noise output as the circuit. For an *electron tube* the resistor is assumed connected in the control grid circuit. Thus the *shot noise* and *partition noise* of the tube are regarded as thermal noise generated in the equivalent noise resistance.

equivalent resistance That value of resistance which can replace a given *two-terminal network* without affecting the voltage or current at the terminals. For a simple network of resistances R_1, R_2, R_3, etc, in the series the equivalent resistance R_{eq} is equal to the sum of the individual resistances thus

$R_{eq} = R_1 + R_2 + R_3 + \text{etc}$

For a simple network of resistances R_1, R_2, R_3, etc, in parallel the reciprocal of the equivalent resistance R_{eq} is equal to the sum of the reciprocals of the individual resistances thus

$$\frac{1}{R_{eq}} = \frac{1}{R_1} + \frac{1}{R_2} + \frac{1}{R_3} + \text{etc.}$$

erasable, programmable, read-only memory (EPROM) A *non-volatile read-only memory* in which the stored data can be repeatedly changed by the user.

Most EPROMs are based in the floating-gate structure illustrated in *Figure E.2*. Programming is achieved by applying a high voltage (e.g. 20 V) to the drain terminal. This injects high-energy electrons from the substrate through the silicon dioxide layer to the floating gate, where the charge is stored, the gate being surrounded by insulating silicon dioxide.

Because of the presence of the stored charge a high voltage must now be applied to the select gate to turn the transistor on; normal operating voltages (e.g. 3 V) have no effect. In the absence of a stored charge the normal low voltage would, of course suffice to turn the transistor on.

To erase the cell requires internal emission of photo-electrons from the floating gate to the select gate and the substrate. This is achieved by exposing the cell to ultra-violet light for several minutes. The light must be applied via a window in the package and this requires the EPROM to be removed from the equipment for erasure. This disadvantage is overcome in the *EEPROM*.

erasing (1) In *magnetic recording* the obliteration of an existing recording to make the medium available for a new recording. (2) In *digital computers* the removal of information from the store leaving the space available for new information.

error In *computers* and *data-processing equipment*, circumstances which prevent normal execution of a *program* through the introduction of conditions which the *central processing unit* cannot handle. Examples are an instruction to divide a number by zero or to perform an operation on a file that does not exist. In such conditions an error message is often given.

error-correcting code A data code in which the signals conform to specific rules of construction to enable departures from the standard form to be automatically detected and corrected. Such codes require more signal elements than are necessary to convey the information. See *error-detecting code, parity check*.

error-detecting code A data code in which the signals conform to specific rules of construction to enable departures from the standard form to be detected. Such codes require more signal elements than are necessary to convey the information. As an example in a digital code it may be arranged that all words contain an even number of pulses. Thus any word with an odd number can be detected by a parity check. See *error-correcting code, parity check*.

Esaki diode Same as *tunnel diode*.

ethernet Form of *Local Area Network* for connecting computers, printers and other equipment within one building. The connection is via twisted wires or coaxial cable and can transmit data at up to 10 M bits/second.

exalted-carrier reception Reception of amplitude-modulated single-, vestigial or double-sideband signals in which the carrier is amplified more than the sidebands with the object of reducing distortion. This method is often used to receive reduced-carrier single-sideband transmissions. See *amplitude modulation, double-sideband transmission, single-sideband transmission, vestigial-sideband transmission*.

except element Same as *exclusive-OR element*.

excitation In general the application of an input signal or of the power supply to an item of equipment or to a stage. For example, the current which energises the field magnets of a DC generator is called the excitation. The input signal applied to the control grid of an *electron tube* is also known as the exciting voltage.

excitron A grid-controlled gas-discharge rectifier with a mercury-pool cathode and a single anode capable of a large current output. The arc is initiated by an igniter electrode and is maintained by a keep-alive electrode fed from an auxiliary DC source.

exclusive-NOR gate A gate with two inputs which gives a logic-0 output when one and only one of the inputs stands at logic-1. It is therefore the complement of an exclusive-OR gate and its performance is identical with that of an *equivalence element*.

exclusive-OR gate A gate with two inputs which gives a logic-1 output when one and only one of the inputs is at logic 1. The graphical symbol for an exclusive-OR gate is given in *Figure E.16*.

Figure E.16 Logic symbol for an exclusive-OR gate

executable code Same as *machine code*.

exit In mainframe computers a point in the *program* at which control can be transferred to a user-defined *subroutine*. It allows users to tailor standard software to meet their precise needs.

expansion Increase in the volume range of a signal by varying the gain to which it is subjected according to the mean volume level, gain being greater for high volume levels than for low volume levels. This is the converse of *compression* and it is used to restore the original volume range of a compressed signal. To achieve exact compensation for compression the laws of the compressor and expander must be the same.

expansion card An electronic circuit board which plugs into the *backplane* in a computer. It might contain an interface for some peripheral, additional memory, a video digitiser, an extra disk drive or some other storage medium.

expert system A computer application that simulates the behaviour of a human expert in some subject. The application stores large quantities of information in a logically organised structure. It has precise rules about the circumstances in which the information is presented to the user. For example, a medical expert system might begin with a menu listing types of symptoms. Each menu item might lead to a further sub-menu of more specific symptoms. These in turn might lead to a set of questions which the user must answer concerning his condition. When this is completed, the application describes possible causes of the condition and methods of treatment. The application has, in fact, followed much the same course that a consultation with a live doctor might have taken.

exploring coil Same as *search coil*.

exponential horn An acoustic horn the cross-sectional area of which increases exponentially with distance from the throat. Such horns were used in acoustic record players and sometimes with loudspeakers, the best example of which is the Voigt corner horn loudspeaker.

export In computing, to extract data from an application in a format that is foreign to that application or to that computer type. For example, most *vector graphics* packages allow users to export the current document in *EPS (Encapsulated PostScript)* format. This is useful because this format can also be *imported* by a wide variety of applications on many computer

126

types. Exporting facilitates the transfer of data between applications and *platforms*.

Extended Binary Coded Decimal Interexchange Code (EBCDIC) The code used to represent alphanumeric and other printable characters and control characters in some mainframe and mini-computers, in contrast with the *ASCII* code used in microcomputers. The two codes are not compatible and applications which transfer textual data between, e.g. mainframe computers and microcomputers, are required to convert the data from one code to the other.

extended character set An *ASCII* character set that includes characters having codes above 127. The additional characters usually included accented letters for most European languages, a selection of currency symbols and a wide range of punctuation marks.

extinction current Of a *gas-filled tube* the minimum current at which the discharge can be maintained.

extinction frequency In *magnetic recording* the frequency at which the output from the *reproducing head* falls to zero as a result of the recorded wavelength becoming comparable with the effective width of the gap in the reproducing head.

extinction voltage Of a *gas-filled tube* with decreasing anode voltage the value of the voltage at which the discharge ceases.

extrinsic semiconductor A *semiconductor*, the conductivity of which is dependent on the presence of an impurity. Most semiconductor materials in their pure state are almost non-conductive but by the introduction of a controlled amount of a suitable impurity the conductivity can be increased to a value at which the material can be used in the manufacture of semiconductor devices such as *diodes* and *transistors*. The proportion of impurity needed for this purpose is very small, typically a few parts in 10^8 so that the initial semiconductor material must be in a very high state of purity. See *intrinsic semiconductor*.

F

facsimile (1) The process of *scanning* a still picture to obtain corresponding electrical signals which can be used locally or remotely to produce a recorded likeness of the picture.

The process is extensively used by news media for the radio transmission of photographs and documents over telephone circuits. The original is clamped to a cylinder which rotates, causing a spot of light to scan the material in a spiral path. The light transmitted through the original falls on a *photocell*, the output of which is sent to the reproducing machine. The image is reproduced on photo-sensitive material clamped to a rotating cylinder and scanned by a light beam the density of which is controlled in effect by the output of the photocell. *Synchronising signals* are used to keep the scanning process at the transmitting and receiving ends in step.

(2) A form of paper-based *electronic mail* using standard telephone lines. A 'fax machine' consists of a *scanner*, a *printer*, a *modem* and a dedicated microcomputer. In the sending instrument the document is scanned and the resulting *bitmap* is *compressed* before transmission over the telephone lines. In the receiving instrument the bitmap is reconstructed and printed. Computer software is available which allows computers connected to suitable modems to send fax messages based on software *documents* which may never have been printed on paper and to receive and display fax messages on screen without any need to print them.

(3) The term is also used to mean an image reproduced by either of the above processes.

fading Accidental or deliberate variation of signal *amplitude*. Random amplitude variations occur in the signal input to a radio receiver as a result of the vagaries of radio-wave propagation and AGC systems are used in *receivers* to minimise the effects of such variations.

Deliberate variations of signal amplitude are employed in sound and vision mixing equipment where programme material is introduced by bringing the signals up to standard amplitude (known as fading up or fading in). Similarly programme material is removed by reducing the signal amplitude to zero (known as fading out).

fader Control used for fade-in and fade-out effects. See *fading*.

failure Cessation of the ability of a component or equipment to carry out the required function.

failure rate The number of failures per unit time.

fall time A measure of the steepness of the trailing edge of a pulse waveform. More specifically it is the time taken for the instantaneous *amplitude* to change from 90% to 10% of the peak value as illustrated in *Figure F.1*.

The ability of a circuit to bring about changes in voltage or current as required in the reproduction of pulses is determined by the high-frequency response of the circuit and the following simple relationship exists between

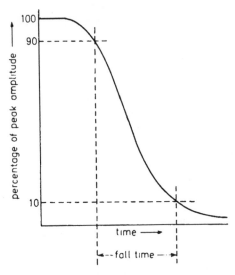

Figure F.1 Fall time of a pulse

fall time and upper frequency limit f_{max}:

$$f_{max} = \frac{1}{2 \,(\text{fall time})}$$

Thus to reproduce a pulse with a 0.1-μs fall time an amplifier requires an upper frequency limit of at least 5 MHz.

In a simple *RC* or *RL* circuit, decreases in voltage or current are exponential in form and for such a change the fall time is simply related to the time constant according to the approximate relationship

fall time = 2.2 × time constant.

fan in In logic circuitry the maximum number of outputs that may be connected in parallel at the input to a logic element whilst still permitting normal operation of that element.

fan out In logic circuitry the maximum number of inputs which can be connected in parallel at the output of a logic element whilst still permitting normal operation of that element.

Farad The unit of capacitance, symbol F. A capacitor is said to have a capacitance of 1 F when 1 coulomb of charge on each plate sets up a potential difference between them of 1 Volt. The farad is an inconveniently large unit and the microfarad (μF) and the picofarad (pF) are more commonly used.

Faraday cage An earthed *electric screen* of wire mesh or parallel conductors which shields equipment within it from the effects of *electric fields* but is transparent to *magnetic fields*.

Faraday's law Of electromagnetic induction states that the EMF induced in a circuit is directly proportional to the rate of change of the magnetic flux linked with the circuit.

fatal error An *error* which necessitates the immediate termination of the current application, but does not necessarily affect other tasks in the computer.

fax Same as *facsimile* (2).

feed Of an *active device* the mean current taken by the device from the DC power supply.

feedback The return of a portion of the signal at any stage in an amplifier to an earlier stage so as to augment or reduce the signal at that stage. If the returned signal is in phase with the signal at the earlier point the feedback is said to be positive and its effect is to increase gain. If the returned signal is in phase opposition to the signal at the earlier point the feedback is said to be negative and its effect is to reduce gain. See *negative feedback, positive feedback*.

feeder In general any *transmission line* carrying electrical power. In particular the type of line used to connect an *antenna* to a transmitter or a receiver.

ferrite A homogeneous non-metallic material with high permeability and high electrical resistance with the general formula MFe_2O_4 where M is a divalent metal such as nickel, cobalt or zinc. Ferrite cores have very low eddy-current losses making them suitable for use in RF inductors and transformers and in the yokes of deflection coils. Ferrites are of ceramic nature and cannot be sawn or drilled. They are produced in the form of cups, rods and rings from which magnetic circuits can be constructed.

ferrite bead store Same as *bead store*.

ferromagnetism A property of certain materials subjected to a magnetising field which causes induced magnetism which greatly aids the applied field. Such materials are strongly attracted to a magnetic pole and have high effective permeabilities which are greatly dependent on the applied magnetising field. Iron, cobalt, nickel and certain alloys are typical examples of ferromagnetic materials. See *diamagnetism, paramagnetism*.

fibre optics The study of the transmission of light along thin fibres of transparent material. Light can be transmitted along such a fibre by a succession of internal reflections and transmission is improved if the fibre material is clad in a material of lower refractive index. Transmission is unaffected if the fibre is bent and thus optical fibres provide a method of conveying light from one point to another and around bends if necessary.

By using an assembly of a very large number of very fine fibres it is possible to convey optical images from one point to another and around bends. This is useful in surgery for internal examinations of the body and there are also television applications.

Because of the enormous *bandwidth* obtainable with optical fibres they are also used as a means of communication over long distances.

fidelity The degree to which a transmission system or part of a system reproduces at its output the essential characteristics of the signal applied to its input. The term is often applied to audio equipment such as amplifiers and loudspeakers.

field In TV the part-picture composed by the lines described in one downward sweep of the scanning agent. In *twin-interlaced scanning*, as used in most TV systems, two vertical sweeps are needed to cover all the lines of the picture and each field is thus a half-picture. If the lines are numbered in sequence from the top to the bottom of the picture, the scanning agent first covers lines 1, 3, 5, etc (this being known as the odd field) and then returns to the top of the picture to cover lines 2, 4, 6, etc (this being known as the even field). The field was formerly known as a *frame*.

field blanking In TV the suppression of the picture signal during the interval between successive fields.

field-effect transistor (FET) A *transistor* consisting essentially of a channel of *semiconductor* material, the resistance of which can be controlled by the voltage applied to one or more input terminals (gates).

A feature of the FET is that the input resistance of the gate is very high so that input current is negligible. FETs are therefore voltage-operated devices, a property they share with the *electron tube*. FETs differ from *bipolar transistors* in that the output current is carried by only one type of charge carrier, e.g. electrons in n-channel devices. See *insulated-gate field-effect transistor, junction-gate field-effect transistor*.

field emission The release of electrons from an unheated surface under the action of a strong magnetic field.

field frequency In TV the number of vertical sweeps made by the scanning beam in one second. For *interlaced scanning* it is equal to the product of the picture frequency and the number of fields per picture. In most TV systems the field frequency is approximately equal to the frequency of the supply mains. In Europe this is 50 Hz.

field sync signal In TV the signal transmitted at the end of each field to initiate vertical *flyback* of the scanning beam in receivers so keeping field scanning at the receiver in step with that at the transmitter.

In most TV systems the field sync signal consists of one or more pulses (each longer than the line sync pulse) and so arranged that the continuity of the line sync pulses is not interrupted. See *Figure F.2*.

edges triggering line time base

time —→

Figure F.2 Waveform of a field sync signal

field time base In TV the circuits responsible for generating the signals causing vertical deflection of the scanning beam.

filament The cathode of an *electron tube* which is heated to give the required electron emission by passing an electric current directly through it.

For low-power tubes such as those used in battery-operated equipment, such cathodes consisted of a fine wire consuming say 0.1 A at 1. 4 V or 2 V and coated with metallic oxides to yield copious electron emission at a low temperature.

For tubes of higher power such as mains rectifiers the filament was often in the form of a tape consuming say 2 A at 5 V. For high-power transmitting tubes the filament is of pure tungsten or thoriated tungsten consuming up to several hundred amps at say 30 V.

file In *computers* and *data-processing equipment* a series of *bytes* given a unique identity (*filename*) for storage in a *filing system*.

fileserver A computer in a *Local Area Network* which incorporates a high-capacity *hard disk* on which storage space is provided for the users of the network. In practice some storage space may be declared 'public'; this contains files which any network user may access. Other storage space is dedicated to individual users of the network, access to these being protected by a password system. The fileserver may also control other shared resources such as a network *printer*.

filing system In computers and data processing equipment a system which stores *programs* and *data* as *files* in an orderly manner which permits their rapid location for retrieval. The storage media most commonly used are magnetic disks.

filter A *network* which passes signals with frequencies within certain bands (*passbands*) with little attenuation but greatly attenuates signals within other bands (stopbands). The block symbol for a filter is given at *Figure F.3*.

Figure F.3 Block symbol for filter

firing voltage The minimum direct voltage applied to a gas-discharge tube which will initiate the discharge.

firmware Software resident in *read-only memory* in a computer system.

first generation computer See *digital computer*.

fixed-point arithmetic, computation or representation In computers a form of presentation in which a number is displayed as a single set of digits, the decimal (binary or other) point being fixed in position with respect to the set.

fixed disk Same as *hard disk*.

fixed storage Same as *read-only memory*.

flag In computers and data-processing equipment a *bit* or *byte* used to signal the occurrence of a particular condition or event, e.g. an *error*.

flare In an optical or TV image, spurious areas caused by scattering of light in the camera or picture tube.

flash arc In an *electron tube* an arc between the *electrodes* causing a violent increase in cathode emission which, if maintained, can destroy the tube. The effect is thought to be caused by irregularities on electrode surfaces which cause local concentrations of electric field sufficient to ionise the residual gas.

flash EPROM A type of EPROM which can be erased and re-used. Unlike EEPROM, however, the whole memory chip must be erased at once. See *erasable, programmable, read-only memory*.

flashing Same as *reactivation*.

flashover *Arc discharge* between two conductors or between a conductor and earth caused by excessive voltage or a breakdown of the insulation. It may occur between the *electrodes* of an *electron tube*.

flash test Application for a brief period of a voltage considerably greater than the working voltage of a component or equipment to test its insulation resistance.

flat random noise A noise signal in which the components have approximately equal amplitudes over the frequency range of interest.

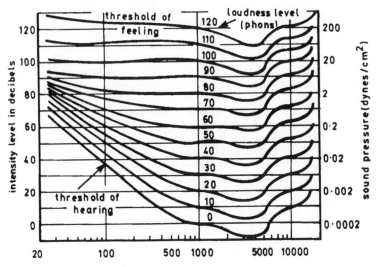

Figure F.4 Fletcher-Munson equal-loudness curves

Fletcher-Munson curves Equal-loudness curves plotted in terms of sound intensity and frequency (*Figure F.4*).

Each curve is an equal-loudness contour, i.e. it shows how the sound intensity necessary to give a particular value of loudness level varies over the audio band. The lower contours show that at low loudness levels considerably more power is needed to make very low frequency signals and very high-frequency signals as loud as signals around 1 and 2 kHz. This is another way of saying that the human ear is most sensitive at frequencies around 1 and 2 kHz.

As loudness level is increased, however, the contours become flatter showing that the sound intensity needed to give constant loudness does not vary greatly with frequency. In some audio amplifiers the volume control is made frequency discriminating to allow for the change in shape of the contours with change in loudness level.

The number of *phons* for each contour is numerically equal to the intensity level at the point where the contour crosses the 1-kHz ordinate; this stems from the definition of the phon.

flicker In TV, unwanted regular variation in the brightness of the reproduced picture. Flicker can be annoying when the *field frequency* is low and, in fact, this consideration sets a lower limit to the field frequency which can be used. It was for this reason that *interlaced scanning* was adopted because this permits a high field frequency (minimising flicker) whilst allowing a low picture frequency (minimising bandwidth).

flicker effect In *electron tubes*, random variations of the output current causing noise which is inversely proportional to frequency.

flip-flop (US) Same as *bistable circuit*.

floating Not connected to any source of potential. For example the potential of the open-circuited grid of an *electron tube* could be described as floating.

floating-carrier modulation A system of *amplitude modulation* in which the carrier is modulated in the normal manner and is simultaneously controlled by the envelope of the *modulating signal* so that the carrier amplitude is at all times just large enough to accommodate the modulation envelope.

Thus the *modulation depth* remains constant at nearly 100% no matter what the amplitude of the modulating signal. Such a system gives more efficient transmitter operation than if the carrier amplitude is kept constant as in conventional amplitude modulation.

floating-point arithmetic, computation or representation In computing a form of presentation in which a number x is displayed as two numbers y and z such that $x = yb^z$ where b is usually 2 or 10. y is known as the fixed-point part, z as the exponent and b is the base. As an example the number 12500000 could be written 125, 5 (representing 125×10^5).

floppy disk (diskette) A *demountable* data storage and distribution medium used in *microcomputers* and *word processors*. The magnetic disk is enclosed in a rectangular plastic sleeve which, on the earlier 8-inch and 5¼-inch disks, was flexible, hence the term *floppy*. The 3½-inch disks now most widely used are housed in rigid plastic cases which are more robust and incorporate a metal shutter over the aperture for the *read/write* heads, affording better protection against grease and dust. Each side of the disk carries 80 tracks spaced at 135 tracks per inch, giving a storage capacity, depending on the operating system and disk quality, of 720 kbytes, 800 kbytes, 1.44 Mbytes, 1.6 Mbytes or 2.88 Mbytes.

floptical disk and drive A magnetic *data storage* system using media which resemble 3½-inch *floppy disks*, but which use an optical track to allow more accurate tracking of the *read/write* heads. Storage capacity is 21 Mbytes per disk. The drive can also read and write to standard 3½-inch floppy disks.

flow angle Same as *angle of current flow*.

Floyd and Steinberg error diffusion See *dithering*.

fluctuation noise Same as *random noise*.

fluorescence The emission of light from materials irradiated by energy of a higher frequency or bombarded by *electrons*. The effect differs from *phosphorescence* in that the light emission lasts only for the duration of the stimulus.

flutter Distortion in sound reproduced from disk, film or tape and caused by undesired rhythmic speed variations on recording or reproduction. It is usually caused by eccentric or unbalanced driving of the film, tape or turntable. The variations occur at a rate exceeding approximately 20 Hz and cause a roughness in the sound quality. See *wow*.

flux A term used to describe a flow of particles, *photons* or the lines of force of an electric or magnetic field.

flux density The number of particles, *photons* or *lines of force* of an electric or magnetic field which pass through unit area of a surface normal to the direction of the beam or field.

flyback In *cathode ray tubes* the rapid return of the *electron beam* to its starting point at the end of each *trace*. In TV there is a horizontal flyback at the end of each *scanning line* and a vertical flyback at the end of each field.

focusing In *electron optics* the process of converging the beam to minimum cross section (ideally to a small spot) on the target of the tube.

foldback A protective system for power suppliers which ensures, in the event of a sustained overload, that the output current is automatically reduced to zero thus reducing the possibility of damage to the supplier and/or to the load.

forbidden band In an *energy-level diagram* a range of electron energies lying between the permitted bands such as the *conduction band* and *the valence band*.

No electron can have an energy corresponding to that of a forbidden band; it must have lower energy and be, for example, in the valence band or higher energy and be in the conduction band. Electron energy is therefore in steps and if sufficient energy can be given to electrons in the valence band to lift them into the conduction band, the element can be made conductive. See *Figure E.12*.

forced-air cooling A method of removing heat from a component or equipment by blowing air against it or against finned radiators in good thermal contact with it.

forced oscillation See *oscillation*.

format (1) In data transmission the way in which the code characters are grouped into blocks. (2) A process known as *formatting* must be applied to magnetic storage media such as *floppy* and *hard disks* before they can be used to store data and programs. The process lays down the pattern of tracks (sometimes called *cylinders*) divided into *sectors* which the *operating system* uses to locate files and to check for errors and also creates the *root directory*.

form factor Of an alternating quantity the ratio of the root mean square value to the mean value of the positive or negative half cycle. For a sinusoidal *waveform* the RMS value is 0.71 ($1/\sqrt{2}$) and the mean value 0.63 of the peak value. The form factor is hence $0.71/0.63 = 1.11$.

Forth See *computer language*.

FORTRAN See *computer language*.

forward automatic gain control An AGC system in which the gain of *transistors* is reduced by use of forward control bias. An essential feature of the circuit is a resistor in the collector circuit (R_1 in *Figure F.5*) which causes the collector voltage to fall as the forward bias increases. The transistors used for this application are so designed that their collector characteristics become more crowded at low collector voltages thus decreasing gain.

forward current In a rectifier or *semiconductor diode* the current which flows through it in the forward direction.

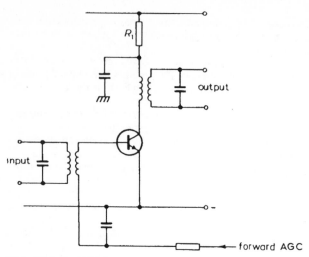

Figure F.5 Essential features of a stage of amplification using forward automatic gain control

forward power dissipation In a *semiconductor diode* the power dissipated within the diode by the *forward current*.

forward recovery time Of a *semiconductor diode* in a specified circuit the time taken for the *forward current* to reach a specified value after the application of forward voltage.

Foster-Seeley discriminator A detector for phase- or frequency-modulated signals in which the centre tap of the tuned secondary winding of a transformer is coupled to the primary winding and in which two *diodes* in series opposition are connected across the secondary winding, their net output giving the modulation-frequency output.

A typical circuit diagram of a Foster-Seeley discriminator is given in

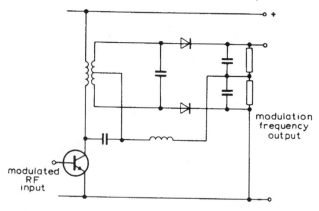

Figure F.6 A typical Foster-Seeley discriminator circuit

136

Figure F.6. The effect of the two types of *coupling* between the primary and the secondary windings is that an increase in input-signal frequency from its unmodulated value causes the voltage at one end of the secondary winding to increase whilst that at the other end decreases. One diode therefore yields a larger output than the other. When the frequency of the input decreases from the unmodulated value this process is reversed and the second diode now gives the larger output and it is of opposite polarity to that of the first diode. Thus a modulation-frequency output is obtained from the detector.

A disadvantage of this form of discriminator is that it responds to amplitude modulation of the input signal. In a receiver therefore it must be preceded by one or more limiter stages to give protection against unwanted noise signals. See **amplitude limiter**.

Foucault current Same as *eddy current*.

Fourier analysis A mathematical method of determining the number, *amplitude, frequency* and *phase* of the components of a given repetitive complex waveform.

The French mathematician Fourier showed that any repetitive waveform could be synthesised by adding together a number of sinusoidal waves of suitable frequency, amplitude and phase. The frequencies of the waves are simple multiples of the repetition frequency (i.e. the fundamental frequency) of the original waveform.

four-quadrant multiplier A multiplier which operates normally irrespective of the sign of the input signal. See *two-quadrant multiplier*.

fourth generation computer See *digital computer*.

Fowler-Norheim tunnelling See *EEPROM*.

fractal An image, usually mathematically generated, in which the detail appears similar at any level of magnification. A fern leaf provides an example from Nature of a fractal pattern. Fractal-based techniques offer exciting possibilities for data *compression*.

frame (US) In TV and cinema film practice, one complete picture.

frame frequency (US) Same as *field frequency*.

frame grabber Same as *video digitiser*.

frame-grid tube An *electron tube* with a *control grid* of fine pitch, wound with fine wire and situated very close to the *cathode* so giving higher values of *mutual conductance* than is possible with grids of more conventional construction. Mutual conductances as high as 15 mA/V can be obtained in a small receiving tube by using frame-grid construction.

framing In digital data transmission systems a procedure used to ensure that the *receiver* can identify the subchannel to which each set of data belongs.

Franklin oscillator An *oscillator* in which the frequency-determining circuit is coupled via very small *capacitances* to the input and output of a two-stage *electron tube* amplifier. Because of the double phase inversion in the amplifier, the output is in phase with its input so giving the positive *feedback* necessary for oscillation. By making the coupling capacitances to the tuned circuit only just large enough to sustain oscillation the effect of the amplifier on the LC circuit can be minimised so giving good frequency stability. A circuit diagram of a Franklin oscillator is given in *Figure F.7*.

free oscillation See *oscillation*.

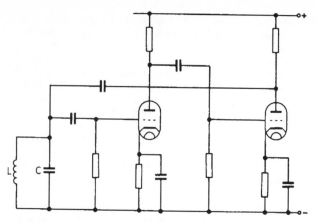

Figure F.7 Circuit diagram of a Franklin oscillator

freeware *Software* which may be freely copied, distributed and used subject to conditions described in a text file that forms part of it. Unlike *public domain software*, its originator retains the copyright and may demand payment if, for instance, a software distributor wishes to incorporate it or a part of it into a commercial package.

freeze frame (1) On a *video tape recorder*, the facility to stop the tape temporarily during playback while the head continues scanning, thus producing a still picture. In practice this facility often has a time limit after which normal playback is resumed; this is to prevent excessive wear of the tape surface. (2) In computing, a system of primitive animation in which successive frames appear too infrequently to give the illusion of true movement.

frequency band In general the range of frequencies between specified upper and lower limits. In particular one of the following frequency ranges which are agreed internationally.

VLF	very low frequency	3–30 kHz
LF	low frequency	30–300 kHz
MF	medium frequency	300 kHz–3 MHz
HF	high frequency	3–30 MHz
VHF	very high frequency	30–300 MHz
UHF	ultra high frequency	300 MHz–3 GHz
SHF	super high frequency	3–30 GHz
EHF	extra high frequency	30–300 GHz

frequency changer A circuit which accepts a modulated signal at a particular carrier frequency and transfers the modulation to a different, usually lower, carrier frequency. Such circuits are used in *superheterodyne receivers* to transfer the modulation of all received signals to the intermediate frequency. Most frequency changers consist of an *oscillator* (at a frequency f_1), the output of which is fed, together with the input signal (frequency f_2), to a mixer stage the output of which contains the

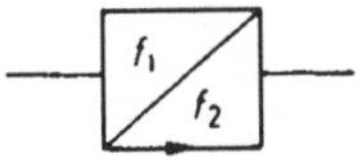

 Figure F.8 Block sumbol for a frequency changer

required difference frequency (f_1-f_2). Many practical frequency changers use separate *active devices* for oscillation and mixing although it is possible to make one device serve both purposes as in the *self-oscillating mixer*. The block symbol for a frequency changer is given in *Figure F.8*.

frequency converter (US) Same as *frequency changer.*

frequency deviation In angle modulation the peak difference between the instantaneous frequency of the modulated wave and the carrier frequency.

frequency discriminator Same as *discriminator.*

frequency divider A circuit which accepts a signal with a frequency f and gives an output at a frequency f/n, where n is an integer. See *counter*.

frequency division multiplex (FDM) The process of transmitting two or more signals along a common path by using different frequency bands for each of them. Each signal modulates a separate carrier and the carrier frequencies are spaced to avoid mutual interference between *sidebands.*

frequency modulation (FM) Methods of modulation in which the frequency of the carrier wave is made to vary in accordance with the instantaneous value of the modulating signal.

(a)   (b)

Figure F.9 (a) An unmodulated carrier wave and (b) the effect of frequency modulation

Figure F.9 (a) illustrates a carrier wave of constant amplitude and constant frequency and *Figure F.9* (b) shows the effect of frequency modulation. The amplitude of the modulating signal determines the extent of the change in carrier frequency and the frequency of the modulating signal determines the number of times per second the carrier frequency is swept above and below its nominal value.

Frequency modulation is used in sound broadcasting in the VHF bands and in most TV broadcasting services for the associated sound channel. An important feature of frequency modulation is that the carrier amplitude is unaffected by modulation: thus by making frequency modulation receivers immune to amplitude changes in the received signal much interference can be eliminated so making frequency-modulated reception noise-free.

frequency multiplier A circuit which accepts a signal with a frequency f and gives an output at a frequency nf, where n is an integer. The commonest examples of frequency multipliers are frequency doublers and frequency triplers.

One form of frequency multiplier is a device with a markedly non-linear input-output characteristic (e.g. an *active device* operating in class C). This generates an output rich in harmonics of the input frequency and the desired component can be selected by a circuit resonant at the required frequency.

frequency pulling The displacement of the frequency of an oscillator towards the frequency of an external applied signal. The effect becomes more marked as the frequency of the applied signal approaches that of the oscillator.

frequency range The range of frequencies over which a component or equipment operates satisfactorily. For example an AF amplifier may have a frequency range of 30 Hz to 15 kHz.

frequency response The variation in output amplitude of a component or equipment as the frequency of a constant-amplitude input is varied over the working frequency range of the equipment. The response is usually displayed as an amplitude/frequency curve and shows up any *attenuation distortion* of the equipment.

frequency shift keying (FSK) A form of *frequency modulation* in which, without interrupting the carrier, its frequency is switched between two predetermined values termed the mark and space frequencies.

frequency stability A measure of the constancy of the frequency of an oscillation. The frequency stability of an oscillator is important in many applications, e.g. when it is used as the carrier source in a *transmitter* which must stay on its assigned carrier frequency within very close tolerance.

frequency standard A highly stable *crystal-controlled oscillator* the output of which can be calibrated against National Standard Frequency radio transmissions.

frequency swing In *angle modulation* the difference between the maximum and minimum values of the instantaneous frequency of the modulated wave.

fringing Distortion of the *electric field* at the edges of a parallel-plate capacitor. The *lines of force* are normal to the plates over most of the area of the plates but at the edges they tend to bulge outwards, a feature that may be important in the design of electrostatic deflecting plates in a *cathode ray tube*.

front end (1) Of an equipment, that section of it which handles the *input* signal(s). (2) Of a computer *application*, the user interface, i.e. the screen display and input facilities with which the user interacts.

front porch In a TV signal the period of *blanking level* immediately preceding the *line sync signal*. See *Figure F.10*.

full adder Same as *adder*.

full motion video Smooth animation in a full-colour picture occupying the whole screen. This contrasts with the jerky *freeze-frame* animation or fractional-screen pictures offered in some computer animation systems.

full radiator An ideal radiator and absorber of radiation. Its radiation in any part of the spectrum is the maximum obtainable from any radiator at the same temperature. The nearest practical form of full radiator is a cavity with opaque walls maintained at a constant temperature and with a small opening for observation. It was formerly known as a black-body radiator.

full-wave rectification Rectification in which power for the load is taken

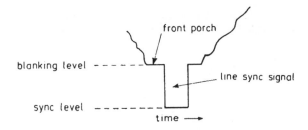

Figure F.10 Front porch of a television signal

from the AC supply during both half-cycles of the supply. One circuit for a full-wave rectifier is given in *Figure F.11*. Diode D_1 conducts during

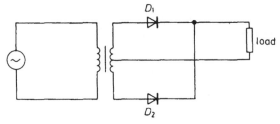

Figure F.11 A full-wave rectifier using a centre-tapped transformer

positive half-cycles of the secondary voltage and diode D_2 during negative half-cycles, the diodes being so connected that current flows through the load in the same direction during both half-cycles. As shown in *Figure F.12* a bridge arrangement of rectifiers is often used for full-wave rectification, this circuit avoiding the need for a centre tap on the transformer secondary winding.

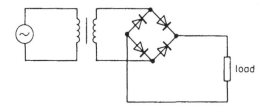

Figure F.12 A full-wave rectifier using a bridge rectifier

function In computing, a *subroutine* which is named and which returns either a numeric value or a character string. The usefulness of functions lies in that, once defined, they can be treated in much the same way as *variables*; for example, an instruction to display a function executes the code in the function and then displays the value it has returned.

141

function keys On a *computer keyboard* the keys normally numbered F1 to F12 (and sometimes to F24). They have no fixed function, but *operating systems* and *programs* may assign temporary functions to them. For example, F1 is often used to call up an *on-line help* system.

fundamental frequency Of a regular complex *waveform* the repetition frequency of the wave.

Fourier analysis of a complex wave yields a series of components the frequencies of which are multiples of the repetition or fundamental frequency of the wave. It is possible for the fundamental frequency component to be missing but in a complex sound signal it is the fundamental frequency, whether present as a discrete component or not, which determines the pitch of the sound.

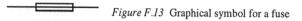

Figure F.13 Graphical symbol for a fuse

fuse A protective device which heats and melts so interrupting a circuit when the current exceeds a certain value which depends on the material and the cross-sectional area of the fuse. The graphical symbol for a fuse is given in *Figure F.13*.

G

gain   In general an indication of the extent to which the amplitude of a signal is increased by its passage through an electronic system or part of it. More specifically the ratio of the output-signal power, voltage or current to the input-signal power, voltage or current. The ratio is usually expressed in decibels as

$$10 \log_{10}\frac{P_{\text{out}}}{P_{\text{in}}};\ 20 \log_{10}\frac{V_{\text{out}}}{V_{\text{in}}};\ \text{or } 20 \log_{10}\frac{I_{\text{out}}}{I_{\text{in}}}$$

gain-bandwidth product   Of an *active device* a figure of merit assessing its usefulness as a wideband amplifier. A given active device can give a particular value of *gain* over a particular value of bandwidth and if the bandwidth is changed the gain is changed in the inverse ratio so that the product remains constant. For an *electron tube* the gain-bandwidth product is proportional to $g_{\text{m}}/(C_{\text{in}} + C_{\text{out}})$, where g_{m} is the mutual conductance, C_{in} the input capacitance and C_{out} the output capacitance. Thus for wideband applications, tubes with high mutual conductance and low input and output capacitances should be chosen.

gallium arsenide (GaAs)   A *semiconductor* material in which the electron mobility is better than that of silicon. This makes the material suitable for use in microwave transistors (MESFETs) and for high-speed operations in computers. GaAs also has the property that the electrons can exist in two states as described under *Gunn effect*. GaAs is also used in LEDs.

galvanometer   An instrument for the detection and measurement of very small electric currents. In one type the current passes through a rotatable coil situated in the field of a permanent magnet. The resultant magnetic interaction causes the coil to rotate against the restoring force of a spring, so moving a pointer or a small mirror which reflects a narrow beam of light across a scale.

gamma (γ)   In TV a factor expressing the relationship between the contrast in the reproduced image and that in the original scene. Quantitatively it is the slope of the curve of output *luminance* plotted against input luminance, both being plotted on logarithmic scales. As shown in *Figure G.1* the overall gamma is given by the slope of the chord connecting two points P and Q which define the length of the characteristic used in the system. Thus overall gamma is given by

$$\gamma = \frac{\log R_2 - \log R_1}{\log S_2 - \log S_1}$$

where R_1 and R_2 are the luminances of the two points in the reproduced image and S_1 and S_2 are the luminances of the corresponding points in the original scene.

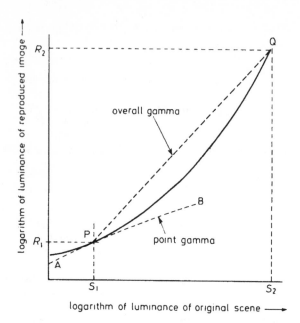

Figure G.1 Distinction between overall gamma and point gamma

For a colour TV system the overall gamma should be unity but for black-and-white television where the contrast due to colours is absent an overall gamma of slightly greater than unity, say, 1.2, is recommended.

Because of the curvature of the characteristic the gamma as measured by the slope of the curve can differ appreciably from the overall value. Thus point gamma or contrast gradient at a given point in the characteristic is measured by the slope of the tangent to the point and it is useful to know its value because it measures the visibility of detail in tonal values in the image corresponding to the point. For example in *Figure G.1* the tangent AB measures the point gamma in low-light areas of the image and clearly this gamma is less than the overall gamma, suggested that perhaps some 'black stretching' may be desirable.

gamma rays *Electromagnetic waves* similar to *X-rays* and of approximately the same frequency emitted from radioactive atoms during disintegration.

ganging Simultaneous adjustment of two or more circuits by a common mechanical control. In a stereo amplifier, for example, operation of the gain control simultaneously adjusts the gain in both channels. In a *superheterodyne receiver* tuning of the oscillator and RF circuits is simultaneously adjusted by a single control.

gas amplification factor Of a gas-filled photocell the ratio of the sensitivity with and without *ionisation* of the gas.

gas current In a *vacuum tube* the flow of positively-charged *ions* to a negatively-charged electrode. The ions are produced by ionisation of residual gas by the electron current of the valve.

144

gas discharge The conduction of a current between two cold electrodes in an ionised gas tube when the potential difference between them is made sufficiently high.

gas-filled rectifier A *gas-filled tube* used as a rectifier. The term is a general one which embraces tubes with a mercury-pool cathode or an indirectly-heated cathode; the anode current may or may not be controlled by a third electrode.

gas-filled tube A tube of which the electrical properties are dependent upon the *ionisation* of gas deliberately included in the envelope.

gas-filled valve Same as *gas-filled tube*.

gas focusing In a *cathode ray tube* focusing of the electron beam by allowing a trace of gas to remain in the envelope. The electron beam ionises the gas and the positive *ions* so released set up a field which counteracts the mutual repulsion between the beam electrons and concentrates the beam into a convergent form.

gas tube Same as *gas-filled tube*.

gate (logic) See *logic gate*.

gate (semiconductor) In a *field-effect transistor* the *electrode* to which the signal voltage is applied to control the effective width and thus the resistance of the *channel*. In a *thyristor* the electrode to which the control signal is applied to switch the device to the on-state (and in some devices to the off-state).

gate array See *Application Specific Integrated Circuit*.

gated-beam tube A *pentode* designed to operate as a self-limiting phase-difference discriminator in FM receivers, the input signals from the primary and secondary windings of the IF transformer being applied to the control grid and suppressor grid. The tube operates in the same manner as a *nonode* and the electrodes are laid out in a manner quite dissimilar to those of a conventional pentode in order to achieve the type of limiting action required at the *grids*.

gate turn-off switch A *semiconductor* device combining the voltage- and current-carrying capacities of the *thyristor* with the control of the transistor. It is a four-layer device which can be switched on and off by signals applied to the gate.

Gaussian distribution For a quantity which varies statistically about a mean value the shape of the curve which shows, for any chosen value, the probability of a sample having that value. If a very large number of observations of the quantity are taken and if, for a number of discrete values, the number of observations having each value are plotted, a curve of the shape of *Figure G.2* is obtained.

This is often known as a probability curve and it shows that the number of observations having a value markedly different from the mean value is very small but that the number increases as the value selected approaches the mean value.

Geiger-Muller tube *Electron tube* used for the detection of alpha or beta particles or *gamma rays*. The tube consists of a fine wire anode surrounded by a co-axial cold-cathode cylinder, the intervening space being filled by gas at a low pressure. The voltage between anode and cathode is adjusted to just below the ionising potential of the gas. When any of the radiations to

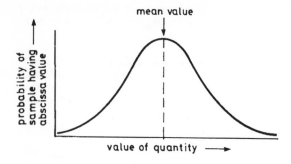

Figure G.2 Gaussian distribution curve

be detected enters the tube it causes momentary ionisation of the gas so that a pulse of current flows through the tube. These pulses are usually transferred to a counter circuit to form a Geiger counter equipment.

geometric distortion In television any displacement of picture elements in a reproduced image from the correct positions, i.e. those of the corresponding elements in the original scene.

Germanium (Ge) A tetravalent crystalline element (atomic number 32) widely used in the manufacture of *semiconductor* devices. In its pure state the element is an insulator but by suitable doping p- or n-type conductivity of a value suitable for semiconductor devices can be obtained. Germanium devices have a significant *leakage current* at normal temperatures making them liable to *thermal runaway* and have to a large extent been superseded by *silicon* devices.

getter A material included in a *vacuum tube* to remove traces of gas remaining after pumping and sealing or which may be released when the tube is put into use. The most commonly-used getter is barium and after the tube is sealed a pellet of the material is vapourised by eddy-current heating and condenses on the walls, combining with the remaining gas as it does so.

glitch Distortion of a pulse waveform in the form of a short-duration disturbance. A typical example is shown in *Figure G.3*.

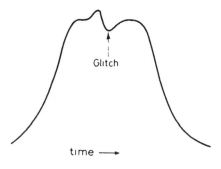

time ⟶

Figure G.3 An example of glitch

glow discharge In a *gas-filled tube* with a cold cathode, conduction as a result of electronic emission caused by bombardment of the *cathode* by positive *ions*. There is usually weak emission of light.

Gouriet oscillator Same as *Clapp oscillator.*

graded-base transistor A *bipolar transistor* in which the impurity concentration in the base region increases smoothly between the *collector* and *emitter* junctions.

Such a graded concentration, which can be produced by *diffusion*, gives an electric field which speeds up the passage of charge carriers across the base region, so reducing *transit time* and improving high-frequency performance.

gramophone Dated term for a disk player.

graphic equaliser An audio *equaliser* in which the audio *spectrum* is divided up into a number of contiguous bands which are individually amplified, adjustment of the gain controls permitting any desired shape of *frequency response* to be obtained. The frequency bands may be one octave or one third octave in *bandwidth* and the gain controls are usually slider types so arranged that a level frequency response is obtained when the controls are in line. When the sliders are moved, their positions give an indication of the shape of the frequency response of the equaliser.

graphical user interface (GUI) A user interface for a computer system in which most routine operations are performed graphically, i.e. using a *mouse* or other *pointing device* to move a *pointer* on to *icons* on the screen instead of typing textual commands. For example, to copy a file from one directory to another typically involves displaying *windows* representing the source and destination directories, selecting the *icon* in the source directory that represents the file to be copied and dragging this into the destination directory.

graphics In computing, matter appearing on a *VDU* or in a *printout* apart from text in the computer's or printer's regular typeface (font). Usually such graphics consist of pictures, diagrams or rules, boxes, borders or other embellishments. There are two principal systems for creating and editing graphics using a computer.

In *bitmap* graphics (also called *raster graphics*) images are stored as a two-dimensional array of *pixels*. Examples include the output from *scanners* and *video digitisers. Applications* for creating and editing bitmaps are sometimes called *painting* software, since ultimately they are concerned only with the colours of areas of pixels. Such applications often feature sophisticated tools for creating lines and shapes in a selected colour and for *image processing*, that is, subjecting the whole or part of the image to some change of colours or dimensions.

The alternative system is *vector graphics*, in which the image is stored effectively as a sequence of co-ordinate-based data describing the lines and shapes of which the image is composed. Applications for creating and editing vector graphics are sometimes called *drawing* software, since it is essentially line-based. Advantages of vector graphics are that the stored image is independent of the output device, it is easy to edit and scale and it may have more than two dimensions. Three-dimensional drawing packages are widely used, for example to create the mathematical models that form the basis of *virtual reality*. Even two-dimensional packages may divide a drawing into a number of superimposed layers which can be turned on and off individually to show different aspects of the subject, such as the various utilities in a map of a city. *Computer-aided design* packages are usually vector-based.

In computers that use a *graphical user interface*, however, it is common practice for *text* to be reproduced on the screen or in printouts in industry-standard typefaces (fonts) that have been stored in the computer as vector graphic images (see *outline fonts*) and are subsequently rendered using graphical routines rather than textual routines.

graphics accelerator An *expansion card* that accelerates a computer's handling of *graphics* on the screen. Some accelerators provide a separate *microprocessor* dedicated to handling screen output, freeing the main processor for the application task. Others use a variety of techniques to circumvent the more processor-intensive aspects of writing to the screen.

Graphics Interface Format (GIF) A *bitmap file* standard originally developed for use in *bulletin board systems*. It incorporates *LZW* compression, substantially reducing most file sizes and therefore also the time taken to transmit the image over costly leased lines. Its disadvantage is its limitation of 8 bits per pixel and it is now being superseded to some extent by *JPEG* which offers 24 bits per pixel and even higher compression factors.

Gray code A binary code in which the codes for any two consecutive numbers differ by only one binary digit.

grid An *electrode* located between the cathode and the anode of an *electron tube* and through the interstices of which the main electron stream of the tube passes.

A tube may have a number of grids (e.g. a *pentode* has three) and each usually has the form of a wire spiral surrounding the *cathode*, the electron stream passing through the meshes of the spiral. The pitch of the spiral is chosen to give the required degree of control over the density of the electron stream.

grid base Of an *electron tube* the range of grid voltage between zero and that which gives cut off of anode current. This is illustrated in *Figure G.4*. The grid base depends on the value of anode and screen-grid voltages and these should be stated when values of grid base are quoted. The value of the grid base gives an indication of the maximum peak-to-peak signal input voltage the tube can accept under *class-A* conditions.

grid bias (GB) Of an *electron tube* the steady voltage applied to the control grid to ensure that signal excursions operate over the required region of the characteristic. The value of the grid bias depends on the anode and screen-grid voltages and these should be stated when values of grid bias are quoted. In *class-A* operation the grid bias voltage is chosen to give an operating point near the centre of the most linear part of the characteristic.

grid characteristic Of an *electron tube* the relationship between grid voltage and grid current exhibited graphically.

grid emission Of an *electron tube* the emission of *electrons* or *ions* from the grid. This can arise as a result of heating of the grid (primary grid emission) or of bombarding it by electrons or ions (secondary grid emission).

148

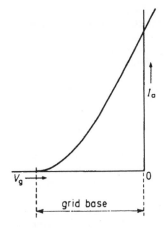

Figure G.4 The grid base of an electron tube

grid leak Dated term for a *resistor*, usually of high value, connected to the control grid of an *electron tube* and used to apply *grid bias*.

grid-leak detector An AM detector in which the control grid and cathode of a triode or multi-grid *electron tube* operate as a *diode detector*, the AF output being amplified by the tube.

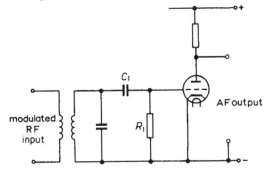

Figure G.5 The circuit diagram of a grid-leak detector

One possible circuit for a grid-leak detector is given in *Figure G.5*. The output of the *diode detector* is the voltage generated across C_1 and this provides the tube with *grid bias* and also with the AF input signal. Clearly only for one particular value of input signal will the direct component across C_1 be the optimum value of grid bias. In general, therefore, the detector gives considerable *harmonic distortion* but with the aid of *reaction* it can be made very sensitive and it was very popular in the early days of radio.

grid stopper See *parasitic stopper*.

ground (US) Same as *earth*.

grounded-anode, -base, -emitter, -grid, etc circuit Literally a circuit in which the anode, base, emitter grid, etc is earthed but the term is generally used to mean a common-anode, common-base, common-emitter, common-grid circuit, etc.

ground loop Same as *earth loop*.

ground return Same as *earth return*.

group code An *error-detecting code* used to verify correct transmission of a group of characters.

group delay Same as *envelope delay*.

group velocity Same as *envelope velocity*.

grown transistor An early method of manufacturing *bipolar transistors* by 'pulling' a *crystal* from the molten *semiconductor* material, the required p- and n-regions being formed in the crystal by introducing pellets of appropriate impurity into the molten material as the crystal is withdrawn.

guard band A *frequency band* between neighbouring *channels* which is left vacant to give a margin of safety against mutual interference.

Gunn effect The splitting of *charge carriers* into domains by a steady electric field. In certain *semiconductors*, typically *gallium arsenide*, electrons can exist in a high-mass low-velocity state as well as their normal low-mass high-velocity state and they can be forced into the high-mass state by a steady electric field of sufficient strength. In this state they form clusters or domains which cross the field at a constant rate so that current flows in a series of pulses. This can be made the basis of a microwave oscillator. See *Gunn-effect diode*.

Gunn-effect diode A semiconductor device used as a microwave oscillator. It is not, strictly speaking, a diode, consisting of an *epitaxial* layer of n-type *gallium arsenide* grown on a GaAs *substrate*. Ohmic contacts are made to the substrate (anode) and the n-layer (cathode). A potential of a few volts between anode and cathode produces an electric field causing electrons to cross the device in clusters or domains (see *Gunn effect*). Thus the current is in the form of pulses with a frequency dependent on the transit time and hence on the thickness of the n-layer. In use the device is mounted in a *cavity resonator* and powers up to 1 W at frequencies between 10 and 30 GHz are possible. Certain other semiconductor materials can also be used in Gunn-effect diodes.

H

hacking Colloquial term for the deliberate gaining of unauthorised access to private computer systems over the telecommunications network. For some hackers this is simply an intellectual challenge; others cause serious problems by tampering with confidential stored data or by making fraudulent use of the data accessed.

halation In a *cathode ray tube*, degradation of the image caused by an area of light surrounding the spot where the *electron beam* strikes the screen. This unwanted area is caused by light from the spot reaching the screen by reflection at the front and rear surface of the faceplate of the tube.

half-adder In logic circuitry a combination of logic elements with two inputs and two outputs so related that one output is the sum of the two inputs and the other is an exclusive-OR function of the two inputs. Two such circuits can be combined to form a *full adder* which can perform binary addition. See *exclusive-OR gate*.

half-duplex operation Of a telegraph system, a *duplex* system permitting operation in either direction, but not in both directions simultaneously.

half-section Of a filter. See *ladder network*.

half-wave rectification Rectification in which power for the load is taken from the AC supply only during alternate half-cycles of the supply. A typical circuit for a half-wave rectifier is given in *Figure H.1*.

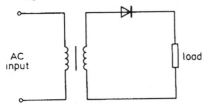

Figure H.1 A simple circuit for half-wave rectification

Hall constant (or **coefficient**) Factor relating the voltage generated as a result of the *Hall effect* with the product of the current and the magnetic field.

Hall effect The production of a transverse EMF in a current-carrying conductor or semiconductor subjected to a *magnetic field*. When a conductor or a semiconductor carrying a current is subjected to a magnetic field perpendicular to the direction of current flow a voltage proportional to the current and to the magnetic field is generated across the faces of the conductor perpendicular to the direction of current flow and to the direction of the magnetic field.

Hamming code In digital transmission a code so designed that errors in signals can be detected and corrected. See *parity check*.

151

Hamming distance Same as *signal distance*.

handset An instrument containing a *microphone* and an *earphone* designed to be hand held, as in the familiar telephone instrument.

hard copy In data processing a copy of a document on paper or similar material, in contrast with a copy on magnetic media such as a floppy disk or in some electronic format. Hard copy is so called because, once committed to paper, it cannot be easily changed; in contrast, a file in a data storage medium, such as a hard or floppy disk, can be readily edited.

hard disk (or hard drive) A data storage system consisting usually of several magnetic disks on a common spindle and sealed with their *read/write* heads in a container. The disks are rigid and double-sided, each surface having its own read/write head. To minimise wear, the heads do not contact the disk surface, but are suspended very close to it, the sealed container preventing the ingress of dust which would have a catastrophic effect on the drive's reliability. Physical drive sizes vary widely. Storage capacities range from 20 Mbytes to 10 Gbytes and data transfer is significantly faster than that available from *floppy disks* or *CD-ROMs*. Hard disks are sometimes called *fixed disks* because, in contrast with floppy disks, they are usually fixed permanently in computers where they are used for the storage of commonly used *software*, such as *operating systems* and *applications*. This software is transferred on to the hard disk from floppy disks, CD-ROMs, tape streamers or over a network. There are, however, some *demountable* hard disks. In these the sealed container housing the disks and heads slides out of the drive assembly containing the motor and control electronics. These are used not only for software storage, but also for the distribution of large collections of software.

hard valve Same as *vacuum tube*.

hardware Term used particularly in computer technology to describe the electronic equipment, its circuits and components. See *software*.

hardware key A device used to protect a *software* package against unauthorised copying. Also known as a *dongle*, the device is plugged into a computer *port*, often the parallel printer port; the printer is connected to a printer port on the hardware key itself. At start-up the protected software checks for the presence of coded data in the hardware key; if it is absent, the application is terminated. Since the software cannot be used without a suitable hardware key, unauthorised copies of it are useless. The hardware key itself is far more difficult to copy than the software it protects.

harmonic analyser Equipment for measuring the amplitude and phase of the *harmonic components* of a complex waveform. It is usually mechanical in nature.

harmonic component Of a complex waveform, one of the components with a frequency equal to a multiple of the fundamental frequency. The analysis of a complex waveform into its harmonic components can be carried out mathematically by *Fourier analysis* or practically by use of a *harmonic analyser*.

harmonic distortion Distortion arising from the non-linearity of the input-output characteristic of a system, equipment or component and resulting in the production of new signals at harmonics of the frequency of a sinusoidal input signal.

harmonic generator Same as *frequency multiplier*.

Harries tetrode Same as *critical-space tetrode*.

Hartley oscillator A sinusoidal oscillator in which the frequency-determining element is a parallel-tuned LC circuit connected between the input and output terminals of a *active device, positive feedback* being obtained by connecting a tapping on the inductor to the common terminal *(cathode, emitter, source)* of the active device. *Figure H.2* gives the circuit diagram of a transistor Hartley oscillator.

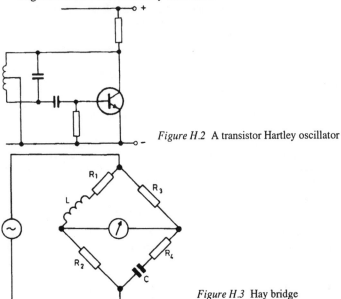

Figure H.2 A transistor Hartley oscillator

Figure H.3 Hay bridge

Hay bridge A bridge circuit generally used for the measurement of inductance in terms of capacitance, resistance and frequency. It differs from the Maxwell bridge in that the capacitor is in series with its associated resistor as shown in Figure H.3. The conditions for balance are:

$$L = R_2 R_3 \frac{C}{1 + \omega^2 C^2 R_4^2}$$

$$R_1 = R_2 R_3 \frac{\omega^2 C^2 R_4}{1 + \omega^2 C^2 R_4^2}$$

head A device which records information on a storage medium, reproduces the information or erases it. The storage medium may be tape, film or disk and the information may be in digital form as in *data-processing* equipment or may be in analogue form as in audio and TV *recording*.

head amplifier An audio or video amplifier incorporated in a *microphone, television camera* or motion picture projector to raise the level of the output signal before it is sent along a cable. This technique is used as a means of improving the *signal-to-noise ratio*.

headphones An electro-acoustic *transducer* designed to feed sound directly into the ear and mounted on a headband. Some headbands carry only one transducer so as to leave one ear free. Others carry two transducers which are connected in series or in parallel for monophonic reproduction or can be connected to the two channels of stereo equipment for stereophonic reproduction.

headset A combination of headphone and microphone mounted on a headband and used for two-way communications by, for example, telephone operators and aircraft pilots.

hearing aid An equipment consisting of a *microphone*, an *audio amplifier* and an *earphone* and powered by an internal battery which allow persons suffering hearing deficiency to take part in normal conversations and to enjoy radio and television broadcasts. Modern hearing aids are extremely small and can be hidden behind or within the ear. Many also feature a pick-up coil which may be selected instead of the microphone for use in a theatre, church or other auditorium fitted with an *induction loop*.

heater In general any *resistor* carrying current and used to supply heat. In particular the resistor which supplies the heat necessary for *thermionic emission* from an indirectly-heated *cathode*. The heater is usually a tungsten wire contained within the cathode cylinder but electrically insulated from it.

heat sink A metal structure arranged to be in intimate thermal contact with a heat-generating component such as a power transistor to aid heat dissipation and so limit temperature rise in the component. To be efficient a heat sink requires adequate thermal capacity (mass × specific heat). It should also be a good conductor of heat and have large surface area. Often heat sinks are finned to increase surface area.

Heising modulation A circuit for *amplitude modulation* in which the *anodes* of a modulation-frequency amplifier and a carrier-frequency amplifier are coupled and fed via a common inductor with a high reactance at modulation frequencies.

If the carrier-frequency amplifier operates in *class C* its output is proportional to its HT supply voltage. If, therefore, the HT voltage is varied by the modulation-frequency amplifier, amplitude modulation can be achieved. When the anode current of one tube is instantaneously high, that of the other is low and in fact the sum of the two currents is constant. The circuit is sometimes called constant-current modulation.

The basic circuit described here is, in practice, modified to enable very deep modulation to be achieved without excessive distortion, e.g. by the use of a transformer in place of a choke.

Helmholtz resonator A cavity open to the external environment via a small aperture in the cavity wall. Such an arrangement resonates at a frequency dependent on the dimensions of the cavity. Helmholtz resonators are used extensively as tuned elements for electromagnetic waves in microwave tubes such as *klystrons*. The vented enclosures used in loudspeaker designs and the tuned absorbers used in acoustic treatment of sound studio walls are also examples of Helmholtz resonators.

Henry The unit of self-inductance or mutual inductance, symbol H. An inductor is said to have an inductance of 1 H when a current in it, changing at the rate of 1 A per second, generates a voltage across it of 1 V. Practical inductors have inductances ranging from many H to fractions of a μH.

heptode An *electron tube* with five grids situated between cathode and anode. It was usually used as a *frequency changer*, the first two grids acting together as the grid and anode of the oscillator. Grids 4 and 5, together with the anode, constitute an RF *tetrode* acting as an RF mixer. Grid 3 is a screen to minimise capacitive interaction between oscillator and mixer sections. Mixing occurs by virtue of the oscillator-frequency component impressed on the electron stream by grids 1 and 2. However the signals on these grids are usually in antiphase and their effects on the electron stream are therefore mutually destructive. To minimise this effect grid 2 is not of conventional construction but may take the form, for example, of a number of rods which can carry sufficient current to sustain *oscillation* but have little modulating effect on the electron stream. Because grids 3 and 5 both function as *electrostatic screens* they are often connected together inside or outside the tube.

The graphical symbol for a heptode is given in *Figure H.4* which also indicates the function of each grid.

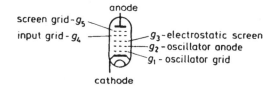

Figure H.4 Graphical symbol for a heptode indicating the function of each grid

heterodyne The process of combining two signals of different frequencies so as to produce an output at the sum or difference frequency. The process is thus equivalent to non-linear additive mixing or multiplicative mixing. The process is extensively used in sound and TV receivers where the received signal is combined with the output of the local oscillator to produce a difference term at the intermediate frequency. Because the difference frequency is above audibility this type of receiver is known as a supersonic heterodyne, abbreviated to *superheterodyne* or superhet.

Hertz The unit of frequency and equal to 1 cycle per second, symbol Hz. The multiple units of kilohertz (kHz), megahertz (MHz) and gigahertz (GHz) are commonly used in electronics.

hexadecimal (hex) A counting scale containing sixteen digits, each digit in a number representing a power of sixteen. Conventionally the digits of the hexadecimal scale are represented by the decimal numbers from 0 to 9 followed by the letters A to F. Thus the decimal equivalent of the hexadecimal number 5A6 can be evaluated in the following manner:

$$5A6 = 5 \times 16^2 + 10 \times 16^1 + 6 \times 16^0$$
$$= 5 \times 256 + 10 \times 16 + 6 \times 1$$
$$= 1280 \quad + 160 \quad + 6$$
$$= 1446$$

Hexadecimal numbers are often used as a shorthand method of writing binary numbers. The base 16 of the hexadecimal scale is 2^4 and this facilitates conversion between the hexadecimal and the binary scales. Each four digits of a binary number can be uniquely represented by one digit of a hexadecimal number. For example an 8-bit byte such as 01101100 can be represented by the two-digit hexadecimal number 6C, both having a decimal equivalent of 108. Similarly a 16-bit binary word can be represented by a four-digit hexadecimal number.

hexode An *electron tube* with four grids between cathode and anode. The tube may be regarded as an RF *tetrode* (grid 4 being the screen grid) with two *control grids* (1 and 3) separated by a *screen grid* (2) to minimise *capacitive coupling* between the two inputs. Grids 2 and 4 may be connected together inside or outside the tube.

The tube was designed for use as an RF mixer, the oscillator and signal-frequency inputs being applied to grids 1 and 3. The tube is often combined with a triode oscillator in a single envelope to form a triode-hexode frequency changer.

The graphical symbol for a hexode is given in *Figure H.5* which also indicates the function of each grid.

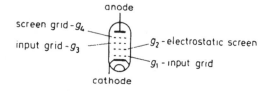

Figure H.5 Graphical symbol for a hexode indicating the function of each grid

hi-fi, high-fidelity A vague term used to describe audio equipment in whose design some effort has been made to ensure that sound is reproduced with minimal distortion.

high-definition television (HDTV) A colour television system providing wide-screen flicker-free pictures with a resolution comparable to that of 35-mm cinematograph film. Considerable research has been invested in the development of such systems using a 16:9 *aspect ratio* and 1125 lines, but there are difficulties in establishing standards that are acceptable worldwide.

high frequency resistance Same as *effective resistance* (2).

high level modulation Same as *high-power modulation*.

high-noise-immunity logic (HNIL) A form of logic similar to *diode-transistor logic* but including a zener diode to increase voltage levels. It is largely obsolete.

Figure H.6 Block symbol for a high-pass filter

high-pass filter A *filter* designed to pass signals at frequencies above a specified *cut-off frequency*. Thus the passband extends from the cut-off frequency to an infinitely-high frequency. In general such filters comprise series capacitors and shunt inductors. The block symbol for a high-pass filter is given in *Figure H.6*.

high-power modulation In an AM transmitter *amplitude modulation* of the carrier by introducing the *modulating signal* into the anode circuit of the final amplifying stage. The power of the modulating signal thus approximates to that at the transmitter output. See *low-power modulation*.

high tension (HT) The high-voltage supply required for the *anodes* and *screen grids* of *electron tubes*. In battery-operated equipment the HT supply was commonly between 45 V and 90 V but in mains-operated equipment supplies for early stages were often about 200 V and for output stages up to 500 V.

high threshold logic (HTL) Same as *high-noise-immunity logic*.

highway Same as *bus*.

H-network A five-element network in which both legs contain two elements in series, a shunt element bridging their junctions as shown in *Figure H.7*. It can be regarded as a balanced form of *T-network*.

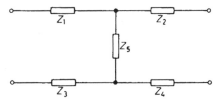 *Figure H.7* An H-network

hold control In early TV receivers the controls which determined the free-running frequency of the line and field *time bases*. The controls were adjusted to bring the time-base frequencies into the range at which the time bases lock at the frequencies of the line and field synchronising signals. Hence receivers had two hold controls: line or horizontal hold; field or vertical hold. Modern receivers do not have hold controls.

hole A deficiency of one *electron* in the atomic structure of an atom in an *extrinsic semiconductor*. As a result of this deficiency the atom has a net positive charge equal to that of an electron and neutralisation of such charges by electrons can give rise to a current in the form of a unidirectional movement of holes as described under *p-type semiconductor*.

hole conduction The process by which current flows through a *p-type semiconductor*.

hole injection The basic process underlying the action of *pnp transistors*. The base-emitter junction is forward-biased and the current which flows between these regions is largely carried by holes moving from emitter to base. These enter the very thin *base region* and are swept into the *collector region* by the collector-base voltage so giving rise to a considerable collector current. It is the holes injected into the base region from the emitter region which are responsible for the collector current.

hole storage The process which causes the collector current of a *pnp transistor* (which has been driven hard into conduction) to continue for a brief period after the emitter current has been cut off. When the transistor has been driven hard on, the emitter injects into the base region more holes than are required to give the collector current and the excess holes are stored ready to be swept into the collector region to prolong the collector current when the emitter current has been cut off. See *carrier storage*.

homodyne A system of *heterodyne* reception in which the received signal is mixed with another at the carrier frequency of the signal and derived from the received signal. Thus there is no local oscillator as in normal heterodyne reception.

horizontal amplifier In oscilloscopes the circuits which amplify the signals responsible for horizontal deflection of the beam. Also known as an *X* amplifier.

horizontal blanking Same as *line blanking*.

horizontal hold See *hold control*.

horizontal polarisation Property of an electromagnetic wave in which the plane of polarisation of the electric field is horizontal.

horizontal time base In a *cathode ray tube* the circuits generating the signals which give horizontal deflection of the beam. In TV this is usually termed the line time base.

host The computer controlling a *network*. More specifically, the *mainframe* or other large computer served by a network of *terminals* or the computer running a *bulletin board* or other communications system to which other computers are temporarily connected for data exchange purposes.

hot carrier diode Same as *Schottky diode*.

hot-electron diode Same as *Schottky diode*.

hot spot (1) In tubes with a *mercury-pool cathode* a small heated area formed on the mercury surface by the ignition electrode and which initiates the discharge. (2) In a high-power *electron tube* a small heated area on the *anode*.

H-parameters Same as *hybrid parameters*.

HSV colour model See *colour model*.

hue That quality of a colour which enables it to be classified as red, green, blue, etc. For colours appearing in the *spectrum*, hue can be quantitatively defined by quoting the wavelength of the colour.

hum An unwanted low-frequency noise originating from mains-driven equipment and comprising harmonics of the mains frequency. The sound may be radiated directly from components such as mains transformers or may be caused by ripple on the DC supply to an amplifier or receiver and heard in the output of *loudspeakers* or *headphones*.

hunting In a controlled system an undesired oscillation, usually at a low frequency, resulting from over-correction in which the controlled quantity fluctuates about the required value.

hybrid coil A *transformer* with three windings and four pairs of terminals,

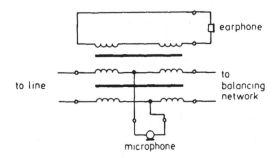

Figure H.8 An application of the hybrid coil. Signals from the microphone can enter the line but do not appear at the earphone. Signals from the line can enter the earphone, but do not enter the microphone

two of the windings being designed for use in a bridge circuit which isolates one pair of terminals from another pair provided the remaining two pairs are correctly terminated. The circuit diagram of one application of a hybrid coil is given in *Figure H.8*.

hybrid computer A computer in which *analogue* and *digital techniques* are used.

hybrid integrated circuit An integrated circuit using *monolithic* and *thin-film* techniques.

hybrid network A network with four pairs of terminals which, when two pairs are correctly terminated, transmits power from the third pair into the terminated pairs but not into the fourth pair.

hybrid parameters Of a *transistor* a method of expressing the electrical characteristics by representing it as a four-terminal *equivalent network*, for which the input voltage and output current are expressed in terms of input current and output voltage. The fundamental equations are:

$$V_{in} = h_i i_{in} + h_r V_{out}$$

$$i_{out} = h_f i_{in} + h_o V_{out}$$

from which h_i has the dimensions of an impedance, h_o of an admittance, whilst h_r and h_f are both pure numbers. The parameters are thus mixed or hybrid in nature. In fact h_f is the current gain of the transistor and h_i is the input impedance (both with the output terminals short-circuited), h_o is the output admittance and h_r the voltage feedback ratio (ratio of input voltage to output voltage) both for open-circuited input terminals. See *Y parameters, Z parameters*.

hypermedia *Interactive multimedia*, that is, a computer *application* in which the user can steer his way through presentations consisting of *hypertext*, still images, video and sound.

hypertext A form of *interactive* text display used in computer *applications* and often forming part of *on-line help* and *hypermedia* systems. Unlike a plain text file through which the user must navigate using scroll bars or other means to control the portion of the document displayed, hypertext uses an active cross-reference system. Certain words in the text are highlighted (often using a different colour). When the user moves the *pointer* over one of these and clicks the mouse button, the screen displays an entry giving further information on that topic. Facilities are usually provided to allow the user to retrace his steps and to jump to an interactive index to the whole document.

hysteresis In general a relationship between a force and its effect in which the effect lags on the force so that the magnitude of the effect depends not only on the present value of the force but also on its previous value.

As a result the magnitude of the effect for a given value of force depends on whether the force is increasing or decreasing and any particular value of effect can be produced by two values of force. If the force is varied cyclically the relationship between force and effect, if shown graphically, is a closed loop, the most familiar example of which is the *B–H* curve for a magnetic material.

hysteresis loop Of a magnetic material, the closed figure obtained by plotting the *magnetic flux density B* against the *magnetising force H* as the magnetising force is varied throughout a complete cycle. An example of this closed loop is shown in *Figure H.9.*

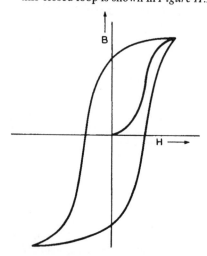

Figure H.9 B–H hysteresis loop for a magnetic material

hysteresis losses In a magnetic material the energy loss whenever varying flux is produced in the material. The loss is proportional to the area of the *hysteresis loop.*

icon Small graphical representation displayed on the screen to illustrate a computer function. For example, a representation of a file folder to indicate a *directory*.

iconoscope The earliest form of TV camera tube in which the optical image of the scene to be televised is focused on a photo-emissive target *mosaic* which is obliquely scanned by a high-velocity *electron beam.*

The mosaic is deposited on one face of a mica sheet which is backed by a conductive signal plate from which the output of the tube is taken. When the optical image is focused on the mosaic, photo-electrons are released from each element in proportion to the light falling on it. Thus a positive charge image is built up on the mosaic surface and this grows with time as the capacitance between element and signal plate is charged. The charge image is discharged by the scanning beam and the resulting voltage change is transferred to the signal plate via the capacitive coupling to the mosaic.

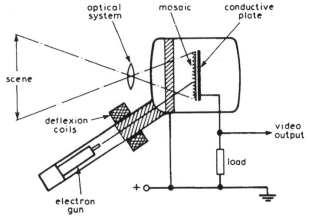

Figure 1.1 The standard emitron television camera tube: an example of an iconoscope

In spite of the charge storage thus achieved the tube is not very sensitive and required high scene illumination for a satisfactory *signal-to-noise ratio*. Moreover secondary emission from the target as a result of bombardment by the high-velocity scanning beam results in spurious signals in the tube output which produce undesirable shading effects in reproduced images. These effects were minimised by mixing with the tube output, sawtooth and parabolic waveforms at line and field frequencies.

One example of an iconoscope is the standard emitron tube, the construction of which is illustrated in *Figure 1.1*; this is the type of tube used when Britain started the world's first regular high-definition TV service from Alexandra Palace in North London towards the end of 1936.

ideal radiator Same as *full radiator*.

identity gate Same as *AND gate*.

IF-THEN gate A gate with two inputs and one output such that the relationship between inputs and outputs is given in the following *truth table*:

input 1	input 2	output
1	1	1
1	0	0
0	1	1
0	0	1

igniter (also known as **ignition electrode**) In a mercury-pool discharge tube a stationary electrode which is in contact with the mercury pool and, when suitably biased, causes a local hot spot on the surface so initiating the main discharge.

ignition voltage Same as *firing voltage*.

ignitron A gas-discharge rectifier tube with a *mercury-pool cathode* and a single *anode* capable of a large current output. The arc is initiated by an ignition electrode which dips into the mercury pool. A positive voltage is applied to the igniter once per cycle of the applied alternating voltage and this causes a small arc at the mercury surface which precipitates the main discharge.

image attenuation coefficient (or **factor**) See *image transfer coefficient*.

image converter An *electron tube* in which an image of a scene focused on a *photo-cathode* gives rise to a corresponding visible image on a fluorescent screen. One advantage of such a tube is that the photo-cathode may be made sensitive to infra-red radiation and can thus give a visible image of a scene invisible to the human eye.

image dissector An early form of TV camera tube in which an optical image of the scene to be televised is focused on a *photo-cathode*, the released *photo-electrons* forming an electron image which is focused on the plane of a defining aperture and is swept over the aperture to effect *scanning*.

The defining aperture was a hole in an *anode* and *electrons* passing through it were either intercepted by a collector electrode or were directed into an *electron multiplier*. The tube lacked sensitivity because of the absence of any form of charge storage but it was used for transmitting cinema film where a large light input could be achieved.

image frequency In superheterodyne reception a frequency as much above (or below) the oscillator frequency as the wanted signal frequency is below (or above) it and which is therefore accepted with the wanted signal by the IF amplifier so causing *interference*.

image iconoscope An early form of TV camera tube consisting of an *iconoscope* with an *image section*.

image impedances Two parameters of a two-terminal-pair *network* such that if the output is terminated with an impedance Z_2 the input impedance is Z_1 and if the input is terminated with a impedance Z_1 the output impedance is Z_2.

162

Networks are used on an image-impedance basis for matching purposes. For example if an amplifier requires a load of impedance Z_1 in order to perform properly and if the load to be fed has an impedance Z_2, then a network with image impedances Z_1 and Z_2 can be used to connect the amplifier to its load. In this way the equipment has an effective load impedance of Z_1 and the load is effectively supplied from a generator of impedance Z_2: in other words correct matching is achieved at the input and the output of the network.

If the network contains inductance and capacitance then it is possible that the image impedances will be correct at only one frequency. A *transformer* can achieve matching over a wide frequency range.

image interference In superheterodyne reception interference from signals on the *image frequency*. The frequency of such interfering signals differs from that of the wanted signal by twice the intermediate frequency and, to minimise image interference, the signal-frequency circuits of a *superheterodyne receiver* are designed to give great attenuation at the image frequency.

image isocon An *image orthicon* in which the output is obtained from an *electron multiplier* into which the electron beam scattered from the *target* is directed. When the scanning beam strikes the target some electrons are absorbed to neutralise the charge image, others return to the electron gun and the remainder are scattered. It is the scattered, not returned, electrons which are captured and directed into the electron multiplier in the image isocon.

image orthicon An *orthicon* TV camera tube with an *image section* and in which the output signal is obtained from an *electron multiplier* into which the return scanning beam is directed.

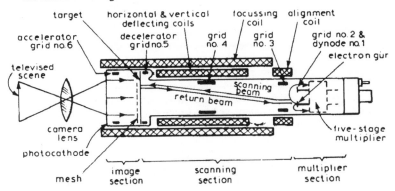

Figure I.2 Simplified diagram of an image orthicon tube

The essential features of the tube are illustrated in *Figure I.2*. An optical image of the scene to be televised is focused on the *photo-cathode* and the *photo-electrons* so released are focused by a combination of electrostatic and magnetic *electron lenses* on the image-section face of the target where they give rise to secondary emission which is collected by the nearby positively-charged mesh.

163

Thus a positive charge image is established on the face of the target. If the tube is directed at a very bright light causing the target potential to exceed that of the mesh, the excess secondary electrons are returned to the target: in this way the mesh keeps the tube stable for all light inputs.

The target is very thin and the charge image is rapidly transferred to the opposite face which is scanned by the low-velocity beam. The beam lands on the target to neutralise the positive charge image and thus the return scanning beam is amplitude modulated by the required picture signal. The return beam is directed into the input of a multi-stage *electron multiplier* which surrounds the *electron gun*.

The tube is extremely sensitive and is capable of high-quality pictures. It is, however, complex and too bulky (typically 15 in long and 3 or 4½ in diameter) to be used in colour cameras where three or four tubes are necessary.

image phase-change coefficient (or **constant**) See *image transfer coefficient*.

image processing The application of digital processing to *bitmapped* images, such as *scanned* photographs or *digitised* video, to change their appearance. Often the processing is to improve the appearance, e.g. by improving the contrast in an underexposed photograph, but sometimes the intention is the reverse, e.g. to reduce the number of colours in the image in order to achieve some special effect.

image section An electron-optical stage included in some television camera tubes to increase sensitivity. As shown in *Figure I.3* the optical image is

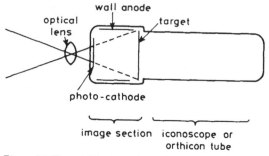

Figure I.3 Essential features of one type of image section for a camera tube

focused on the *photo-cathode* in the image stage and the liberated *photo-electrons* are focused on the target to form a charge image by secondary emission from the target. The use of an image section thus separates the functions of photo-emission (now carried out by the photo-cathode) and secondary emission (carried out by the *target*). In the *iconoscope* the target is required to carry out both functions.

image transfer coefficient (or **constant**) For a *network* terminated in its image impedances at both ends, one half the natural logarithm of the complex ratio of the steady-state volt-amps entering the network to the volt-amps leaving the network.

The real part of the image transfer coefficient is known as the image attenuation coefficient and the imaginary part as the image phase-change coefficient or, more simply, the image phase coefficient.

When networks are connected in *cascade* on an image basis the image attenuation coefficient of the group is equal to the sum of the image attenuation coefficients of the individual networks and the image phase coefficient of the group is equal to the sum of the sum of the individual phase coefficients.

For a symmetrical network the real and imaginary parts of the image transfer coefficient are equal to the real and imaginary parts of the *propagation coefficient*.

immersion lens In a *cathode ray tube*, an electrostatic electron lens designed to concentrate the *electrons* liberated from the *cathode* into a beam. Because these electrons have very low velocities the lens is situated very close to the cathode, so close in fact that the cathode may be regarded as immersed in the lens. The lens usually consists of two plates containing apertures and which may have cylindrical extensions.

immitance A term which can mean *impedance* or *admittance*. It is used in network theory where the distinction between impedance and admittance is irrelevant.

impact dot-matrix printer See *printer*.

impact ionisation The liberation of orbital electrons from an atom in a crystal lattice as a result of a high-energy collision.

impatt diode A four-layer semiconductor device of pnin construction used as a microwave oscillator. A reverse bias causes *avalanche breakdown* at the pn junction. Electrons cross the i-region which acts as a *drift space* and the *transit time* is arranged to be one half the period of the required oscillation. The device then has a *negative resistance* and, given a suitable resonant load, can give up to 50 W output at 10 GHz.

impedance (Z) In general an indication of the opposition offered by a circuit to a flow of alternating current in it. More specifically it is the ratio of the alternating EMF applied to the circuit to the resultant current flowing in it. As there is normally a phase difference between the voltage and the current, the impedance Z is complex and can be written

$$Z = R + jX$$

where R, the real component of the impedance is the *resistance* of the circuit and X, the imaginary component is the *reactance* of the circuit. The numerical value of the impedance is given by

$$|Z| = \sqrt{(R^2 + X^2)}$$

and can be calculated by dividing the RMS applied voltage by the resulting RMS current. The impedance concept is useful in solving problems where components are connected in series because in such circuits resistances can simply be added to give the total or effective resistance and reactances are added or subtracted depending on their sign.

impedance matching The process of ensuring that two *impedances* are equal. This is important for two reasons: (1) To ensure that maximum power is transferred from a generator to a load, the generator and load impedances must be matched. To do this over a wide *frequency range* the resistive and reactive components of the load impedance must in effect be made to equal those of the generator, e.g. by the use of a *transformer*. If, however, *matching* is necessary at only one frequency then maximum power transfer occurs when the load impedance is in effect made to equal the conjugate of the generator impedance, i.e. the resistive components are equal, the reactive components are equal in magnitude but opposite in sign. (2) To ensure that there is no reflection at the termination of a *transmission line* or *filter* network the termination must in effect be equal to the characteristic impedance of the line or the iterative impedance of the network.

imperfection Of a crystal, any difference in its structure from that of the ideal crystal. As an example an atom may be missing from a site or a site may be occupied by a foreign atom. Imperfections are responsible for hole and electron conduction.

import In computing to introduce data into an application in a format that is foreign to that application. For example, to introduce a *bitmap* into a *vector graphics* application.

impulse Same as *pulse*.

impurity In *semiconductor* technology a foreign element added in minute but controlled quantities to a semiconductor element to give it the required *p-type* or *n-type* conductivity. In semiconductor compounds an excess or deficiency of an element belonging to the compound.

impurity diffusion See *diffusion*.

inclusion gate Same as *IF-THEN gate*.

inclusive-OR gate Same as *OR gate*.

incremental permeability The ratio of the change in *magnetic flux density* to the small change in *magnetising force* which gives rise to it when this is superimposed on a steady magnetising force.

independent sideband transmission (ISB) A system of *amplitude modulation* in which one *sideband* is produced by modulation by one signal and the other sideband is produced by modulation by a different signal. The system is not greatly used largely because of the difficulty of separating the two signals at the *receiver*.

indirectly-heated cathode A *cathode* of an *electron tube* for which the heat necessary for the thermionic emission is supplied by an independent heater contained within the cathode but insulated from it. The cathode is usually a nickel cylinder coated with electron-emitting material and the heater is a tungsten spiral or hairpin within it but insulated from the cylinder by, for example, aluminium oxide.

induced current A current flowing in a circuit as a result of an *induced EMF*.

induced EMF The EMF generated in a circuit as a result of *electromagnetic induction*.

inductance (L) In general that property of a current-carrying circuit which enables it to generate an EMF in itself (or in a nearby circuit) as a result of changes in the current and therefore in the associated magnetic field. The direction of the induced EMF is such as to oppose the change in current which gave rise to it and thus inductance may be regarded as the electrical analogue of inertia. The EMF induced in the circuit itself is more properly ascribed to self-inductance and that induced in a nearby circuit to *mutual inductance*. Clearly the inductance of a conductor can be increased by coiling it so as to increase linkage with the magnetic field and the introduction of a *magnetic core* increases it further. The practical unit of inductance is the Henry (H) and a circuit is said to have an inductance of one Henry if a current, changing at the rate of one ampere per second, induces an EMF of one volt in it.

induction See *electromagnetic induction, magnetic flux density*.

induction loop In general any looped conductor used to transfer energy to another by *mutual inductance*. More particularly, a form of localised wireless broadcasting used in churches, theatres, museums and other public buildings to assist those hard of hearing or to provide a commentary. They are also sometimes installed in the homes of partially deaf people to help them to hear radio and television programmes. The induction loop surrounds the area to be served and is connected to the output of an *audio amplifier* reproducing the required audio programme. Receivers are either *hearing aids* fitted with suitable pick-up coils (selected on the so-called 'T'-switch) or dedicated receivers usually fitted with an *earphone* that the user holds against his ear.

inductive coupling Coupling of two circuits by virtue of a common inductor or by *mutual inductance*. The common inductor may be a series-connected component as in *Figure I.4* (a) or a shunt component as in (b). *Figure I.5* shows coupling by mutual inductance *M*.

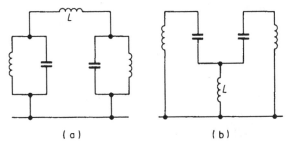

(a)   (b)

Figure I.4 Coupling by series inductances is shown at (a) and by shunt inductance at (b)

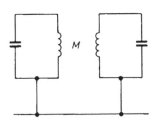

Figure I.5 Coupling between two circuits by mutual inductance M

inductor A component used because of its *inductance*. A wide variety of types of inductor are used in electronics. At low and audio frequencies inductors of many Henries consists of windings with laminated ferromagnetic cores. At higher frequencies *eddy-current* losses become important and non-conductive ferrite cores are used in RF conductors and *transformers*.

inertance That property of an acoustical vibrating system which enables it to resist changes in velocity. It is a function of the mass of the medium in the system, being given by m/A^2, where m is the mass and A is the cross-sectional area over which the driving pressure acts. Inertance is the acoustical analogue of *inductance*.

infinite baffle A loudspeaker cabinet which is totally enclosed at the rear so that sound radiated from the rear of the diaphragm cannot interfere with forward-radiated sound. The cabinet therefore behaves as a baffle of infinite dimensions. In a practical design provision must be made to absorb the rear radiation from the diaphragm and allowance must be made for the inevitable raising of the fundamental resonance frequency of the *loudspeaker* caused by the stiffness of the air trapped in the cabinet. Nevertheless some very successful designs have been produced, some so small that the cabinet can be readily accommodated on a bookshelf or a mantlepiece.

infinite-impedance detector An *anode-bend* detector in which the load is connected in the *cathode* circuit. A typical circuit diagram for the detector is shown in *Figure I.6*. The load R_1 and associated capacitor C_1 behave as in a diode detector circuit, i.e. C_1 is charged on positive-going half-cycles of carrier input and discharges through R_1 during negative-going half-cycles, so developing the AF waveform across the combination. The reactance of C_1 is very high at AF and thus the tube has 100% negative feedback. No grid current flows and the input of the tube is very high, imposing negligible *damping* on the *tuned circuit*.

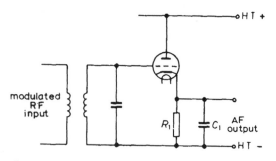

Figure I.6 Circuit diagram of an infinite-impedance detector

information superhighway A vague term denoting any *network* over which information in a variety of formats is accessible to a wide public. It is sometimes used to refer to the *Internet* although equally applicable to many other networks including interactive *video on demand* systems.

Information Technology (IT) A vague term embracing all electronic equipment that is used in the processing and presentation of *data*.

infra red *Electromagnetic waves* with frequencies extending from the red end of the visible *spectrum* to the microwave region (see *Figure E.7*). The radiation is invisible but gives a sensation of heat and is used in cooking and industrial heating applications as well as in remote control systems.

inhibiting signal A signal which prevents a particular action occurring. For example it may close a gate to prevent any output which might otherwise occur. An inhibiting signal is a negated *enabling signal*.

ink-jet printer See *printer*.

in-line colour picture tube See *precision-in-line colour picture tube*.

input (1) Of a device or equipment, the signal applied to it and upon which it performs some function, e.g. amplification. (2) The terminals to which the input signal is applied.

input impedance Of a device or equipment, the *impedance* it presents to the source of input signal.

insertion gain (or loss) The gain resulting from the insertion of a *network* between a generator and its load. It is given by the ratio (usually expressed in *decibels*) of the power (voltage or current) delivered to the load before insertion to the power (voltage or current) delivered after insertion.

insertion signal In TV a signal inserted into one of the line periods during the *field blanking* period. The signal is not seen on the screens of viewers' receivers and is used by the transmitting authority to transmit information such as the source of programme or control data. The signal is also used for test purposes to give information on the performance of the television links. Of a television waveform, a signal introduced during flyback periods to carry information for engineering purposes or for display on receiver screens, e.g. *teletext*.

instability Generation of unwanted and sustained *oscillations*.

instrument transformer A transformer giving at its secondary winding a voltage or current which is precisely related in magnitude and phase with that at the primary winding and is therefore suitable for application to a measuring instrument or a control or protective device.

insulated-gate field-effect transistor (IGFET) A *field-effect transistor* in which the input electrode is capacitively coupled to the channel. A cross section of an IGFET is given in *Figure I.7*. It consists of a base layer of p-type silicon with diffused n-regions at each end to which source and drain connections are made. During manufacture by the *planar process*, the device is sealed by a layer of silicon dioxide obtained by heating it to 1200°

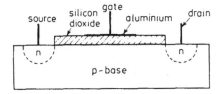

Figure I.7 Simplified diagram showing the structure of an n-channel insulated-gate field-effect transistor

in an atmosphere of water vapour or oxygen. A thin layer of aluminium is then deposited on the device to provide a gate connection, the aluminium, silicon dioxide and the p-layer forming a capacitor.

With no voltage applied to the gate the only current which flows between source and drain connections is the negligibly-small leakage current of the pn junctions. If, however, the gate is biased positively (with respect to the source) electrons are attracted to the surface either from thermal breakdown of the p-layer or from the n-region and these provide a n-type conducting channel between source and drain permitting a longitudinal current flow. Increase in the positive bias increases channel conductivity and drain current. Because current is zero for zero gate bias and increases with increase in forward bias this device is said to operate in the enhancement mode.

It is however possible in the manufacture of IGFETs to provide an n-layer on the p-base so that there is conductivity in the channel and drain current can flow even with zero gate bias. For such IGFETs negative bias on the gate cuts off the channel as in JUGFETs and for such values of bias the device operates in the depletion mode. Positive gate bias will still increase drain current giving enhancement mode operation as before. IGFETs may have more than one gate and dual-gate IGFETs are sometimes termed *tetrode* FETs.

The distinction between enhancement and depletion types of FET is indicated in the graphical symbols by showing the channel of an enhancement device as a broken line. This is illustrated in *Figure I.8* which gives the graphical symbols for a dual-gate depletion-type IGFET and a single-gate enhancement-type IGFET, both with n-type channels.

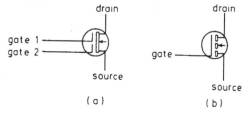

Figure I.8 Graphical symbols for (a) a dual-gate depletion-mode IGFET and (b) a single-gate enhancement-mode IGFET, both with n-channels

insulation-displacement connector A form of connector for terminating *ribbon cables* which clamps on the cable causing contacts in the connector to pierce the insulation of the cable to make connection with the conductors. It makes a quick and effective method of termination which avoids the need for soldering and cable stripping. A typical application for an insulation-displacement connector is for interconnecting printed-circuit or printed-wiring boards.

integrated circuit (IC) See *hybrid integrated circuit* or *monolithic integrated circuit*.

integrating amplifier An *operational amplifier* the output of which is equal to the time integral of the input waveform.

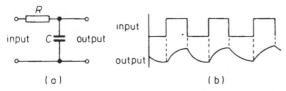

Figure 1.9 A simple integrating circuit (a) and typical input and output waveforms (b)

integrating circuit A circuit the output of which is approximately proportional to the time integral of the input signal. A common example is a circuit comprising a series resistor followed by a shunt capacitor as shown in *Figure 1.9*. The *time constant RC* must be long compared with the period of the input pulses.

integrator Any circuit, network or transducer yielding an output waveform substantially similar to the time integral of the input waveform.

intelligent terminal A computer *terminal* which, in contrast with a *dumb terminal*, allows some local data processing and storage facilities. It may run an application which accesses the *host* in a more efficient or *user-friendly* manner. Often an intelligent terminal is a general-purpose microcomputer running terminal emulation software. See *programmable workstation*.

intensifier electrode An electrode used to increase the electron-beam velocity in a *cathode ray tube* after the beam has been deflected.

intensity modulation Variation of the density of an *electron beam* in accordance with the instantaneous value of the *modulating signal*. An obvious example of intensity modulation occurs in the reproduction of TV images by a picture tube in which the electron-beam density is controlled by the *video signal* so as to produce the variations of light intensity on the screen necessary to make up the picture.

interactive Responding instantly to signals from the user. Many computer *applications*, e.g. word processing, are interactive in that the user provides input to the system via the *keyboard*, *mouse* or other device and this affects the execution of the program. Conventional television and radio broadcasting, in contrast, are not interactive since the viewer or listener has no immediate control over the material being transmitted.

inter-carrier reception In a TV receiver a method of sound reception in which the FM sound signal is derived from the vision detector or a post-detector stage as an FM signal on a carrier frequency equal to the difference between the vision and sound carrier frequencies. The method has the advantage that the centre frequency of the sound signal is unaffected by drift of the local oscillator.

interface A boundary between two pieces of equipment with different functions or between systems in which data are expressed in different forms.

interference (1) In radio or TV reception any unwanted signal, natural or man-made, which adversely affects reception of the wanted signal.

Natural interference signals can arise from lightning flashes, and man-made interference from signals on nearby frequency channels or image channels, from electrical equipment and from car ignition systems.

Interference signals can be picked up on the receiving *antenna* or can reach the receiver via the supply mains. Some reduction in interference may be possible by using a directional antenna, by siting the antenna in an 'electrically-quiet' spot, or by the use of RF filters in the receiver mains supply. The best method is to prevent the radiation of interfering signals by fitting suppressors to the offending equipment. (2) In optics the effects observed when two sources of light of the same frequency are superimposed. Areas where the two waves are in phase are illuminated more brightly than those where the waves are in phase opposition and thus one of the most familiar effects is the formation of interference bands or rings.

interlaced scanning In TV a system in which the scanning agent, during each of its vertical sweeps, scans the image in a series of equidistantly-spaced lines, the lines of each sweep being slightly displaced vertically from those of the previous sweep, so that two or more sweeps are necessary to scan the whole of the picture area. In *twin-interlaced scanning* the lines of each vertical sweep fall midway between those of the previous sweep.

intermediate frequency (IF) In a *superheterodyne receiver* the carrier frequency to which the modulation of all received signals is transferred by the frequency changer. The signal from the frequency changer is accepted by the IF amplifier which is responsible for most of the gain and the selectivity of a superheterodyne receiver.

intermodulation Interaction between the components of two or more complex signals in a non-linear system leading to the production of new components with frequencies equal to the sums and differences of those of the components of the waves (*combination frequencies*).

intermodulation distortion Distortion arising from the non-linearity of the input-output characteristics of a system, equipment or component and resulting in the generation of new signals at *combination frequencies* of the two or more sinusoidal input signals. It is the generation of these new signals some of which are not harmonically related to the frequencies of the input signals which is responsible for the harsh sound of an overloaded audio amplifier.

Internet An autonomous global computer *network* originally set up in 1969 to allow academic and governmental bodies to exchange data. Interest in the Internet became fashionable in the early 1990s when it began to be used by a diversity of individuals and organisations for the exchange of scientific, cultural, religious, commercial and other information.

interpretation See *computer language*.

interrogator A combination of radio transmitter and receiver designed to radiate signals to a *transponder* and to display the response received.

interrupted continuous wave (ICW) Type of wave used for radio telegraphy in which an audio-frequency modulating wave or the audio-frequency modulated wave is keyed on and off.

interval In *acoustics* the difference in frequency or *pitch* between two sounds.

intrinsic diode Same as *pin diode*.

intrinsic semiconductor A semiconductor material in which the concentration of donor and acceptor impurities is equal so that there is no resultant excess of *holes* or *electrons* to act as *charge carriers*. Such material is known as i-type to distinguish it from n-type and p-type material.

inverse amplifier   Same as *inverter*.

inverse feedback   Same as *negative feedback*.

inverse impedances   Two impedances the product of which is independent of frequency. Two simple examples of inverse impedances are a purely-inductive and a purely-capacitive impedance. The series and shunt elements in a *constant-k filter* are inverse impedances.

inverse networks   Two two-terminal networks the impedance of which are inverse, i.e. their product is independent of frequency.

inverse voltage   Of a diode in a rectifying circuit, the voltage appearing across the diode when it is non-conductive. See *peak inverse voltage*.

inverter   (1) Equipment for converting DC into AC. (2) In logic circuitry an amplifier the output of which is equal in amplitude but opposite in polarity to the input signal. Such an amplifier can be used as a negator to convert a logic-1 signal to a logic-0 and vice versa.

ion   A charged particle formed from an atom or a molecule by the loss or gain of *valence electrons*. If electrons have been lost, the resultant ion has a positive charge and if electrons have been gained, the ion has a negative charge. As matter is normally uncharged, ions are formed in pairs, one with a positive and the other with a negative charge. In gases the negatively-charged ions may be electrons and the positively-charged ions comprise the remaining parts of the atom or molecule where most of the mass resides.

ion burn   An area of reduced luminosity on the screen of a *cathode ray tube* caused by partial destruction of the phosphor by bombardment by heavy negative *ions* which are liberated from the cathode or are formed by *ionisation* of the residual gas. The area is usually at the centre of the screen because the heavy ions are not deflected to the same extent as electrons by the deflecting fields. See *aluminised screen, ion trap*.

ion gun   A device similar to an *electron gun* but in which the charged particles are *ions*.

ion implantation   In semiconductor-device manufacture a technique whereby a required amount of an *impurity* can be introduced into the semiconductor material by bombarding it with ions of the impurity (usually boron or phosphorus). The ions can be released by an RF discharge or by use of a heated filament and are focused and accelerated by electrostatic means, using voltages up to 500 kV. The process has the advantage over *diffusion* of giving more precise control over the concentration and depth of penetration of the impurity.

ionisation   The splitting of electrically-neutral matter into positively-charged and negatively-charged *ions*. Electronics is particularly concerned with ionisation of gases and there are a number of ways in which this can be achieved. For example the gas can be subjected to radiation such as ultra-violet light, *X-rays* or *gamma rays*. This occurs in the upper atmosphere where air is ionised by ultra-violet radiation from the sun to form the ionosphere.

A second method is ionisation by collision: if a voltage is applied between two cold electrodes in a gas tube the few positive and negative ions formed by ultra-violet light are collected at the *electrodes*. If the

173

voltage is increased sufficiently all such ions will be collected but, in moving rapidly to the electrodes, some ions and electrons collide with gas atoms and split them into further ions. These, too, may liberate ion-pairs and the process results in a rapid build-up of ions. The presence of an electron stream from a heated cathode greatly assists ionisation by collision.

ionisation voltage The minimum voltage applied to a gas-filled tube which will cause ionisation. The ionisation voltage depends on the gas: for example for mercury vapour it is 10.4 V and for helium 24.5 V.

ion trap Method of avoiding ion burn of the screen of a *cathode ray tube*. In one method the *electron gun* is aimed at the neck of the tube and an external permanent magnet is used to deflect the electron beam to the axis of the tube. Heavy negative ions liberated from the cathode or by ionisation of the residual gas which are responsible for ion burn are deflected to a smaller extent than the electrons and continue to bombard the tube neck. The use of an *aluminised screen* has now rendered the use of ion traps unnecessary.

isolate (1) To disconnect equipment completely from all sources of power. (2) To provide a degree of physical protection by making an object or circuit inaccessible unless special measures are taken. (3) To use a *buffer stage* to prevent interaction between stages in electronic equipment.

isotropic Having the same properties in all directions. See also *anisotropic*.

iterative impedance That value of impedance which, if used to terminate a *network*, gives an input impedance of the same value. The virtue of this

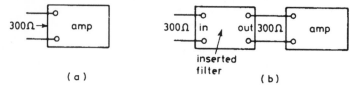

Figure 1.10 The use of a two-terminal-pair network on an iterative-impedance basis

concept is that networks used on an iterative-impedance basis can be inserted into circuits without affecting impedance levels. For example if the input impedance of an amplifier is 300 Ω (*Figure 1.10*(a)) and if a filter with an iterative impedance of 300 Ω is inserted before the amplifier the input to the filter is also 300 Ω as shown in *Figure 1.10*(b).

J

jack A socket containing a number of spring-loaded contacts which mate with the corresponding contacts of a plug inserted in the jack. The contacts within the jack can be arranged to make or break particular circuits when the plug is inserted as shown in the graphical symbols in *Figure J.1*. Plugs and jacks are extensively used in telephone switchboards and in broadcasting to permit rapid connections to be made to equipment or circuits.

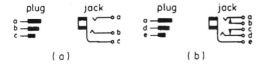

Figure J.1 (a) A three-pole plug and jack connecting two wires and earth and (b) a break-jack. At (b) the contact between a and b and between c and d is broken when the plug is inserted

jamming Deliberate transmission of radio signals to interfere with the reception of other signals.

jitter General term for sudden irregular departures from the ideal value of a parameter such as the *phase, amplitude* or *pulse duration* of a signal. In TV signals jitter can cause errors in synchronising and these can lead to erratic movements in the displayed picture.

JK bistable A *bistable circuit* with two inputs labelled J and K. When the J input is at its logic-1 state, the output (see *Figure J.2*) takes up its logic-1 state. When the J input is at its 0-state it has no effect on the output. When the K input is at its logic-1 state the output takes up its 0-state, but when the K input is at its logic-0 state it has no effect on the output.

The behaviour of the J and K inputs is thus similar to that of the R and S inputs of an *RS bistable* but the JK bistable has the advantage that if logic-1 signals are applied simultaneously to the two inputs the output is predictable. In fact every time the simultaneous logic-1 inputs are applied the output of the bistable goes to the complementary state.

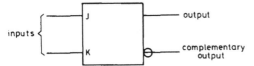

Figure J.2 Logic symbol for a JK bistable

Johnson noise Same as *thermal noise*.

Joint Photographic Experts Group (JPEG) A standard for the storage of *true-colour bitmaps* such as those obtained from colour photographs using a *scanner*. The JPEG standard uses Fourier Transform technology to sacrifice fine detail selectively in order to reduce the storage requirements. The degree of *compression*, however, is controlled by the JPEG file creator who chooses a quality factor expressed as a percentage. At high settings there is little detail loss, although worthwhile space savings may be made; at lower settings deterioration in image quality is more visible and there are greater space savings. Results vary with subject matter, but for many photographs a quality setting of 87 per cent results in a JPEG file about 10 per cent of the size of the original uncompressed bitmap with imperceptible deterioration in image quality at normal magnifications. A purpose-designed *integrated circuit* is available which decodes JPEG images considerably faster than a computer can. See also *MPEG*.

Joule effect The heating effect generated by an electric current in a conductor by virtue of its resistance. Joule's law states that the rate of heat generation is proportional to the square of the current.

joystick A control capable of movement in two mutually-perpendicular directions. Such a control may be used to adjust the position of the cursor of a VDU.

jump In computing, the transfer of control to a different area in memory. This contrasts with the processor's normal sequential execution of the instructions held in memory.

junction In a *semiconductor* device a transition region between two regions of different types of conductivity, e.g. a pn junction is the transition region between a p-region and an n-region.

When a pn junction is forward biased as indicated in *Figure J.3*(a), the p-region being positive with respect to the n-region, a considerable current crosses the junction. This current is carried by majority carriers, i.e. holes from the p-region (which move to the right) and electrons from the n-region (which move to the left). This forward current increases rapidly with increase in applied voltage.

If, however, the battery is reversed the majority carriers move away from each other leaving the junction region deficient in carriers: it is then known as a depletion layer. The only current which crosses a reverse-biased junction is a very small one composed of minority carriers, electrons from the p-region (which move to the right) and holes from the n-region (which move to the left). This small leakage current is largely independent of the applied reverse voltage.

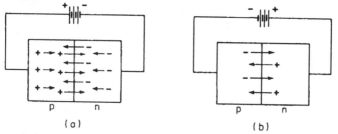

Figure J.3 (a) Movements of majority carriers in a forward-biased pn junction and (b) movement of minority carriers in a reverse-biased junction

Figure J.4 Graphical symbol for a diode. The end marked
+ is that which goes positive when the diode conducts

junction diode A pn junction employed because of its unilateral *conductivity*. Such diodes are extensively used for rectification, detection and in digital circuitry. The graphical symbol for a junction diode is given in *Figure J.4*.

junction-gate field-effect transistor (JUGFET) A *field-effect transistor* in which the gate forms a pn junction with the channel. A simplified structure of an n-channel JUGFET is given in *Figure J.5*. It consists of a crystal of n-type silicon with ohmic contacts provided by highly-doped (n+) regions near the two ends: these provide the source and drain connections to the external circuit. A region of p-type conductivity is formed (e.g. by diffusion) between the ohmic contacts: this provides the gate connection.

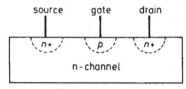

Figure J.5 Simplified diagram showing the structure of an n-channel junction-gate field-effect transistor

Current flowing longitudinally between the n+ regions is carried by the free electrons of the n-type channel. The number of electrons available to act as charge carriers is, however, dependent on the gate bias. The more negative (with respect to the source voltage) the gate is made the greater is the area of the depletion layer surrounding the p-region and the smaller the number of electrons available to carry the longitudinal current. In fact the channel can be cut off completely by a sufficiently-great reverse bias on the gate, the depletion layer then occupying the whole of the cross section of the channel.

Because the pn junction is reverse-biased during normal operation of the FET the transistor has a very high input resistance. The channel of a JUGFET may have more than one gate. A connection to the substrate can also provide a second input terminal although it is not usually as sensitive as the normal gate terminal. JUGFETs operate in the depletion mode and the graphical symbol for an n-channel device is given in *Figure J.6*.

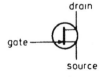

Figure J.6 Graphical symbol for an n-channel depletion-mode JUGFET

junction transistor A *transistor* containing a *pn junction* in its structure. This definition includes a *junction-gate field-effect transistor* but the term is generally used to mean a *bipolar transistor* as opposed to a *field-effect transistor*.

177

K

keep-alive electrode An *electrode* in a mercury-arc rectifier which maintains the vapour in a partially-ionised state.

Kell factor In TV a factor expressing the ratio of horizontal to vertical definition. Suppose each of the n lines of a TV picture is made up of alternate black and white elements. If these elements are assumed to be square-which is equivalent to regarding horizontal definition as equal to vertical definition-then there are $4n/3$ elements in each line, the aspect ratio being 4:3. The lowest video frequency which will enable such a row of elements to be resolved is one in which one half-cycle represents a white element and the other half-cycle a black element. This is the upper video-frequency limit for the system and it gives values appreciably higher than are used in practice. The lower practical values are justified by the assumption that the horizontal definition need only be, say, 0.7 times the vertical definition: in other words they are based on a Kell factor of 0.7.

Kelvin bridge A modification of the *Wheatstone bridge* used for comparing very low values of resistance. As shown in *Figure K.1* the Kelvin bridge

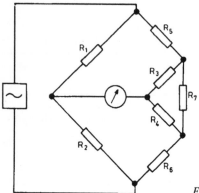

Figure K.1 Kelvin bridge

has an additional pair of ratio arms R_3 R_4 which have the same ratio as R_1 and R_2. R_5 and R_6 are the two low-value resistances to be compared and R_7 is an unknown resistance. At balance the following relationship applies:

$$R_5 = R_6 . \left(\frac{R_1}{R_2}\right)$$

kenotron A hot-cathode high-vacuum diode used for high-voltage low-current rectification in industrial applications, e.g. X-ray equipment.

Kerr cell A light modulator consisting of a liquid cell containing two parallel plane electrodes and situated between crossed polariods. Normally no light emerges because the polariods are crossed. Signals applied between the electrodes cause the plane of *polarisation* of the light to rotate (Kerr effect) so allowing light to pass.

key (1) Type of manually-operated switch used in telephone and broadcasting equipment. One example is a Morse key, a single-pole two-way switch designed for rapid operation to generate Morse telegraph signals. A telephone key is a multi-pole single-way or two-way switch operated by a lever or push button. (2) In a *database* the criterion against which data are retrieved. For example, if a doctor wishes to retrieve the names of all patients aged 70 or over, the age criterion is said to be the *key*.

keyboard An arrangement of buttons or keys by which the user enters data manually into equipment. More specifically, the alphanumeric keyboards used with *computers* or the musical keyboards used with *synthesisers*.

keypad A small *keyboard* often containing only numeric keys and a few others for special functions. Keypads are used in telephones and electronic calculators.

keystone distortion In TV, geometric distortion of the image causing a rectangle to be reproduced as a trapezium or keystone. The effect can be caused by interaction between line-scanning and field-scanning circuits but occurred in the early days of television as a result of oblique scanning of the target by the electron beam in *iconoscope* tubes. It was necessary to introduce correction to obtain an accurately rectangular reproduced image.

keyword In computing a string of characters that is recognised by the *operating system* or by a *computer language* or by *applications software* and which initiates an activity.

kinescope (US) Same as *cathode ray tube*.

Kirchoff's laws (1) In any *network* the sum of the currents which meet at a point is zero. (2) In any closed path in a network the algebraic sum of the products of resistance and current in the branches is equal to the sum of the EMFs acting in the path.

These two fundamental laws apply to instantaneous currents and are of great value in theoretical analyses of the behaviour of electronic circuits.

klystron An *electron tube* in which the *electron beam* is velocity modulated to generate or amplify microwaves. A simplified diagram showing the construction of a typical klystron is given in *Figure K.2*. The electron beam

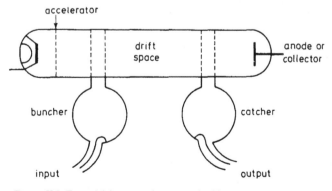

Figure K.2 Essential features of a two-cavity klystron

179

from the cathode passes between the grids of a *cavity resonator* known as a buncher. The input signal is applied to this resonator and the resulting potentials between the grids causes velocity modulation of the beam. Bunching occurs in the drift space and the bunches, in passing through the catcher grids, induce an amplified output signal in the catcher resonator. The beam is finally collected at the *anode* (or collector).

This is in essence the theory of the two-cavity klystron amplifier. The graphical symbol for the tube is given in *Figure K.3*. By coupling the buncher and catcher resonators, the tube can be made to generate microwaves. An alternative form of klystron oscillator is the *reflex klystron*.

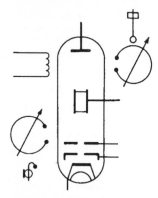

Figure K.3 Graphical symbol for a two-cavity klystron. The symbol shows an indirectly-heated cathode, intensity-modulating electrode, two focusing electrodes and collector. The tunable input resonator is fed via a coaxial line and the tunable output resonator is window-coupled to a waveguide. An external coil is shown to produce the axial magnetic field.

k rating In TV a factor derived from objective measurements of the distortion of a test waveform which expresses the subjective effect of the distortion on reproduced images. The measurement is made on an *oscilloscope* fitted with a graticule using a *pulse and bar test signal*.

180

L

ladder network A four-terminal *network* consisting of alternate series and shunt elements as shown in *Figure L.1*. As shown in the diagram, if each shunt element is replaced by two elements in parallel and if each series element is replaced by two elements in series, the ladder network can be regarded as a succession of *L-networks* (*Figure L.2*). Networks 1 and 2, 3 and 4, etc can be combined to form *pi-networks*. Alternatively networks 2 and 3, 4 and 5, etc could be combined to form *T-networks*. Thus a ladder network can be regarded as a succession of T-networks or pi-networks. These constituent networks are known as sections and the L-networks of which they are composed, as half-sections. See *constant-k filter, filter, m-derivation*.

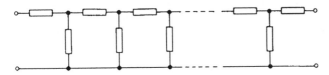

Figure L.1 A ladder network

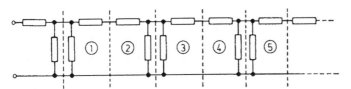

Figure L.2 The ladder network of *Figure L.1* redrawn as a succession of L-networks

lag (1) See *phase angle*. (2) In *photocells* and *camera tubes* the time which elapses between a change in light input and the corresponding change in electrical output. Lag in camera tubes tends to produce blurred images of objects which move rapidly across the field of view. Keeping lag to an acceptable level is one of the difficulties in the design of photoconductive targets.

landline In a communications system a link between two points that uses hard cables, e.g. telephone lines, rather than a radio link.

laptop computer A self-contained *microcomputer* which can be operated from internal batteries and is small enough to be carried in a briefcase and used anywhere, e.g. while travelling.

large-scale integration (LSI) See *monolithic integrated circuit*.

laser (light amplification by stimulated emission of radiation) A device for light amplification which relies for its action on the *radiation* emitted by certain atoms when transitions occur between discrete energy levels. In practice positive feedback is usually applied to the amplifiers (by use of mirrors at the input and the output) to make them oscillate. They then become generators of coherent light in the form of a narrow and sharply-defined beam of good spectral purity and frequency stability.

Certain atoms emit radiation as they change from one energy level to a lower energy level and they can be stimulated into emission by placing them in a radiation field with the same frequency as they would normally emit (a technique known as pumping). Amplification is possible if more atoms leave the higher level than absorb energy at the lower level. Suitable materials are therefore those in which atoms at the upper energy level have a longer lifetime than those at the lower level. A number of materials have this property including some gases, liquids, solids and semiconductors. Thus a number of different types of laser have been developed and these are capable of oscillation at a large number of frequencies between ultra-violet and infra-red.

Gas lasers normally employ a helium-neon mixture, argon or carbon dioxide contained in a tube between 1 and 2 mm in diameter and 20 to 100 cm long. Pumping is achieved by passing a discharge through the gas and the carbon-dioxide laser can give several kW of output, adequate for machining operations.

Liquid lasers use solutions of fluorescent materials and are pumped by light from an argon laser or an electronic flash tube. Ruby is the solid most often used in lasers although neodymium glass or yttrium aluminium garnet are also employed. These are optically pumped using light from a flash tube, a krypton arc or a tungsten iodine lamp. Outputs up to 1 kW are possible using a number of cascaded rods of the solid.

Semiconductor lasers are different in nature from the types just described. In pn junctions using certain semiconductors such as gallium arsenide and indium arsenide, radiation is emitted when electrons and holes recombine and by using very high doping it is possible to achieve the condition where the higher-level charge carriers have a longer lifetime than those at the lower level. A high-density forward current across the junction can then lead to laser action and oscillation is possible by using a structure in which the crystal is cleaved at right angles to the plane of the junction, the cleaved surfaces acting as the mirrors which give positive feedback. In early semiconductor lasers of this type, intermittent operation was necessary to avoid overheating but continuous operation is possible in modern multi-layer structures.

Because of the small dimensions possible, semiconductor lasers can be designed to feed light into optical fibres. They are also used to replay *compact disks*. The block symbol for a laser is given in *Figure L.3*.

Figure L.3 General block symbol for a laser

laser printer See *printer*.

latch In general a two-state device which changes state on receipt of an input signal and then remains in that state, ignoring all subsequent input signals, until reset. Latching may be mechanical or electronic.

latch up In *computers* or *data-processing equipment* a phenomenon in which, as a result of a fault, a particular section of the equipment is prevented from operating normally.

lattice network A two-terminal-pair *network* consisting of two series elements, one in each leg, and two shunt elements each bridging an input and output terminal. The lattice network is shown in its usual form in *Figure L.4* but it can be redrawn in the form shown in *Figure L.5* which is generally known as a bridge network.

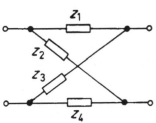

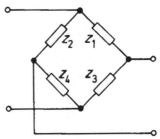

Figure L.4 A lattice network

Figure L.5 The lattice network of *Figure L.4* redrawn in the form of a bridge network

lead-acid cell (accumulator) A *secondary cell* in which the positive plate is of lead dioxide, the negative plate is of lead (in sponge form) and the electrolyte is dilute sulphuric acid. During discharge the acid reacts with both plates to form lead sulphate and thus becomes more dilute. On recharging these reactions are reversed and the acid regains its strength. The cell has an EMF of about 2 V on load and can supply currents up to 100 A. The cells are extensively used for power supply and are most tolerant of variations in charging current. Their chief disadvantage is that the EMF and capacity both fall as temperature is lowered.

leakage current In general any current which flows via an undesired, usually high-resistance path. For example a leakage current may flow across the surface of an insulator. In particular in a *bipolar transistor* a component of the collector current which is not controlled by the input current. It is generated by thermal breakdown of covalent bonds in the collector-base junction and increases rapidly with increase in temperature. In germanium transistors leakage current can be large enough to seriously limit the performance of the transistor and it sets a limit to the temperature at which such transistors may be used. In silicon transistors leakage current is negligible at normal temperatures.

leakage flux Of a magnetic circuit that part of the *flux* which does not flow in the main (intended) path. For example the leakage flux in a transformer is that which does not link with the secondary winding and does not therefore contribute to the secondary voltage.

leaky-grid detector Same as *grid-leak detector*.

Leclanché cell A primary voltaic cell consisting of a carbon electrode (positive) and a zinc electrode (negative) immersed in an electrolyte of ammonium chloride solution. In its dry form this is the most widely used of primary cells. The carbon electrode is a rod surrounded by a paste of ammonium chloride and manganese dioxide (depolarising agent) within a zinc container acting as the negative electrode. See *depolarisation*.

leddicon A *camera tube* with a photoconductive target of lead oxide.

Lempel-Ziv-Welsh (LZW) compression A form of data *compression* widely used in computing, especially for the storage of *bitmaps*.

level The amplitude of a signal particularly when compared with a reference amplitude. The level may be expressed in decibels relative to the reference level which is taken as 0 dB. For example if zero level is taken as 1 V, a voltage level of 10 mV can be expressed as −40 dB.

life test A test carried out on a component or device to determine its probable life under normal working conditions.

lifetime In *semiconductor* theory the average time between the formation of a *minority carrier* and its recombination.

lift In TV a *pedestal* of adjustable height.

light-emitting diode (LED) A pn *junction* which emits visible radiation when forward biased. When a pn junction is forward biased electrons are driven into the p-region and holes into the n-region as shown in *Figure J.3(a)*. Some of these *charge carriers* combine in the junction area and with suitable choice of semiconductor material this is accompanied by the emission of visible light. It is possible to obtain light of a number of different colours but maximum electrical-optical efficiency is obtained when the light is red. LEDs are widely used as indicator lamps in electronic equipment and as seven-segment arrays in numerical displays. The graphical symbol for a LED is given in *Figure L.6*.

Figure L.6 Graphical symbol for a light-emitting diode

lighthouse tube An electron tube made suitable for operating at UHF in which the closely-spaced electrodes are extended as annular disks which project through the glass envelope to locate with the ends of the coaxial

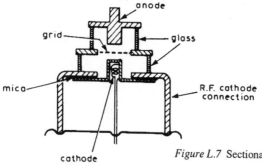

Figure L.7 Sectional view of a lighthouse tube

lines used as tuning elements. The stepped sizes of the disks gave the tube the appearance of a lighthouse as shown in *Figure L.7*. The close spacing of the electrodes is necessary to minimise transit time and so raise the limiting frequency of oscillation and amplification. This is an American version of the British *disk-seal tube*.

light pen A device used to produce a visible image on the screen of a *visual display unit*. By passing the pen over the surface of the unit the operator can communicate with a stored programme and can thus modify the display. Light pens are extensively used in *computer-aided design*.

light valve A device which enables light transmission to be varied in accordance with the instantaneous value of the applied electrical signal. An example of a light valve is the *Kerr cell* used with polarised light.

limiter Same as *amplitude limiter*.

line (1) Abbreviation for *transmission line*. (2) Abbreviation for *scanning line*.

linear distortion Any form of *distortion* which is independent of the amplitude of the signal. Thus distortion which results from variation of a parameter with frequency, e.g. attenuation distortion is an example of linear distortion. See *non-linear distortion*.

linearity control In early TV receivers a control which varied the scanning speed of the *electron beam* during the forward trace so as to correct geometric distortion of the image. Such control might be applied to the horizontal and vertical movement of the beam so that there were two linearity controls: line linearity and field linearity.

linear network See *network*.

line blanking In TV the suppression of the picture signal during the interval between two successive *scanning lines*.

line frequency In TV the number of horizontal sweeps made by the scanning beam in one second. It is equal to the product of the number of lines per picture and the picture frequency. In a twin-interlaced system such as used by most TV services the picture frequency is one half the field frequency. For example in the British 625-line system there are 50 fields per second and thus the line frequency is $625 \times 50/2 = 15.625\,\text{kHz}$.

line microphone A directional microphone in which the *transducer* is fed with sound via a long tube with spaced or distributed acoustic elements so designed that only sound waves travelling along the tube axis or at small angles to it are accepted.

lines of force Lines representing the direction of an electric or magnetic field. The direction of the line at any point in it indicates the direction of the force of attraction or repulsion on a small positive charge or small north pole placed at that point, and the number of lines crossing a unit area at right angles to the line is a measure of the field strength.

line printer See *printer*.

line sync signal In TV the signal transmitted at the end of each *scanning line* to initiate horizontal flyback of the scanning beam in receivers, so keeping the scanning at the receiver in step with that at the transmitter. In most TV systems the signal consists of a single pulse from blanking level to sync level as shown in *Figure L.8*, the leading edge of which locks the receiver *line time base*.

185

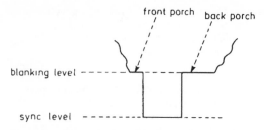

front porch   back porch

blanking level

sync level

Figure L.8 Waveform of television line sync signal

line time base In TV the circuits responsible for generating the signals causing horizontal deflection of the *scanning beam*. In modern TV receivers the line output stage generates, in addition to the line scanning current, a direct voltage to boost the supply to the output stage, the heater supply for the picture tube, the EHT supply for the picture tube and possibly a low-voltage supply for early stages in the receiver.

liquid crystal display (LCD) A display system consisting essentially of a very thin layer of liquid sandwiched between two conducting glass plates between which the control voltage is applied. One way in which the applied voltage controls the light transmission of the device is by varying the light scattering in the liquid which is specially chosen because of its long-molecule construction. The conducting areas of the plates are such that, by applying voltages to certain of the leads, specified areas of the display can be illuminated by light transmitted through the device or reflected at the rear glass plate. Thus a seven-segment pattern can be used to give a numerical display.

Liquid crystal displays consume very little electrical power compared with *light-emitting diode* displays and have superseded LEDS in battery-operated calculators and digital watches.

Lissajous figures Patterns generated on the screen of a *cathode ray tube* when the *electron beam* is defected horizontally and vertically by sinusoidal signals with a simple relationship between their frequencies. Some typical patterns are shown in *Figure L.9*.

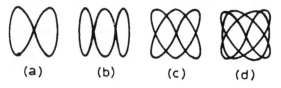

(a)   (b)   (c)   (d)

Figure L.9 Some typical Lissajous figures

Litz wire See *skin effect*.

L-network A network consisting of one series and one shunt element. See *Figure L.10*.

load (1) Of any signal source the circuit connected across its terminals and into which it delivers power. See *anode load*. (2) In computers and data

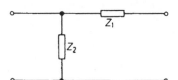

Figure L.10 An L-network

processing to transfer *data* or *programs* into the system's *memory* from a *storage* medium.

load line   A line drawn on a family of characteristics of an active device to illustrate the relationship between the voltage across and the current in the load when an input signal is applied to the device. *Figure L.11* shows an example of a load line on a set of I_c–V_c characteristics for a *bipolar transistor*. AB is the load line and its slope is equal to the reciprocal of the load impedance. P is the operating point where the load line meets the

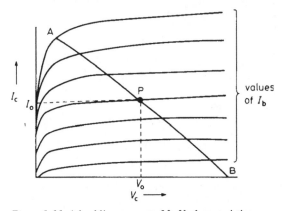

Figure L.11   A load line on a set of I_c–V_c characteristics

characteristic for the chosen value of base bias: P therefore represents the quiescent or no-signal values of voltage (V_o) and current (I_o) for the load. When a signal is applied to the base the operating point oscillates about P along the load line to an extent dependent on the input-signal amplitude.

Local Area Network (LAN)   A *network* of microcomputers within a limited area such as one building or even one office within a building. LANs allow several microcomputer users to share *peripherals* that may be too expensive to be allocated to each individual user. Usually one computer is set aside as a *fileserver*; its *hard drive* provides *storage* space for individual network users and may also include a 'public' area in which all network users may store and access files and may use this facility to transfer files amongst themselves. Other peripherals, such as laser printers, modems or CD-ROM drives connected to the fileserver will also be accessible to all users of the network. Facilities are usually provided for network users to send messages to other network users.

local oscillator In a *superheterodyne receiver* the oscillator the output of which is mixed with the received signal to give an output at the *intermediate frequency*.

locking The process by which an oscillator can be synchronised at the frequency of a signal applied to it.

lock out See *latch up*.

logarithmic decrement A measure of the rate of decrease of amplitude in a damped oscillation. It is equal to the natural logarithm of the ratio of one amplitude to the next in the same direction.

It is the natural logarithm of the *damping coefficient* or factor. In an electrical resonant circuit the logarithmic decrement is related to the circuit constants, being given by $R/2f_0L$, where R is the series resistance in the circuit, L is the inductance and f_0 the resonance frequency. This is equal to $\pi R\sqrt{(C/L)}$, where C is the capacitance in the circuit.

logic In a *computer* or *data-processing equipment* the operations chosen to carry out the functions required of the equipment. Also the physical form of the *logic elements* chosen for the equipment.

logical In computer *hardware* or *software* the *logical* appearance of a system or an operation is the way in which it is perceived by the user. This may contrast with the actual configuration or method of operation. See *physical*.

logic element In *computers* and *data-processing equipment* a device used to perform a specific logic function. There are two main types: combinational elements usually called *logic gates* and sequential elements of which the *bistable* is the best-known example.

logic gate A device with a number of input terminals and one output terminal in which the state of the output signal depends in a particular manner on the states of all the input signals. For example the output may take up the 1-state only when all the inputs stand at their 1-state: this is true of an *AND gate*.

In an electronic gate the 1-state may be the more positive or the more negative of the two voltage values selected as *logic levels*. The choice is important because the behaviour of the gate depends on it. A gate which functions as an *AND gate* for *positive logic* behaves as an *OR gate* for *negative logic*. Thus the behaviour of a particular gate cannot be predicted until the logic convention has been decided.

logic level In *digital computers* and *data-processing equipment* one of the two values of voltage or current which represent the only two possible significant states permissible on the signal lines in the equipment. One value is designated the 0 state and the other the 1 state.

log in or **log on** In computing, to establish contact with a *host* computer, which may be a mainframe machine serving many terminals or the *fileserver* of a *local area network*. Logging on (or in) usually involves giving a user identity and a password. The identity determines which personal storage areas the user will be able to access. The password reduces the risk of unauthorised access to confidential data. A user must successfully log on (or in) before he is allowed access to the host.

log off or **log out** In computing, to terminate a session on the *host* computer. See *log in*.

long-persistence tube A *cathode ray tube* of which the screen is coated with

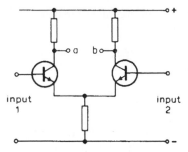

Figure L.12 A long-tailed pair using two bipolar transistors

phosphorescent material as well as *fluorescent* so that the visible trace persists for several seconds.

long-tailed pair A phase-splitting circuit in which push-pull signals are obtained from the output circuits of two similar *active devices* with a common resistance in their emitter (cathode or source) circuits. The circuit (an example is shown in *Figure L.12*) has applications other than phase-splitting: for example if two equal steady signals are applied to the bases, no resulting output appears between a and b. As another example if two equal alternating signals are applied to the bases there is no output provided that the inputs are in phase. Any phase difference results in an output. Thus the circuit can be used as the basis of an FM or PM discriminator.

loss In general an indication of the extent to which the amplitude of a signal is decreased by its passage through an electronic system. More specifically the ratio of the output-signal power, voltage or current to the input-signal power, voltage or current. The ratio is usually expressed in decibels as $10 \log_{10} P_{out}/P_{in}$, $20 \log_{10} V_{out}/V_{in}$ or $20 \log_{10} I_{out}/I_{in}$.

loudness The subjective assessment of the power of a sound.

loudness level A numerical assessment of loudness. The loudness level of a sound in *phons* is equal to the intensity level in decibels of a 1-kHz tone which sounds, to the human ear, equally as loud as the sound, zero intensity level being taken as a sound pressure of $2 \times 10^{-5} N/m^2$. This is a subjective method of measuring loudness level. It can also he measured objectively by the use of a *microphone*, amplifier and level meter, the circuit including a weighting network to simulate the frequency response of the ear.

loudspeaker A *transducer* for converting electrical audio-frequency signals into corresponding sound signals and capable of radiating enough audio-frequency power for domestic and auditorium purposes.

There is a very wide range of types of loudspeaker varying from the miniature single-unit transducers used in portable radio receivers to the large multi-unit transducer capable of high-fidelity reproduction. Loudspeakers utilise a number of different principles. See *crystal loudspeaker, electrostatic loudspeaker, moving-coil loudspeaker*.

low-level modulation Same as *low-power modulation*.

low-pass filter A *filter* designed to pass signals at frequencies below a specified cut-off frequency. Thus the passband begins at zero frequency

Figure L.13 Block symbol for a low-pass filter

189

and extends to the cut-off frequency. In general such filters comprise series inductors and shunt capacitors. The block symbol for a low-pass filter is given in *Figure L.13*.

low-power modulation In an AM transmitter modulation of an RF stage prior to the final high-power stage. This system has the advantage that the modulator need only be of low power but the stage following the modulated amplifier must be linear to avoid distortion of the modulation envelope. See *high-power modulation*.

low tension (LT) The low voltage used to supply power for the heaters or filaments of electron tubes. For tubes intended for operation from batteries the low tension was usually 1.4 V or 2 V. For indirectly-heated tubes normally operated from an AC supply the low tension was usually 6.3 V.

luminance In general a measure of the light flux emitted or reflected by unit area of a source. More specifically the luminance in a given direction of a surface emitting or reflecting light is the quotient of the light flux measured in that direction and the area of the surface projected on a plane perpendicular to the direction. The subjective assessment of luminance is *brightness*.

luminance signal (Y) In colour TV the signal which conveys information about the *luminance* of the elements constituting the picture. Information about the colours of the elements is contained in the *chrominance signal*. It is the luminance signal to which a black-and-white receiver responds and which enables it to reproduce a black-and-white version of the colour picture transmitted.

luminescence General term for the emission of light from a phosphor when suitably excited. See *fluorescence* and *phosphorescence*.

luminophore A luminescent material. See *luminescence*.

lumped circuit constant *Capacitance* (*inductance* or *resistance*) of a circuit which can be treated for circuit analysis as though it were a single capacitor (inductor or resistor) connected at a point in the circuit. See *distributed constants*.

M

machine code or **machine language**   See *computer language*.

magic-eye tuning indicator   Same as *cathode-ray tuning indicator*.

magnetic amplifier   A device consisting of one or more ferromagnetic cores with windings so arranged that alternating current in one winding can be amplified by virtue of the non-linearity arising from *magnetic saturation* of the core. See *saturable reactor*.

magnetic axis   Of a coil carrying current, the line of symmetry of the *magnetic flux* pattern.

magnetic bubble   In a thin layer of permeable material magnetised across its thickness. a domain in which the magnetism is in the opposite direction. A bubble can be produced by passing a current pulse through a looped conductor. The diameter of a bubble depends on the thickness and properties of the layer and the externally-applied field and can be made very small as required in *magnetic bubble stores*.

magnetic bubble store   A *non-volatile random-access memory* in which data can be stored in binary form, the two states being the presence or absence of a magnetic bubble. By forming the bubbles in a epitaxial layer of magnetic material on a semiconductor substrate the bubble diameter can be reduced to a few μm so that very great packing density is possible and such stores offer great possibilities for the future. The bubble store is similar to the CCD store in that it is basically a serial-in/serial-out device which can replace magnetic tape or disk but the magnetic-bubble store has the advantage over CCD that it is non-volatile.

magnetic circuit   The closed path of the *magnetic flux* set up by a permanent magnet or a winding carrying a current. The path is usually confined to permeable material with possibly a small air gap. It is termed a circuit because the flux is determined by the quotient of the magnetomotive force and the reluctance of the magnetic circuit just as the current in an electric circuit is given by the quotient of EMF and resistance. The reluctance of a magnetic circuit is determined by the dimensions and the permeability of the material forming the closed path.

magnetic core   In an *inductor, transformer* or similar device the ferromagnetic circuit which is provided to concentrate the *magnetic flux* within the region where it is required. In magnetic storage devices the core is a device of magnetic material which is required to retain two or more levels of magnetisation. In other magnetic devices, such as *magnetic amplifiers* and *saturable reactors*, the core is needed to provide the non-linear relationship between magnetising force and flux essential to their action.

magnetic deflection   Same as *electromagnetic deflection*.

magnetic dipole   An elementary radiator of electromagnetic waves consisting of a single infinitesimally-small loop.

magnetic drum   In *computers* and *data-processing equipment* a rotating cylinder the surface of which is coated with magnetic material on which

191

data can be stored by selective magnetisation of the surface. Normally the data is in binary form and the recording is by the orientation or the polarity of the local magnetisation of the surface.

magnetic field A region near a permanent magnet or a conductor carrying an electric current in which a magnetic pole experiences a mechanical force caused by the magnet or current.

magnetic field strength (H) At a point in a *magnetic field* a vector representing the magnitude and direction of the mechanical force on a unit north pole at that point. It is equal to the magnetomotive force per unit length measured along a closed path in the magnetic field.

magnetic flux The lines of force constituting a *magnetic field*.

magnetic flux density (B) The magnetic flux per unit area normal to the direction of the lines of force.

magnetic flux leakage Same as *leakage flux*.

magnetic focusing Focusing of the electron beam in a *cathode ray tube* by the use of an *electromagnetic lens*.

magnetic hysteresis Same as *hysteresis* and *hysteresis loop*.

magnetic induction Same as *magnetic flux density*.

magnetic intensity Same as *magnetic field strength*.

magnetic lens Same as *electromagnetic lens*.

magnetic memory Same as *magnetic store*.

magnetic moment (1) Of a magnet the product of the pole strength and the distance between the poles. (2) Of a *magnetic dipole* a vector equal to the product of the current in the loop and its area. The direction is perpendicular to the plane of the loop.

magnetic recording In general the process of impressing the characteristics of a *signal* on a moving magnetic medium. In particular the impressing of an audio or video signal on a magnetic tape or disk as it moves under a recording head. The recording is made by varying the intensity of magnetisation of the medium in accordance with the instantaneous amplitude of the signal.

magnetic saturation The state of a ferromagnetic material subjected to a large magnetising force in which increase in the force gives no increase in *magnetic flux*.

magnetic screen A screen of magnetically-permeable material used to reduce the penetration of a magnetic field into a particular region.

magnetic shield (US) Same as *magnetic screen*.

magnetic shunt A piece of magnetic material connected across a magnetic circuit to reduce the flux in that circuit.

magnetic storage A method of data storage using the magnetic properties of materials in the form of cores, films or plates, or as coatings on tapes, disks or drums. See *floppy disk, hard disk, magnetic bubble, magnetic bubble store, magnetic core, magnetic drum, magnetic tape*.

magnetic susceptibility See *susceptibility*.

magnetic tape A tape used for the storage of data or the recording of audio or video signals and consisting of magnetic material or coated with such material.

magnetisation In general the process of producing a temporary or permanent magnetic field in a material. More specifically the magnetisation or intensity of magnetisation of a material is the magnetic moment per unit volume or the pole strength per unit area.

magnetising force Same as *magnetic field strength*.

magnetism The name given to a number of phenomena arising from the special properties of magnetic materials. For example these materials can be magnetised and then always come to rest in the same position when freely suspended. They can magnetise other materials. Two magnetised materials can attract or repel each other.

magnetomotive force (MMF) The force which gives rise to a *magnetic field*. It is equal to the line integral of the magnetising force around a closed path in a magnetic field and it determines the magnetic flux according to the relationship

$$\text{magnetic flux} = \frac{\text{magnetomotive force}}{\text{reluctance of the magnetic circuit}}$$

which compares with Ohm's law for electrical current. Thus magnetomotive force is the magnetic analogue of electromotive force.

magneto-optical disk and **drive** A *demountable* data *storage* system which offers many of the advantages of *Compact Disc* while also erasable and recordable. To record, the appropriate spot on the disk surface is heated by a laser while a localised magnetic field changes its crystal structure, affecting its reflective properties. The disk is read using a laser as in Compact Disc; indeed many 5-inch magneto-optical drives can also read Compact Discs. The Sony *Mini-Disc*, intended as a *digital audio* recording system for portable and automotive equipment, uses 2.5-inch disks holding 125 Mbytes which, with real-time *companding*, allows about 75 minutes of stereo audio. 5-inch disks for computer use hold a nominal 650 Mbytes.

magnetostriction The change in dimensions of a piece of magnetic material when it is magnetised. The effect is most marked in nickel and is exploited in magnetostriction loudspeakers and oscillators.

magnetron A diode tube in which the electron stream is subjected to a *magnetic field* at right angles to the electric field. In its simplest form the magnetron consists of a diode with a cylindrical *anode*, the *cathode* lying along the axis of the cylinder. The anode is split into two or more segments. A strong magnetic field is set up, parallel to the axis of the cylinder, by an external magnet. The anode is positively charged to attract electrons liberated from the cathode but the magnetic field causes the electrons to pursue curved paths around the cathode and, depending on the relative strengths of the electric and magnetic fields, they may not meet the anode at all and return to the cathode. If tuned circuits are connected between neighbouring segments of the anode, the electron stream becomes velocity modulated as it moves around the cathode and in this way bunching and oscillation can occur.

magnification factor (1) See *amplification factor*. (2) See *Q-factor*.

mailmerge The process of merging data from a *database* repeatedly into copies of a generic *document* for printing. Often the document is a standard letter; the data being merged consists of the necessary personal details for the individual recipients. Each letter printed will therefore carry the appropriate details for its intended recipient.

mainframe computer See *digital computer*.

mains hum See *hum*.

main storage Of a *digital computer* the internal store associated with the *central processor*.

mains unit An assembly of components designed to derive from a mains input outputs suitable for operating electronic equipment. For example a mains unit for transistor equipment requires basically a mains transformer, rectifier and smoothing capacitor suitable for the required direct voltage. For electron-tube equipment in addition to the HT (direct) supply, a low-voltage AC supply was required for the tube heaters and this was obtained from a winding on the mains transformer.

maintaining voltage Of a *gas-filled tube* the voltage between the *electrodes* when the tube is operating normally. The maintaining voltage is independent of the current taken by the tube over a considerable range of current.

maintenance test A test carried out regularly during the life of an equipment or component to check that its performance is still within tolerance.

majority carriers In an *extrinsic semiconductor* the type of charge carrier which outnumbers the other type. For example in *n-type semiconductors* the majority carriers are electrons and in *p-type semiconductors* they are holes.

majority gate A *logic gate* with several inputs and one output which gives a logic-1 output only when more than half the inputs stand at logic-1.

Mallory cell Same as *mercury cell*.

marginal relay A relay which operates when a predetermined change in the coil current or voltage occurs.

maser (microwave amplification by stimulated emission of radiation). A microwave amplifier which makes use of the radiation emitted by certain atoms or molecules when transitions occur from a higher to a lower energy state. A feature of the amplifier is direct interaction between electromagnetic waves and atoms: if the waves are of the right frequency they can stimulate high-energy electrons to release their energy in the form of electromagnetic waves thus amplifying the stimulating signal. Amplification can be almost noise-free particularly if the device is operated at a very low temperature. The block symbol for a maser is shown in *Figure M.1*.

Figure M.1 General block symbol for a maser

masking (1) In acoustics the apparent suppression of a weak sound by a strong one. More specifically it is the amount (in dB) by which the threshold of audibility of a sound is raised by the presence of the other (masking) sound. (2) In colour TV, alteration of the colour rendering by deliberate *cross-coupling* between the three primary colour channels. Masking is used to improve the colour rendering of cinema film when transmitted by colour television.

mass In a mechanical oscillating system a property of a body in motion which enables it to resist changes in velocity and is measured by the ratio of force to rate of change of velocity, i.e. the ratio of force to acceleration. Mass is analogous to inductance in an electrical oscillating system. A moving mass possesses kinetic energy just as an inductor carrying current stores energy in the associated magnetic field.

mass storage Same as *bulk storage*.

master clock Same as *clock*.

master oscillator In a radio transmitter an oscillator used to determine the frequency of the *carrier wave*. Such an oscillator requires great frequency stability to maintain the carrier wave on its allotted frequency.

master-slave operation In general a system of two or more devices in which one (the master) controls the operation of the others (the slaves).

match gate Same as equivalence element.

matching (1) The process of ensuring that maximum power is transferred from a generator to its load. As an example an *L-network* can be used to match the generator impedance to the load impedance at a particular frequency. Where matching is necessary over a wide frequency band a transformer can be used to couple generator to load. Such matching is used to connect an electron tube to a *loudspeaker*. (2) The selection of two devices with approximately-equal characteristics. For example two matched transistors are essential in a *class-B* stage to minimise distortion.

matrix Mathematically a two-dimensional *array* of quantities manipulated in accordance with the rules of matrix algebra. In electronics the term is loosely used to mean any coder or decoder. More specifically in computers a rectangular logic array of intersections between a number of input leads (arranged as rows) and a number of output leads (arranged as columns). *Logic elements* are situated at some of the intersections to enable a single input to produce a number of outputs (decoder use) or a number of inputs to give a single output (coder use).

Maxwell bridge An AC bridge network in which one arm consists of *inductance* and resistance in series, the opposite arm consists of *capacitance* and resistance in parallel, the remaining two arms being resistive.

The bridge is illustrated in *Figure M.2* and is used for the measurement

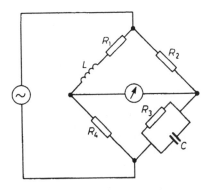

Figure M.2 A Maxwell bridge

of inductance (in terms of a known capacitance) or capacitance (in terms of a known inductance), the relationship at balance being

$$\frac{L}{C} = R_2R_4 = R_1R_3$$

***m*-derivation** A method of modifying the prototype L half section of a *filter* so as to give a sharper cut-off at the edges of the passbands and a better impedance characteristic. The reactance of the series arm (S) of the L section is reduced to mS (m being less than unity) and that of the shunt arm (P) is increased to P/m; in addition a new reactance is added to the half section to give infinite attenuation at a frequency beyond the *cut-off frequency*.

In series *m*-derivation the new reactance has a value of $S(1-m^2)/m$ and is connected in series with the shunt element P/m to form an acceptor tuned circuit across the network. In shunt *m*-derivation the new reactance has a value of $Pm/(1-m^2)$ and is connected in parallel with the series reactance mS to form a rejector tuned circuit. These modifications to the prototype half section (illustrated in *Figure M.3*) do not affect the cut-off frequency of the filter or its characteristic impedance. The frequency of infinite attenuation can be placed anywhere within the stopband by suitable choice of the value of m: the smaller m is made, the nearer is this frequency to the cut-off frequency.

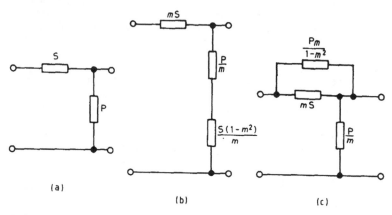

Figure M.3 A prototype L-network (a) and the modifications for series *m*-derivation (b) and shunt *m*-derivation (c)

It is common practice in filter design to employ *m*-derivation in the final half-section of a constant-k filter, which can then be satisfactorily terminated in a pure resistance. The *m*-derived half-section provides a satisfactory termination for the constant-k sections over most of the passband if m is made equal to 0.6. The *m*-derived half section also improves the steepness of attenuation outside the passband.

196

mean time between failures (MTBF)   In general a measure of the reliability of an equipment. More specifically it is the time to the first failure or, if the equipment can be repaired, the average time between failures. The MTBF of equipment is a statistical estimate based on the MTBFs of the components in it.

mechanical impedance   The complex ratio of the alternating force applied to a vibrating system to the resulting alternating velocity in the direction of the force. By analogy with electrical impedance, mechanical impedance is made up of mechanical resistance (e.g. friction) and mechanical reactance, and the latter can arise from mass (mechanical analogue of inductance) or from compliance (mechanical analogue of capacitance).

medium-scale integration (MSI)   See *monolithic integrated circuit*.

megatron   Same as *lighthouse tube*.

Meissner oscillator   An oscillator in which the *LC* circuit which determines the frequency of oscillation is coupled to the input and output circuits of the *active device* by separate coupling coils. There is therefore no physical connection between the frequency-determining circuit and the active device. The circuit diagram of a Meissner oscillator is given in *Figure M.4*.

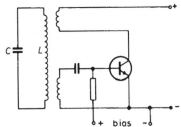

Figure M.4 Circuit diagram of a Meissner oscillator

melt-back transistor   A method of manufacturing *bipolar transistors* in which the junctions are formed by allowing molten *semiconductor* material containing donor and acceptor impurities to solidify. The distribution of the impurities changes on solidification and it is thus possible to form the required npn or pnp structure.

memory   A deprecated term for *store* but one which is likely to persist in such phrases as *random-access memory* and *read-only memory*.

memory cell   The basic unit of a *store*. In a binary system the cell has two states representing the two binary digits. In a semiconductor *dynamic RAM* the cell may consist of a single capacitor which represents logic 1 when charged and logic 0 when discharged but in a practical cell there are other active components which provide external connections to the capacitor for writing and reading purposes. Alternatively a *bistable circuit* consisting of two bipolar or two field-effect transistors can provide the basic storage unit as in a *static RAM*. Again additional components are required to provide loads for the transistors and to provide external connections so that the total number of transistors may be six per cell.

menu   In computers and microprocessor-controlled equipment, a list of options or commands that appears in the display. Using *keyboard* buttons or a *pointing device* such as a *mouse*, the user positions a pointer over or

197

adjacent to the required menu option to select it and then presses a button to initiate the selected action.

mercury arc rectifier (MAR) A rectifier in which the rectification occurs via an arc between an *anode* and a mercury pool acting as *cathode*. The discharge occurs in ionised mercury vapour and the means adopted to initiate the discharge depends on the type of rectifier. See *ignitron* and *excitron*. Very little voltage drop occurs across a mercury arc rectifier which is therefore very efficient.

mercury cell A primary voltaic cell in which the *cathode* is a mixture of mercuric oxide and graphite (usually deposited on a metal base), the anode is of zinc and the electrolyte is a solution of potassium hydroxide.

The cell has an EMF of 1.25 V which remains remarkably constant throughout its life. For a given volume the cell has much greater capacity than the *Leclanché cell*: expressed differently the mercury cell can be made very small whilst giving a useful life. This makes it useful in applications where small size is important, e.g. in cameras and watches. Such cells are also used to power portable electronic equipment.

mercury-pool cathode A cathode in the form of a pool of mercury in an arc discharge tube such as a mercury arc rectifier. Emission from such a *cathode* occurs as a result of bombardment of the mercury surface by positive ions resulting from ionisation of the mercury vapour within the tube and from thermionic emission from hot spots on the surface caused by this bombardment.

mesa transistor A diffused *bipolar transistor* in which unwanted regions of semiconductor are etched away leaving the base and emitter regions as plateaux above the collector region. *Figure M.5* gives a typical cross section of a mesa transistor.

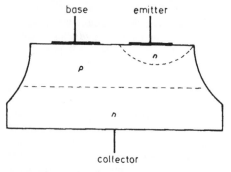

Figure M.5 Simplified diagram illustrating the structure of a mesa transistor

mesh In a *network* those elements which form a simple closed loop. For example in the network shown in *Figure M.6* there are two meshes as indicated by the arrows.

meta-data *Data* which describes data. See *repository*.

metal-film resistor A resistor made by screening a paste containing metals in a pattern on a ceramic base and firing it at a high temperature. See *thick-film circuit, thin-film circuit*.

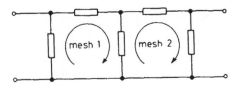

Figure M.6 An example of a two-mesh network

metal-oxide-semiconductor transistor (MOST or MOSFET) Same as *insulated-gate field-effect transistor.*

metal rectifier A rectifier in which *unilateral conductivity* occurs between the inner surface of a coating on a metal and the metal itself. Such a junction constitutes a single rectifier cell and in practical rectifiers the cells are stacked in series to provide the required voltage rating.

Examples of metal rectifiers are the copper-oxide rectifier in which the coating is of copper oxide on copper and the selenium rectifier in which the coating is a selenium alloy on iron. Both have now been superseded by silicon semiconductor rectifiers which are much smaller and more efficient.

metal-semiconductor field-effect transistor (MESFET) A microwave *field-effect transistor* with *gallium arsenide* as the semiconductor and with a reverse-biased *Schottky* junction as the gate. The high mobility of electrons in gallium arsenide and the use of the *unipolar* principle gives such transistors a better performance at microwave frequencies than silicon bipolar or field-effect transistors.

micro-alloy diffused transistor (MADT) A *micro-alloy bipolar transistor* in which the base wafer is initially given a graded impurity concentration by diffusion.

micro-alloy transistor A *bipolar transistor* manufactured by electrolytically etching a wafer of semiconductor material to produce the required thickness of base region and then electrodepositing and subsequently alloying a trivalent or pentavalent element on opposite faces to form the emitter and collector regions.

microcircuit Same as *monolithic integrated circuit.*

microcomputer A *computer* in which the *central processing unit* is a *microprocessor.*

micro-electronics A branch of electronics concerned with the design, manufacture and applications of devices of very small dimensions such as integrated circuits.

microphone An electro-acoustic *transducer* for converting sound waves into corresponding electrical signals. Microphones can respond to the pressure of the sound wave or the particle velocity. See *pressure operation, pressure-gradient operation.*

In general the electro-acoustic energy conversion does not occur in one step. Usually the sound waves strike a diaphragm causing it to vibrate (so creating mechanical energy), the vibrations being communicated to some form of generator which delivers the electrical output. A number of different types of generator are used in microphones. See *crystal microphone, electrostatic microphone, moving-coil microphone.*

microphonics (US) Same as *microphony*.

microphony Undesired modulation of the output current of an *electron tube* caused by vibration of the *electrodes* as a result of mechanical shock. The most familiar example is the 'pong' heard from a *loudspeaker* when a microphonic tube in the amplifier is struck. In an extreme form the tube may respond to sound waves from the loudspeaker so giving rise to acoustic feedback and possibly sustained oscillation.

microprocessor A small *digital computer* consisting essentially of a *central processing unit* with a limited amount of storage and some interface connectors. Microprocessors are now extensively used wherever mathematical and/or logic operations are required to be carried out in sequence and/or under the control of one or more input signals, e.g. in controlling automatic washing machines or machine tools.

It is now possible to include all the components necessary in a microprocessor in a single *monolithic integrated circuit*.

microwaves Electromagnetic waves with a wavelength smaller than approximately 30 cm, i.e. with a frequency above approximately 1000 MHz.

midicomputer See *digital computer*.

Miller effect Increase in the effective input capacitance of an *electron tube* caused by *feedback* from the anode circuit via the internal capacitance between *anode* and *grid*. Such feedback can make the input capacitance many times greater than the physical grid-cathode capacitance.

Miller integrator A circuit in which an external capacitor is connected between *grid* and *anode* of a *pentode* to increase *Miller effect*. The *feedback* introduced by this capacitor has the effect of generating a linear fall of anode potential when anode current is started, e.g. by a positive pulse applied to the suppressor grid. A rectangular pulse applied to the suppressor grid thus generates a sawtooth wave at the anode. Mathematically a *sawtooth* is the time integral of a rectangular wave and this accounts for the name of the circuit.

Miller time base See *Miller transitron*.

Miller transitron A *Miller integrator* circuit made astable by capacitively coupling the screen grid to the suppressor grid. With suitable potentials applied to the electrodes the circuit (*Figure M.7*) generates a linear

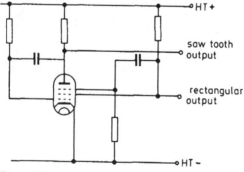

Figure M.7 Basic circuit for a Miller transitron

sawtooth at the *anode* and a rectangular pulse at the screen grid. Thus the circuit is suitable as an oscilloscope time base, the sawtooth providing horizontal deflection voltages and the screen grid giving flyback blanking pulses. The circuit can readily be synchronised at the frequency of a regular signal applied to the suppressor grid.

minicomputer See *digital computer.*

Mini-Disc See *magneto-optical disk.*

minimum phase-frequency characteristic Of the infinite number of phase-frequency characteristics which can be associated with a given attenuation-frequency characteristic, the particular characteristic for which the phase shift at each frequency is the minimum value possible.

minimum phase network A *network* for which the attenuation-frequency characteristic is associated with its *minimum phase-frequency characteristic.*

minority carriers In an *extrinsic semiconductor* the type of *charge carrier* which is outnumbered by the other type. For example in an n-type semiconductor the minority carriers are holes and in a p-type semiconductor they are *electrons.*

mismatch factor (or ratio) At the junction of a source of power and a load, and at a particular frequency, the ratio of the current delivered to the load to the current which would be delivered to the load if it were perfectly matched to the generator. The ratio is given by

$$k = \frac{2\sqrt{(Z_1 Z_2)}}{Z_1 + Z_2}$$

where Z_1 is the generator impedance and Z_2 is the load impedance. See *reflection loss.*

mixer (1) In radio or TV equipment which can combine two or more *signals* and permits their relative levels to be adjusted to form a composite signal giving the desired sound or visual effect. The term is also used to mean the operator of the mixing equipment.

(2) A device which accepts two inputs at different frequencies and generates an output at the combination frequencies. In particular the RF mixer in a *superheterodyne receiver* accepts an input from the antenna (or an RF amplifier) and from the local oscillator and generates an output at the sum or difference frequency which is equal to the intermediate frequency.

mnemonic In general anything that is deliberately made easy to remember. In computing, abbreviations for the various operations which a *central processing unit* can execute. They often consist of just three letters, such as ADC for 'Add with carry' and JMP for 'Jump'. These mnemonics and their *arguments* are used to build up machine-code programs in assembly language. See *computer language.*

modem (contracted form of **modulator/demodulator**) An item of electronic equipment used to enable *digital* signals to be transmitted over *analogue* voice-frequency circuits. A modem operating as a *modulator* is required at the transmitting end and a second, acting as demodulator, at the receiving end. See *modulation, demodulation.*

modulated amplifier In a transmitter a carrier-wave amplifier in which the modulation is introduced. In Heising modulation the amplifier is an *electron tube*, the *carrier wave* being applied to the grid and the modulating wave to the anode.

modulating signal The signal which causes a characteristic of the *carrier wave* to be varied so effecting modulation of the carrier.

modulation The process by which a characteristic of one signal is varied in accordance with another signal. In radio transmission the amplitude or frequency of the *carrier wave* is varied in accordance with the signal it is required to transmit (the *modulating signal*).

modulation depth Extent to which the amplitude of a *carrier wave* is modulated. If expressed quantitatively the modulation depth is equal to the *modulation factor*.

modulation envelope Of an amplitude-modulated wave the curve formed by joining up the peak values of the modulated wave. In *Figure M.8* the dashed line gives the shape of the envelope for a sinusoidal modulating signal.

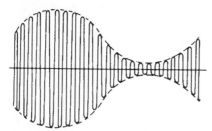

Figure M.8 An amplitude-modulated wave: the dashed line indicates the modulation envelope

modulation factor For an amplitude-modulated wave the ratio of the departure from the unmodulated carrier amplitude to the carrier amplitude. Thus in *Figure M.9* the modulation factor is a/A. When $a = A$, the carrier

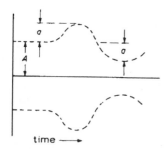

time ⟶

Figure M.9 The modulation envelope of an amplitude-modulated wave: the modulation factor is given by a/A

amplitude swings between twice its unmodulated value and zero. Attempts to increase the *modulation depth* beyond this value cause the carrier to be reduced to zero for appreciable periods. This leads to distortion of the *modulation envelope* and hence of the received signal. Thus the maximum practical value of the modulation factor is unity and smaller values are usually expressed as percentages.

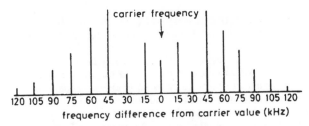

Figure M.10 Spectrum diagram for a frequency-modulated wave with a modulation index of 5

modulation index In *angle modulation* the ratio of the frequency deviation to the modulating frequency. The modulation index determines the side-frequency structure of the modulated wave. As an example for a modulation index of 5 the structure is as shown in *Figure M.10*. This applies, for example, when the frequency deviation is 75 kHz (the rated maximum in most sound-broadcasting frequency-modulated services) and the modulating frequency is 15 kHz (the upper limit in high-quality broadcasting).

modulation percentage The *modulation factor* expressed as a percentage. 100% modulation is equivalent to a modulation factor of unity.

modulator In general any device for achieving *modulation*. The term has two different specific meanings:

　(1) In a transmitter it is the amplifier which applies the modulating signal to the modulated amplifier. (See *Figure M.11*).

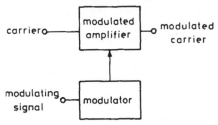

Figure M.11 Block diagram illustrating the terminology used in radio transmitters

　(2) In telephony it is a device which accepts two input signals and produces an output consisting of one signal modulated by the other. See *balanced modulator, ring modulator*.

modulo 2 Same as *OR gate*.

moiré In TV an effect caused by interference between two regular patterned structures. For example a moiré effect can be caused by beats between a pattern in the original scene (say a check suit) and the line structure of the reproduced picture.

monaural Listening with one ear only. Monaural should not be confused with *monophonic*. See *binaural*.

monitor Any device or routine used to observe, supervise, control or check the operation of equipment.

203

monitoring In general the checking of a signal whilst it is being transmitted or recorded to observe some characteristic of the signal, e.g. its information content, times of start or finish, its level (signal strength), technical quality. For quality checking, high-grade monitoring equipment (amplifiers, receivers, loudspeakers, picture-display tubes etc) is essential. Monitoring can be carried out by human observers or by equipment designed to give warnings or take action if the monitored characteristic necessitates it.

monochrome In TV a system in which the transmitted information is confined to the *luminance* of the scene. No colour information is transmitted. A monochrome system is therefore a black-and-white system. The term is unfortunate because monochrome means literally light of a single wavelength whereas white light contains a range of wavelengths. In this dictionary therefore the term 'black-and-white' is preferred.

monolithic Term applied to an *integrated circuit* in which all the elements are formed *in situ* within a single semiconductor chip.

monolithic integrated circuit A combination of interconnected active and passive devices fabricated on a continuous substrate, usually of silicon. As an example, an integrated circuit may contain several transistors and resistors interconnected to form an amplifier (linear IC) or a collection of logic gates (digital IC). Integrated circuits are manufactured by the *planar process* and there is little limit to the degree of complexity and miniaturisation possible by the process.

Although attempts were made to produce integrated circuits in the 1950s it was not until the planar process was developed in 1960 that monolithic structures became possible. These early ICs were digital circuits (RTL, TFL, ECL, etc) composed largely of bipolar transistors and resistors. The number of components per chip was less than 100—a density known as small-scale integration (SSI).

In 1966 MOS transistors were introduced. These had the advantage of requiring fewer diffusions in manufacture and occupied less area on the chip. On the other hand they were slower in operation than bipolar transistors. It was now possible to increase the component density to a few hundreds per chip—a level called medium-scale integration (MSI).

The demand from the computer industry was for still greater component density and by 1969 a single IC containing more than 10000 MOS transistors was available. This stage of development became known as large-scale integration (LSI) and by 1975 component densities were approaching 100000 per chip.

Since 1980 still further miniaturisation has been achieved and component densities of over a million components per chip are possible. This is termed very large scale integration (VLSI). The point has now been reached where integrated circuits require more input and output leads than can be accommodated on a reasonably-sized *dual-in-line package*. Alternative methods of packaging such as the *chip carrier* are being introduced.

There is no doubt that the small space occupied by the modern computer and its inherent reliability are attributable to the use of monolithic integrated circuits.

So far only digital ICs have been mentioned but there have also been significant developments in ICs for linear equipments. The first linear ICs were *operational amplifiers* and towards the end of 1960 ICs were developed for use in radio and television receivers. There is little limit to the complexity of ICs which can therefore be designed to perform many of the functions in receivers, high-fidelity equipment and video cassette recorders to improve their performance, reliability and ease of operation.

monophonic Single-channel sound signal. This is the system used in long-, medium- and short-wave broadcasting. The term is also used to describe the single-channel audio signal derived by a monophonic receiver from a stereo transmission. In the pilot-tone system used for stereo transmission the monophonic signal is the sum of the left and right channels.

monoscope An electron tube containing a target on which a pattern or photograph is printed and which, when scanned by an *electron beam*, generates a picture signal corresponding to the printed image. The tube is generally a high-velocity type and the pattern is printed in a pigment which modifies the *secondary-emission ratio* of the target. Such tubes are useful in TV services because they can replace a complete camera channel when a stationary pattern is to be transmitted.

monostable circuit A circuit with two possible states, one stable and the other unstable. The circuit is normally in its stable state but can be switched to the unstable state by an external signal. It cannot remain in the unstable state, however, and automatically reverts to the stable state, where it remains unless triggered by another external signal.

The circuit takes an appreciable time to return to the stable state and this period is equal to the duration to the unstable state. It can be given any desired value by suitable choice of component values. For this reason monostable circuits are used as delay generators, i.e. generators of pulses of a predetermined duration, the leading edge of which is coincident with that of the trigger pulse. See *multivibrator*.

mosaic Of a *camera tube*, an *electrode* consisting of a very large number of individually-insulated photo-emissive globules on which the optical image is focused. *Photo-electrons* released from the globules in accordance with the amount of light falling on them leave a *charge image* on the mosaic surface.

motherboard The main circuit board in a *computer* on which the *central processing unit* is mounted; most other boards containing *memory* and essential *interface* circuitry plug into edge connectors on the motherboard.

Motion Picture Experts Group (MPEG) A standard for the *storage* and transmission of digitised *full motion video* in a highly *compressed* format. The compressed video is transmitted over telephone lines or read from a storage medium such as compact disk and expanded to full screen size in real time. Individual frames are compressed using the JPEG process (see *Joint Photographic Experts Group*); in frame sequences only the differences between successive frames are stored.

motional impedance Of a *transducer*, the complex impedance obtained by subtracting its impedance when the moving parts are held still from its impedance when the parts are free to move.

motorboating In an audio amplifier self-oscillation at a very low frequency caused by *instability*. The sound of such an oscillation is similar to that of a motorboat engine.

mount In computing, to initialise a *drive* or the storage medium e.g. floppy disk, installed in it. This usually involves *cataloguing* the *root directory*.

mouse A small device moved over a desk top which causes a pointer to move across the screen of a visual display unit in the corresponding direction. The mouse may be connected to the computer by a flexible lead (tail) or via an infra-red link. The mouse greatly simplifies and speeds up computer operations and is much favoured in CAD.

moving-coil loudspeaker A loudspeaker in which the diaphragm is attached to a coil flexibly mounted within the magnetic field of a permanent or electromagnet and which moves in sympathy with the audio signal applied to it.

This is the most popular type of loudspeaker and its construction is illustrated in *Figure M.12*. The magnet system is arranged to give a radial field within a small annular gap in which the speech coil is situated.

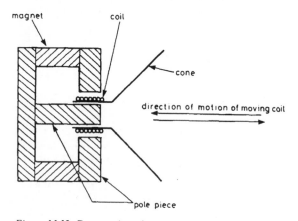

Figure M.12 Construction of a moving coil loudspeaker

moving-coil microphone A microphone in which the diaphragm is attached to a coil flexibly mounted within the magnetic field of a permanent magnet and which moves when sound waves strike it so inducing corresponding EMFs in the coil. The basic form of the microphone is thus similar to that of a *moving-coil loudspeaker* but the diaphragm is made very small so that the microphone has negligible effect on the sound field to which it responds.

moving-coil pickup A *pickup* in which the movements of the reproducing stylus are communicated to a coil flexibly mounted within the field of a small permanent magnet so giving rise to corresponding EMFs in the coil.

moving-iron loudspeaker An early form of loudspeaker in which the diaphragm is attached to a soft-iron armature flexibly mounted between the pole pieces of a permanent magnet and within the field of a stationary winding carrying the audio signal to be reproduced. The magnetic poles induced in the armature by the current in the coil react with those of the

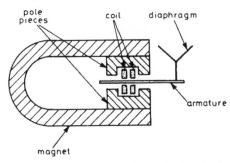

pole
pieces   coil   diaphragm

armature

magnet

Figure M.13 Construction of a moving-iron loudspeaker

pole pieces to cause vibration of the armature and corresponding movement of the diaphragm. See *Figure M.13*.

moving-iron pickup A *pickup* in which the movements of the reproducing stylus are communicated to a small armature flexibly mounted below the pole pieces of a small permanent magnet and within a stationary coil. The construction is similar in principle to that of the *moving-iron loudspeaker*. Movement of the armature causes induced magnetism which links with the coil to give corresponding induced EMFs.

moving-magnet pickup A *pickup* in which the movements of the reproducing stylus are communicated to a small permanent magnet situated within a stationary coil. The field of the magnet embraces the coil and movement of the field causes corresponding EMFs to be induced in the coil.

m signal In stereophonic sound broadcasting one half of the *sum signal*.

multi-cavity klystron A *klystron* in which the bunched *electron beam* excites a number of catcher resonators. By stagger-tuning the *catchers*, i.e. tuning them to different frequencies it is possible to give the klystron a wide bandwidth such as, for example, that required in a television transmitter. Four or five resonators are commonly used and such klystrons can deliver 50 kW output for 1 W input power over a bandwidth of 5 MHz between frequencies of 500 and 1000 MHz.

multi-cavity magnetron See *cavity magnetron*.

multimedia A programme in which information or entertainment is presented through several media such as video, still pictures, text and sound.

multimedia PC An IBM *PC-compatible* computer equipped for multimedia applications. In practice this means that it is fitted with a *soundblaster* or similar sound generation system and suitable speakers and a *CD-ROM* drive. It also requires certain minimum requirements for memory, hard disk capacity and microprocessor.

multipath distortion In FM radio reception distortion of the detector output caused by simultaneous pickup of the wanted signal via two or more paths.

The wavelength used for conventional VHF broadcasting is approximately 3 metres (10 feet). Any object with dimensions comparable to or larger than this, such as a hill, a building or even a car, can reflect such waves. Consequently a mobile FM receiver such as a car radio or a portable radio fitted with an omnidirectional antenna may pick up the required transmission from a number of paths—one directly from the transmitter and the others after one or more reflections. Moreover, the pattern of reflection changes as the receiver moves. Because of their longer path length, the reflected signals are delayed with respect to the direct signal and interference between the direct and delayed signals can cause serious harmonic distortion in the demodulated audio signal.

multiple electron tube A tube in which a number of electrode systems are contained within a single envelope. The systems may share a common *cathode* or may have separate cathodes. Examples of such tubes are double-diode-triodes, double-triodes and triode-hexodes.

multiplexing (MUX) The simultaneous transmission of two or more *signals* along a common path. The signals may share the path on a time-division basis or on a frequency-division basis. See *frequency-division multiplex, time-division multiplex.*

multiplier A device with two inputs, the output of which is substantially equal to the product of the input signals. See also *electron multiplier, frequency multiplier.*

multitasking Of a computer, capable apparently of performing several tasks simultaneously. Such an ability is apparent rather than real since most *central processing units* are capable of executing only one instruction at a time. In a multitasking operating environment either the processor divides its time between tasks on a rota basis or, more commonly, one task is given priority, the others being serviced as *background tasks* when the processor can be spared.

multivibrator A circuit comprising two *active devices*, the output of each being coupled to the input of the other. As a result of the large degree of *positive feedback* thus caused the circuit has only two possible states. If the active devices are electron tubes or transistors of the same type, then in one state device 1 is driven hard into conduction whilst device 2 is cut off and in the alternative state device 1 is cut off whilst device 2 is conductive. If complementary transistors are used, then both are cut off in one state of the circuit and both are conductive in the other. If the inter-device couplings are direct, as in *Figure M.14*, both states are stable and the resulting circuit is bistable. Triggering signals are thus needed to switch the states making the circuit useful as a binary counter or as an element in a store.

If one inter-device coupling is direct and the other capacitive, then one state is stable and the other unstable. The circuit is now *monostable* and will always revert to the stable state having been put into the unstable state

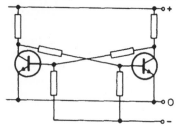

Figure M.14 A bistable multivibrator circuit

by a triggering signal. Such a circuit can be used as a delay generator.

Finally if both inter-device couplings are capacitive both states are unstable and the circuit is *astable*, generating pulses continuously without the need of triggering pulses. It can, however, readily be synchronised at the frequency of any recurrent signal applied to it.

mu-metal A nickel alloy of high *permeability* and low *hysteresis* loss used for the construction of *magnetic screens*.

Musical Instrument Digital Interface (MIDI) A standard that allows electronic musical instruments (such as *synthesisers*) to be connected to each other or to a computer. Via MIDI one instrument may control another or a computer may control one or more instruments or may record what is being played for subsequent editing or playback.

muting The suppression of the audio channel of a device. FM receivers are sometimes equipped with a muting facility which suppresses the audio output unless received signals exceed a certain strength. Thus signals emerge from a silent background when the tuning is adjusted.

mutual conductance (g_m) Of an *active device* the ratio of a small change in the current in one electrode to the small change in voltage at another electrode which gives rise to it. Its value is usually quoted for a specific operating point. *Figure M.15* gives an example of the drain current-gate voltage characteristic for a *field-effect transistor*. The mutual conductance at the operating point P is given by the slope of the tangent to the curve at that point, i.e. by SQ/QR. It is usually quoted in milliamps per volt and is taken as a figure of merit for the transistor.

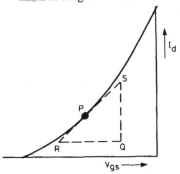

Figure M.15 The mutual conductance of an FET is given by the slope of the tangent to the I_d–V_{gs} characteristic curve at the operating point

mutual impedance Between any two pairs of terminals in a *network* the ratio of the open-circuit voltage between either pair to the current at the other pair, all other terminals being open-circuited.

mutual inductance A property of two circuits such that a change of current in either of them causes an EMF to be induced in the other. The induced EMF is given by $M di/dt$, where M is the mutual inductance and i is the current in one of the circuits. The most common application of mutual inductance is in the transformer in which two windings are wound on a common magnetic core to maximise mutual inductance.

N

NAND gate (NOT-AND gate) A *logic gate* which gives a logic-0 output when, and only when, all the input signals are at logic 1. It may be regarded as an *AND gate* followed by a NOT or logic negating stage as suggested by the graphical symbol shown in *Figure N.1*. See *logic level*.

Figure N.1 Graphical symbol for a NAND gate

natural air cooling Same as *convection cooling*.

natural frequency Of an oscillatory system, the frequency of free oscillations. Thus it is the frequency at which the system oscillates when shock-excited. A system can have more than one natural frequency and the natural frequency of an antenna is the lowest frequency at which it resonates without the addition of inductance or capacitance.

natural period The reciprocal of the *natural frequency*.

n-channel metal-oxide-semiconductor (NMOS) A form of construction for *logic monolithic integrated circuits* using *n-channel field-effect transistors*. These have the advantage over PMOS of higher speed of operation (the charge carriers, being *electrons*, have higher mobility than *holes*) and have largely replaced PMOS.

Near Instantaneous Compression and Audio Multiplex (NICAM) A digital coding system for the inclusion of sound or data signals within a television broadcast transmission. The signal, typically two-channel stereo, is *quantised* to 14 bits and *compressed* to 10 bits. Left- and right-channel data is sent alternately in blocks of 32 11-bit (10 data bits plus one parity bit) words, each lasting 1 ms and containing, with control and framing codes, 728 bits. The NICAM carrier is *quadrature modulated* by the interleaved and scrambled binary data and transmitted at low level adjacent to the normal f.m. sound carrier.

negate In logic to perform the NOT operation. Thus an AND gate followed by a negating operation becomes a NOT-AND or *NAND gate*. The symbol for negation is shown in *Figure N.2*.

—O— *Figure N.2* Graphical symbol for negation

negative AND gate Same as *NAND gate*.

negative feedback (NFB) *Feedback* in which the signal returned to the earlier stage in the amplifier opposes the input signal at that point. Negative feedback has the effect of reducing the gain of that portion of the amplifier included within the feedback loop. It also improves the linearity and makes the frequency response more level.

Negative feedback has an effect on the output impedance of that stage of the amplifier from which the signal is derived and on the input impedance of the stage at which it is re-introduced. The effect on output impedance depends on whether the feedback signal is proportional to the output voltage (see *voltage feedback*) or to the output current (see *current feedback*). The effect on input impedance depends on the way in which feedback signal is re-introduced into the amplifier: if it is connected in series with the input signal it increases input impedance; if it is connected in parallel with the input signal it reduces the input impedance.

Negative feedback is extensively used in the design of electronic equipment. It can be used to give an amplifier a required value of gain, a required value of linearity, a required shape of frequency response (by making the feedback path frequency-discriminating) and required values of input and output impedance. These properties are obtained by suitable choice of *passive components* in the feedback circuit and they are therefore independent of the properties of the active devices contained within the feedback loop—a very useful feature in the mass production of electronic equipment.

negative image An image in which white areas are reproduced as black and black areas as white as in a negative photographic film. The effect can occur in TV images, when, for example, the picture tube has low emission.

negative logic A *computer* or *data processing* system in which the more negative of the two voltages selected as the logic levels is designated the 1 state.

negative modulation In TV transmission *amplitude modulation* in which increased picture brightness results in decreased carrier amplitude. This is the system of modulation used in the British 625-line and in most of the world's television systems.

negative OR gate Same as *NOR gate*.

negative resistance Property of a device in which an increase in applied voltage brings about a decrease in the current through it. The current-voltage characteristic for such a device has a negative slope as shown in *Figure N.3*. Electron tubes can be operated so as to produce a region of negative resistance (see *dynatron*). A semiconductor example with a similar region is the *tunnel diode*. A most useful property of a negative-resistance device is that a simple oscillator can be made by connecting a parallel resonant circuit across it.

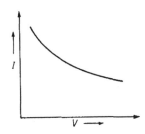

Figure N.3 A current-voltage characteristic with a negative slope

negator Same as *NOT gate*.
NEITHER-NOR gate Same as *NOR gate*.

Figure N.4 Graphical symbol for a neon tube

neon tube A gas discharge tube containing two *electrodes* and filled with neon at low pressure. When a voltage exceeding the *ionisation voltage* is applied between the electrodes a discharge occurs giving a characteristic red glow. Such tubes are used in neon signs and as indicators in electronic equipment. When the discharge occurs the voltage across the electrodes remains constant and the tube can therefore be used as a voltage reference. The graphical symbol for a neon tube is given in *Figure N.4*.

neper The basic unit of a logarithmic scale used to express ratios of voltages, currents and analogous quantities. The neper is based on the use of natural logarithms and is named after their inventor, John Napier. Two voltages V_1 and V_2 are related by n nepers when $\log_e(V_1/V_2) = n$. The number of nepers is simply the natural logarithm of the voltage ratio.

In conditions of equal impedance (as is usual for input and output quantities in telecommunications) a voltage or current ratio is equal to the square root of the corresponding power ratio. Hence

$$n = \log_e \left(\frac{P_1}{P_2} \right)^{1/2} = \tfrac{1}{2} \log_e \frac{P_1}{P_2}$$

By extension, this formula may be applied when the power ratio is not the square of the ratio of the corresponding voltages or currents. It is then important to specify the convention adopted and to indicate the sphere of validity of the usage. The neper may be used to express levels of quantities such as voltage and current in the same way as the *decibel*.

network (1) A system of inter-connected components chosen and arranged to give specific electrical characteristics. A network containing no source of power is termed passive and one containing one or more sources is known as an *active network*. If the values of the components forming the network are dependent on the current in them the network is termed non-linear but if the values are independent of current the network is linear.

(2) In *computing* two or more computers connected together to facilitate the exchange of data. Computer networks are generally classified as *local area networks* if the computers concerned are located on the same premises or as *wide area networks* if they are on separate premises and the connection requires the use of telecommunications lines. See *local area network*, *wide area network*.

network transfer constant Same as *image transfer constant*.

neutralisation Use of a balancing network to counteract the *feedback* between the input and output terminals of an *active device* which occurs via the internal capacitance. The internal feedback causes instability when input and output circuits are tuned to the same frequency and neutralisation is used to enable, for example, transistors in radio receivers to give stable IF amplification. A typical circuit is given in *Figure N.5*: the neutralising capacitor C_n is adjusted to cancel the effect of the collector-base capacitance.

neutrodyne A method of neutralisation developed by Hazeltine in the USA.

212

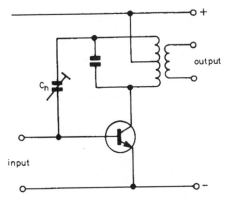

Figure N.5 Simplified diagram illustrating one method of neutralising a transistor IF amplifier

neutron An elementary atomic particle with no charge and similar mass to the *proton*. It is a constituent of all atomic nuclei except that of hydrogen.

n-gate thyristor A thyristor which is switched to the on-state by a negative signal applied between *gate* and *anode*, the gate being a connection to the n-region nearest the anode. It is normally a reverse-blocking thyristor, i.e. one which cannot be switched to the on-state when the anode is negative with respect to the cathode. The graphical symbol for the n-gate thyristor is given in *Figure N.6*.

Figure N.6 Graphical symbol for an n-gate thyristor

nickel-cadmium cell A secondary cell in which the anode consists of a mixture of nickel oxide and hydroxide, the *cathode* is of cadmium and the electrolyte is of potassium hydroxide. Its properties are similar to those of the *nickel-iron cell*. It is widely used in portable electronic equipments.

nickel hydride cell A type of dry battery used in some portable equipment, which offers approximately twice the capacity of *nickel cadmium* cells.

nickel-iron cell A secondary cell in which the anode consists of a mixture of nickel oxide and hydroxide, the *cathode* is of iron and the electrolyte is of potassium hydroxide. This is an alternative to the *lead-acid cell* and has the advantage that the nickel-iron cell will operate satisfactorily at temperatures as low as −30°C. On the other hand it is more expensive than the lead-acid cell and its EMF is only 1.2 V.

Nipkow disk Mechanical scanning device used in early experiments in TV. It consists of a disk containing a number of apertures arranged on a spiral path. The scene to be televised is illuminated by light which passes through these apertures so that, as the disk revolves, the scene is scanned. Light reflected from the scene is picked up by *photocells* the output of which is the required picture signal. The system was used in the BBC 30-line television experiments in the late 1920s.

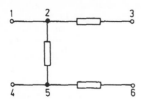

Figure N.7 A C-network has six nodes, numbers 1, 3, 4 and 6 being external, 2 and 5 internal

node (1) In a network a point where three or more elements are connected (internal node) or where signals are applied to or withdrawn from an element (external node). For example the network shown in *Figure N.7* has six nodes, two internal and four external. External nodes are alternatively known as *terminals*.

(2) Of a *standing wave*, a point or plane at which a specified variable has a minimum value. Thus nodes on a *transmission line* are points at which voltage or current is a minimum. For a bowed string nodes are points at which the amplitude of vibration is a minimum. See *antinode*.

noise In general any unwanted signals within the useful frequency band of a communications system which tend to obscure the wanted signal. This is a wide definition and embraces a variety of man-made noises such as mains hum, signals from other channels arising from inadequate selectivity or poor image rejection, cross modulation, etc; it also includes naturally-generated signals arising from lightning flashes. Signals from such causes are better termed *interference* so that the term noise is reserved for signals which arise in components and *active devices* and are known as random noise. Such signals are termed noise because they are heard as undesired sound when reproduced by an acoustic transducer but the term is retained for signals which cause spurious detail on the screen of a *cathode ray tube*.

Random noise from components is known as *thermal noise* or Johnson noise and from active devices is known as *shot noise* and *partition noise*. Even if all sources of man-made noise can be eliminated there still remains random noise and the level of this noise determines the amplitude of the weakest signal which can be successfully transmitted through a communications system. Wanted signals must exceed the noise level by a margin which ensures that the noise does not impair the intelligibility of the signal unduly or the enjoyment of the programme. Thus the noise level and the *signal-to-noise ratio* are two important properties of a communications system.

noise factor (or **noise figure**) Of a device, a measure of the noise it introduces. More specifically it is the ratio of the *signal-to-noise ratio* at the output of the device to the signal-to-noise ratio at the input and is usually expressed in dB. An ideal, i.e. noise-free, device thus has a noise factor of unity (0 dB).

noise generator A device for generating a standard noise signal which can be used for the measurement of *noise factor*. In such generators the *shot noise* in a saturated diode is often employed as the primary source of noise.

noise limiter In a radio or TV receiver a circuit used to reduce the effect of impulsive noise signals on the sound or picture output. Typically the circuit

214

includes a series diode which is situated at the output of the detector and is so biased that it remains conductive for all normal amplitudes of sound or video signals but becomes non-conductive, so interrupting the wanted signal, for the duration of any noise pulse exceeding a pre-determined amplitude. The resulting interruptions are normally so brief that their subjective effect is negligible.

noise temperature At a given pair of terminals and at a given frequency the temperature of a resistance which gives the same noise power per unit bandwidth as the noise to be measured. The standard reference temperature for noise measurement is 290°K.

non-composite colour-picture signal (US) In TV the signal which comprises full colour-picture information including the colour burst but excluding the synchronising signals.

non-composite signal (US) Same as *picture signal*.

non-conjunction gate Same as *NAND gate*.

non-equivalence gate Same as *exclusive-OR gate*.

non-linear circuit element A circuit element for which the significant property is dependent on the current flowing through it or the voltage across it. An example is a *non-linear resistance*.

non-linearity distortion General term embracing all the forms of distortion which result from non-linearity of the input-output characteristics of a system, equipment or component, e.g. *amplitude distortion*, *harmonic distortion* and *intermodulation distortion*. The significant feature of non-linearity distortion is that it is dependent on signal amplitude.

non-linear network See *network*.

non-linear resistance A resistance for which the current is not directly proportional to the applied voltage. Such a resistance may have a current-voltage characteristic of the form shown in *Figure N.8*. For comparison the characteristic for a linear resistor (i.e. one which obeys Ohm's law) is shown as a dashed line.

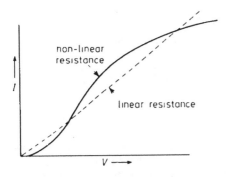

Figure N.8 Current-voltage characteristic for linear and non-linear resistances

nonode An electron tube with seven grids between *cathode* and *anode*. The first grid (nearest the cathode) is a control grid, grids 2, 4 and 6 are screen

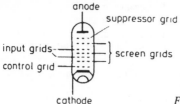

anode

suppressor grid

input grids

control grid

screen grids

cathode

Figure N.9 Graphical symbol for a nonode

grids 3, 5 and 7 are suppressor grids. The arrangement is shown in *Figure N.9*. The tube was used in FM receivers as a self-limiting phase-difference *discriminator* and grids 3 and 5 are fed with FM signals from the primary and secondary windings of an IF transformer. The control grid is given a steady bias and suppressor grid 7 is connected to cathode to suppress *tetrode kink*. The three screen grids are commoned and connected to a low voltage such as 30 V. The limiting action stems from the use of suppressor grids as input electrodes. The signals swing the anode current between zero and the value set by the control-grid bias. Increase in the amplitude of the input signals can produce no corresponding increase in anode-current swing.

non-volatile A term used to describe any form of memory which retains the data it is storing when the equipment is switched off or loses its power supply for some other reason. Under these circumstances the contents of normal RAM is lost. Examples of non-volatile memory include battery-backed RAM and various forms of *EPROM*.

NOR gate (NOT-OR gate) A *logic gate* which gives a logic-0 output when one or more of the input signals is at logic 1. It may be regarded as an OR gate followed by a negating stage as shown by the graphical symbol in *Figure N.10*. See *logic level*.

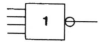

Figure N.10 Graphical symbol for a NOR gate

Norton's theorem If an admittance Y' is connected between any two points of a linear network the voltage across it is given by $I/(Y + Y')$, where I is the short-circuit current and Y is the admittance measured between the points before Y' was connected.

NOT-AND gate Same as *NAND gate*.

NOT-BOTH gate Same as *NAND gate*.

notch filter A bandstop filter for which the stop band is extremely narrow. Such filters are used for reducing interference which is confined to a single frequency (e.g. heterodyne interference) or to a narrow band of frequencies. The frequency response for a notch filter is illustrated in *Figure N.11*.

NOT gate A *logic gate* with a single input and a single output, which gives a logic-0 output when the input is at logic 1, and a logic-1 output when the input is at logic 0. It is often termed a logic negator or *inverter*.

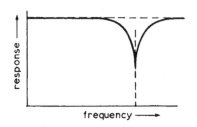

Figure N.11 Frequency response of a notch filter

npin transistor A type of *bipolar transistor* in which the base and collector regions are separated by a region of intrinsic material.

npn transistor See *bipolar transistor*.

NTSC colour television system A compatible colour TV system pioneered by the National Television Systems Committee in USA in which the *luminance signal* is transmitted by *amplitude modulation* of the *vision carrier* and two *chrominance signals* are transmitted simultaneously by *quadrature modulation* of a subcarrier, the total *bandwidth* being the same as that of a black-and-white system using the same line standards. A *colour burst* is included in the *back porch* to facilitate decoding of the colour subcarrier.

This is the system used in USA and a number of other countries. The PAL system used in UK and a number of other countries is a development of the NTSC system. See *PAL* and *SECAM* systems.

n-type semiconductor An *extrinsic semiconductor* in which the free-electron density exceeds that of the hole density. Thus the negative charge carriers outnumber the positive charge carriers and current flowing through such materials is due to movement of negative charge carriers.

nuvistor Miniature *electron tube* of rigid construction with small *electrodes* and electrode clearances designed for high reliability.

Nyquist diagram Of a feedback amplifier a graph in rectangular co-ordinates showing the real and imaginary components of the ratio μ/β, where μ is the gain of the amplifier and β is the fraction of the output voltage fed back. Such a diagram, if plotted for a very wide frequency range, can be used to predict whether the amplifier will be stable.

O

object In computing, any self-contained set of data. For example, in a *filing system* the objects are *files* and *directories*, in a *database* the objects are the various items of data that are stored and in *DTP* or *vector graphics* the objects are the various textual items (commonly called *stories*) and individual graphics of which the *document* is composed.

object code The form taken by a *computer program* after it has been translated from the source code by compilation or assembly. See *computer language*.

object linking and embedding (OLE) In *computing*, the practice of storing pointers to *files* instead of the file contents themselves. For example, in a *Desktop Publishing document* which includes a half-tone illustration as a *bitmap* the conventional approach would be to copy the entire bitmap into the document file. This would occupy appreciable memory and in some circumstances its presence might affect the computer's performance. Using OLE, however, only the name and address of the bitmap file is stored in the document. The bitmap itself is loaded into memory only when it is displayed on the *VDU* or printed. OLE makes it possible for a computer to handle longer and more complex documents. It does, however, increase the time taken to display or print the document and it introduces complications if the document is transmitted elsewhere, e.g. for phototypesetting.

object-oriented programming A philosophy of computer programming that divides data into separate units called *objects* which may be of various types, each handled independently.

Occam See *computer language*.

occlusion Property of certain solids, particularly some metals, which enables them to absorb gases without chemically combining with them. The final stage in the evacuation of *electron tubes* is the removal of the last traces of gases occluded within the electrodes. See *degassing*.

octal scale A counting scale containing eight digits, each digit in a number representing a power of eight. Thus the decimal equivalent of the octal number 352 can be evaluated as follows:

$$352 = 3 \times 8^2 + 5 \times 8^1 + 2 \times 8^0$$
$$= 3 \times 64 + 5 \times 8 + 2 \times 1$$
$$= 192 + 40 + 2$$
$$= 234$$

Octal numbers are often used as a shorthand method of writing binary numbers. Because 8 is 2^3 it is easy to convert between the two counting scales. Each three digits of a binary number can be uniquely represented by one of the octal digits, e.g. the binary number 101010110 can be represented by the octal number 526, each having a decimal equivalent of 342.

octave The interval between two frequencies with a ratio of 2:1. Two sound waves with such frequencies blend harmoniously because one is precisely one octave higher than the other.

octode An electron tube with six grids between *cathode* and *anode*. It is normally used as a frequency changer and may be regarded as a *heptode* to which a suppressor grid has been added to eliminated the *tetrode kink*. As in the heptode, grids 3 and 5 may be connected together either inside or outside the envelope. The graphical symbol for an octode is given in *Figure O.1* which also indicates the functions of each grid.

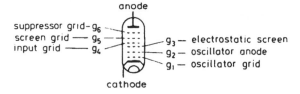

Figure O.1 Graphical symbol for an octode indicating the function of each grid

off-line In computers and data-processing equipment, a term used to describe the peripheral equipment which is not currently under the direct control of the *central processor*.

off-line storage A storage device or medium not under the direct control of a *central processor*.

offset current For a *differential amplifier* the difference between the two bias currents. It is necessary to minimise offset current and this can be achieved by arranging for the source impedances of the two input signals to be equal.

offset voltage For a *differential amplifier* the output voltage arising from inherent DC unbalance of the amplifier. Usually offset voltage can be reduced to zero by adjustment of an external potentiometer but it is often temperature-dependent necessitating a compromise adjustment giving best results over the temperature range to be expected.

ohmic contact A purely-resistive contact between two surfaces such that the voltage drop across the contact is directly proportional to the current through it.

Ohm The unit of electrical resistance, symbol Ω although some standards use R. Practical values of resistance range from fractions of an ohm, through kilohms ($K\Omega$ or simply K) to several megohms ($M\Omega$ or simply M).

Ohm's law The current (I) in a conductor is directly proportional to the EMF (E) applied to it and is inversely proportional to its resistance (R) provided the temperature remains constant. This may be expressed mathematically as

$$ I = \frac{E}{R} \qquad R = \frac{E}{I} \qquad E = IR $$

which enable the third quantity to be calculated when the other two are known. This law is probably the most fundamental in electronics.

one-shot circuit (US) Same as *monostable circuit*.

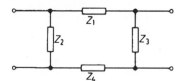

Figure O.2 An O-network

O network A *network* consisting of four elements connected to form a square, the input being applied across one element, and the output being taken from across the opposite element. The network is illustrated in *Figure O.2* and can be regarded as a balanced form of the *pi-network*.

on line In *computers* and *data-processing equipment*, a term used to describe peripheral equipment which is directly under the control of the *central processor*. It is also used to describe operation in which a user can interact directly with a computer, i.e. in which the results of a particular operation can be made immediately available to the user.

on-line help In a computer *application* a series of screens containing text (and possibly also graphics) designed to help the user to understand the application. It is usually invoked by selecting an option from a Help menu or by pressing a *function key*.

open circuit A circuit which is broken and in which current cannot therefore flow.

open-circuit impedance Of a two-terminal-pair *network* the impedance measured at one pair of terminals when the other pair is open-circuited.

open-collector (drain) output An integrated-circuit logic element in which the collector (drain) circuit of a transistor is completed externally to the integrated circuit. By parallelling such outputs for a number of elements it is possible to achieve a logic operation at the common output point without need for a specific logic element at that point. See *distributed connection*.

open system A system that conforms to accepted standards so that it can be extended readily using compatible components.

operating point In general a point on the current-voltage characteristic of an *active device* which represents the direct voltage and current in the output circuit. The term is used with a number of distinct meanings:

(1) To represent the output current and voltage in the absence of an input signal: this is perhaps better termed the *quiescent point*.

(2) To represent the average current and voltage when an input is applied to the active device. For a symmetrical input signal and for a device with ideal characteristics this point coincides with the quiescent point but for an asymmetrical signal and for practical characteristics the two points do not coincide.

(3) To represent the instantaneous current and voltage in the output circuit. Thus for a purely-resistive load and a sinusoidal input signal the operating point moves up and down the load line making one complete oscillation about the mean position for each cycle of input signal as shown in *Figure O.3*.

operating system The resident *software* routines used by a computer to control access to the *hardware* features such as the keyboard, VDU and disk drives. In most computers these routines are stored on disk and loaded

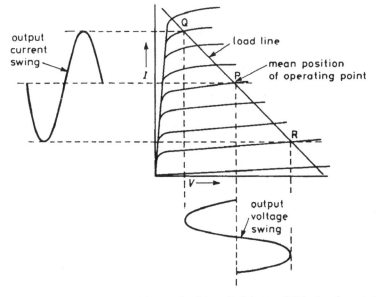

Figure O.3 When a sinusoidal input signal is applied to an active device, the operating point P oscillates about its mean position between the limits of Q and R so generating the output-current and output-voltage waveforms indicated

into RAM as required; in some computers the operating system is permanently resident in ROM.

operational amplifier A high-gain direct-coupled integrated-circuit amplifier with external circuitry designed to perform a specific arithmetical or mathematical operation. The external circuitry usually provides overall *feedback* suitable for the intended function. For example in *Figure O.4* in which all resistors are equal the output of the amplifier is equal to the sum of the two input signals. The operational amplifier is used here as a summing amplifier.

Operational amplifiers are extensively used in *analogue computers*.

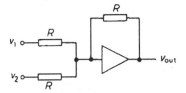

Figure O.4 An operational amplifier used as a summing amplifier

optical amplifier A device which amplifies an optical signal without converting it to an electronic signal, e.g. a *semiconductor laser*.

optical character recognition (OCR) A computer *application* which recognises the shapes of alphanumeric characters and punctuation marks in a *bitmap*, for example that obtained by *scanning* a printed document, and converts these to a *text* file.

optical disk Any disk-based data storage system using optical technology. See *compact disk, magneto-optical disk, phase-change disk, video disk*.

optical fibre See *fibre optics*.

opto coupler or **isolator** A photo-emissive source such as a *light-emitting diode* encapsulated with a photo-sensitive device such as a *photo-transistor* designed to respond well to the radiation (often infra-red) from the source. Such a combination gives an electrical output dependent on the electrical input and can thus be used to couple two circuits. There is, however, no electrical connection between the input and the output of the opto coupler making this a useful device where electrical isolation is required between the coupled circuits. The graphical symbol for an opto coupler is given in *Figure O.5*.

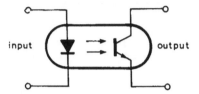

input   output

Figure O.5 Graphical symbol for opto coupler or opto isolator

optoelectronics The study and application of electronic devices that emit light (such as *lasers* and *light-emitting diodes*), amplify light (such as *optical amplifiers*), detect light (such as *photodiodes* and *phototransistors*) and control light (such as *Kerr cells*).

Oracle The *teletext* service of the Independent Broadcasting Authority.

OR-gate A *logic gate* which gives a logic-1 output when one or more of the input signals are at logic 1. The graphical symbol for an OR-gate is given in *Figure O.6*.

Figure O.6 Graphical symbol for an OR gate

orthicon An early form of TV camera tube in which the optical image of the scene to be televised is focused on one face of a photo-emissive target, the reverse face of which is orthogonally scanned by a low-velocity *electron beam*.

The use of a low-velocity scanning beam avoids secondary emission and thus the generation of shading signals as in the *iconoscope*: it also ensures that the target remains stabilised near the electron-gun cathode potential (hence its British name cathode-potential stabilised emitron). The essential features of the tube are illustrated in *Figure O.7*. The target consists of a layer of mica carrying a transparent signal plate on one face and a mosaic of individually-insulated photo-emissive elements on the other.

When an optical image is focused on the target the elements lose electrons in proportion to the amount of light falling on them and so build up a charge image over the field period. This image is neutralised by the scanning beam which restores to the elements the electrons lost by *photo-emission*. Control of the low-velocity beam is obtained by use of an axial

222

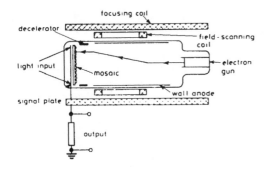

Figure O.7 Essential features of the orthicon tube

magnetic field from the solenoid entirely surrounding the tube. The output of the tube is taken from the signal plate but it could also be taken from the return scanning beam as in the *image orthicon*. See *orthogonal scanning.*

orthogonal scanning Scanning in which the *electron beam* always approaches the target at normal incidence. This is necessary in low-velocity camera tubes such as *orthicons* to achieve cathode-potential stabilisation of the whole area of the scanned face of the target. If conventional electrostatic or magnetic methods of deflection are used with a low-velocity beam, the beam approaches the centre of the target at normal incidence but all other points at an angle to the normal. As a result the whole area of the target is not cathode-potential stabilised and shading signals are present in the tube output. This can be avoided and the whole area of the target stabilised by arranging that the beam approaches all points of the target at normal incidence. This can be achieved by use of an axial magnetic field in conjunction with magnetic deflection of the scanning beam. The axial field can also be used as a long magnetic lens to focus the beam on the target.

oscillation In electronics the generation of alternating signals in a circuit containing *inductance* or *resistance* and *capacitance*. As an example if a closed *LC* circuit is shock excited, oscillations are set up at the resonance frequency of the circuit. The energy which keeps the alternating current going is stored at one moment in the charged capacitor and at another in the magnetic field of the inductor. As this energy is dissipated in the inevitable resistance of the circuit the oscillation amplitude falls exponentially with time. Such oscillations are termed free and their frequency is determined solely by the constants of the circuit.

If the lost energy can be restored by connecting an *active device* to the LC circuit, then sustained oscillation at constant amplitude can be produced: this is the basis of all *LC* oscillators. If an alternating EMF is applied to an LC circuit, then oscillation occurs in the circuit but at the frequency of the applied EMF: this is an example of forced oscillation.

Oscillation can also occur in a circuit containing resistance, capacitance and some active devices and here the frequency of oscillation is governed by the *time constants* of the *RC* combinations: an astable multivibrator is an example of such an oscillator.

223

Oscillation can also occur in a mechanical system and is manifested by the rhythmic motion of part of it. The quantities analogous to inductance and capacitance in a mechanical system are the mass of the moving parts and the elasticity which returns them to the equilibrium position. The energy which keeps oscillation going alternates between the kinetic energy of the moving parts and the potential energy stored in the elastic element.

oscillatory circuit A circuit which consists primarily of *inductance* and *capacitance* and is capable of producing a current which reverses at least once in direction when the circuit is shock excited, e.g. by the momentary application of a voltage. Practical oscillatory circuits inevitably contain resistance and this, if large enough, can damp the circuit to such an extent that it is rendered non-oscillatory. See *critical damping*.

oscillograph Same as *cathode ray oscillograph*.

oscilloscope Same as *cathode ray oscilloscope*.

outline font A typeface stored in a *computer* as a series of line drawings (*vector graphics*). Outline fonts are used in *Desktop Publishing* and graphic design applications where they have the advantage over *bitmapped* fonts that one set of drawings can be scaled to any required size by multiplying the co-ordinate data by the appropriate constant. Text in outline fonts can also be converted readily to vector graphics for more complex manipulations. Such text is usually printed as graphics; a printer intended solely for use with a computer system using outline fonts strictly needs no resident fonts at all and such cut-down *laser printers* are available very cheaply.

outphasing system Same as *Chireix system*.

output (1) Of a device or equipment, the signal power it produces. (2) The terminals from which the output power is delivered.

output impedance Of a device or equipment, the *impedance* presented by the output terminals to the load.

overcoupled circuits Two circuits resonant at the same frequency and between which the degree of coupling is intentionally greater than the critical value. The effect of overcoupling is to give a double-humped frequency response symmetrically disposed about the resonance frequency.

overloading (1) Condition of a source of electrical power when more power is drawn from it than the source can supply continuously without overheating or damage. For example if *voltaic cells* are short-circuited the resulting overload current can do irreparable damage to them. (2) Of an equipment, *active device* or component the application of an input signal with an excessive amplitude. In analogue equipment such an input exceeds the extent of the linear part of the input-output characteristic and leads to non-linearity distortion.

overshoot Form of transient distortion of a step or pulse signal in which the response temporarily exceeds the final value. This is illustrated in *Figure*

Figure O.8 Overshoot on the leading edge of a pulse

224

0.8 and the magnitude of the distortion is usually given as the ratio of the overshoot to the step amplitude expressed as a percentage. Overshoot can also occur, of course, on the trailing edge of a pulse.

overtone See *harmonic component*.

P

packet In *data* communications a set of data in digital format accompanied by control information (such as destination and origin) and *error* correction bits.

pad Same as *attenuator*.

padder or **padding capacitor** A capacitor connected in series with the oscillator tuning capacitor in a *superheterodyne receiver* to maintain the correct frequency difference between the oscillator and signal-frequency tuned circuits. If the oscillator inductor is fixed, the padding capacitor is a preset component which is adjusted for optimum tracking. Normally, however, the padding capacitor is a fixed component and the oscillator inductance is adjusted for optimum tracking. See *tracking, trimming*.

page printer A *printer* which prints a whole page at a time, such as a laser printer.

pager A compact portable radio receiver carried by personnel and used to contact them within a limited area surrounding the *base station*. A pager may reproduce speech communication, may give a visual display of a message or may simply emit an audible tone. Pagers are used mostly in large buildings such as hospitals and airports.

painting software Software for creating or editing *bitmaps*.

PAL colour television system (Phase Alternation, Line) A compatible colour television system similar to the *NTSC system* but in which one of the two *chrominance signals* is reversed in phase for the duration of alternate scanning lines. The phase reversal enables more consistent colour rendering to be achieved in receivers by averaging the chrominance signals for two successive lines before displaying them on the picture tube. This technique minimises the effect of any hue errors due to phase shift of the chrominance signals. This is the colour television system used in UK and many other countries.

palette A list of the colours available in a *VDU* screen or in a *bitmap* with their composition in terms of primaries.

panoramic potentiometer (pan pot) In stereo sound equipment a *potentiometer* which can be used to divide a signal in any desired ratio between the

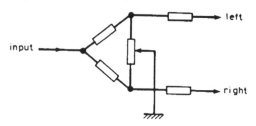

Figure P.1 One possible circuit arrangement for a panoramic potentiometer

left and right channels. Thus the signal can be panned or steered to any desired area between the extreme left and extreme right. *Figure P.1* gives a circuit diagram of one possible way in which a panoramic potentiometer can be connected.

panotrope (US) Disc-reproducing equipment in which the sound is reproduced by loudspeakers driven by amplifiers fed from an electrical pickup.

parafeed Same as *parallel feed*.

parallel connection A method of connecting components or circuits so that they share the same voltage, the current dividing between the circuits depending on their impedance. See *Figure P.2*. The equivalent impedance Z_{eq} of a number of impedances Z_1, Z_2, Z_3, etc is given by

$$\frac{1}{Z_{eq}} = \frac{1}{Z_1} + \frac{1}{Z_2} + \frac{1}{Z_3} + \text{etc.}$$

but problems on impedances in parallel are probably best solved by use of the admittance concept, the equivalent admittance being the sum of the individual admittances thus:

$Y_{eq} = Y_1 + Y_2 + Y_3 + \text{etc.}$

See *series connection*.

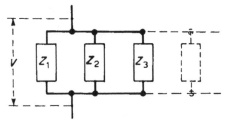

Figure P.2 Components connected in parallel have a common signal voltage V

parallel feed Circuit in which the signal output from an *active device* is taken via a series capacitor, so separating the alternating and direct components of the output current. The direct circuit is completed by a resistor or inductor. The circuit is sometimes used to feed transformers as shown in *Figure P.3* and has the advantage of avoiding *polarisation* of the

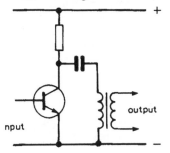

Figure P.3 A parallel-fed transformer

227

transformer core by the direct component of the preceding device, thus enabling a smaller and cheaper transformer to be used.

parallel processing In a computer with multiple microprocessors the simultaneous carrying out of two or more sequences of instructions.

parallel resonance Resonance in which the applied signal is connected to an inductor and capacitor in parallel as shown in *Figure P.4*. Resonance is indicated by the impedance of the *LC* circuit which is a maximum at the resonance frequency given by

$$f = \frac{1}{2\pi\sqrt{(LC)}}$$

as for **series resonance**.

Figure P.4 A parallel resonant circuit

parallel storage (access) A device in which all the bits of a binary word can be stored simultaneously. Cf *serial storage*.

parallel T-network A circuit comprising two *T-networks* in parallel as shown in *Figure P.5*. It is often used as an *RC* filter, as in the diagram, and is capable of giving infinite attenuation at a particular frequency depending on the values of *R* and *C*.

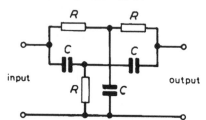

Figure P.5 A parallel-T network used as an *RC* filter

parallel transmission Simultaneous transmission of the elements of a signal over separate channels. See *serial transmission*.

paralysis Effect of applying a large input signal to an electron-tube amplifier causing the tube to be cut off and the amplifier rendered ineffective for a short period. The signal causes grid current in the tube and this charges the inter-tube coupling capacitor to a voltage (the negative plate being connected to the grid) large enough to bias the tube beyond anode-current cut off. The state of paralysis lasts until the charge on the capacitor has leaked away through the grid resistor sufficiently for the tube to begin taking anode current again. This effect is exploited in the *blocking oscillator*.

paramagnetism A property of certain materials subjected to a magnetising field which causes induced magnetism which slightly aids the magnetising field. Such materials are attracted very feebly by a magnetic pole and the effective permeability is slightly greater than unity, being almost independent of the magnetising field. See *diamagnetism, ferromagnetism*.

parameter Strictly one of a number of variables used to express the behaviour of an *active device* or *network* and which is maintained constant whilst the relationship between other variables is being investigated. For example the I_c–V_c characteristics of a *bipolar transistor* are normally displayed as a family of curves, each plotted for a different constant value of base current, which is therefore a parameter. The term is also loosely employed to mean any of the variables used to express electrical behaviour. For example the iterative impedance and cut-off frequency are termed parameters of a network.

parametric amplifier (or **converter**) An amplifier (or converter) which depends for its action on a device, a parameter of which is varied sinusoidally (pumped) at a suitable frequency. The parameter is often the reactance of a semiconductor device such as a *varactor* and parametric amplifiers are used for RF amplification at microwave frequencies where they can give a good *signal-to-noise ratio*.

paraphase Same as *push-pull*.

parasitic oscillation Unwanted sustained oscillation in an *active device* or equipment usually at a frequency well above the passband. It is generated by coupling between stages of the equipment or between the electrodes of an active device. For example if the input and output leads of a *transistor* are not screened from each other, the capacitance between them, even though very small, can provide the positive feedback necessary for oscillation. The leads act as the conductors of a transmission line and their dimensions determine the frequency of oscillation which can be very high indeed when the leads are very short. Such oscillation impairs the performance of equipment by introducing noise and reducing power output. *Parasitic stoppers* are used to suppress such oscillation.

parasitic stopper Component employed in electronic equipment to suppress *parasitic oscillation*. Low-value resistors (e.g. $47\,\Omega$), small inductors and ferrite beads are commonly used for this purpose, the components being included in the leads to active devices as close to the device as possible.

parity bit In digital transmission a binary digit added to each binary code group to make the sum of all the binary digits, including the parity bit, always odd or always even. The technique is used to detect errors in binary code groups.

parity check In digital transmission a technique for detecting that an error has occurred in code groups by use of a parity bit. The binary digits in each code group (including the parity bit) are added to check if the sum is odd or even.

partial A component of a complex sound wave the frequency of which is not necessarily an integral multiple of the fundamental frequency. The sound of a struck bell contains a number of partials. See *harmonic component*.

partition A division introduced in a *hard disk drive* so that it behaves as though it were several separate *logical* drives. There are many reasons for introducing partitions: (1) Separate *operating systems* may reside in different partitions. (2) Some operating systems cannot access hard drives having capacities over a certain threshold, but by introducing partitions such an otherwise unacceptable drive can be used. (3) Some disc formats use a small drive more efficiently than a large one, because the *sectors* are smaller; by partitioning a large drive, its effective storage capacity may be increased.

partition noise Noise arising from random variations in the division of the current from the cathode of an *electron tube* among the various electrodes. Although the division ratios are constant when measured over an appreciable period there are momentary surfeits and deficits at any given electrode and their variation with time constitutes partition noise.

Such noise arises in *tetrodes* and more complex tubes where the cathode current is shared between the anode and the screen grids and makes such tubes about three to five times as noisy as a *triode* which has no partition noise. This was the reason for the choice of triodes for the first stage of amplifiers with very small inputs such as microphone and TV-camera head amplifiers.

Pascal See *computer language*.

passband See *filter*.

passive component A component which does not contain a source of energy. The chief types of passive component in electronics are inductors, capacitors and resistors. See *active device*.

passive network See *network*.

patch (1) A temporary lash-up of equipment typically using jumper leads in order to provide a service at short notice or in other adverse circumstances.

(2) An additional file in a *software* package which modifies the main files in order to overcome some *bug* that was found shortly before release.

path (1) The route taken by a stream of electrons through a component or a *signal* through an equipment. (2) The route taken by radio waves between the transmitting antenna and the receiving antenna. (3) In *computing* the complete address of a file in a storage medium, starting with the drive identity and including all intermediate directories. (4) In *vector graphics* the course followed in plotting a line or the outline of a shape.

pattern generator A *signal generator* which produces on the screen of a TV receiver a regular geometric pattern which can be used to facilitate the adjustment of linearity and convergence.

PC-compatible Of a computer, capable of running software intended for the IBM *PC* and its successors. Of *software*, capable of running on the IBM PC, its successors and compatible machines.

p-channel metal-oxide-semiconductor (PMOS) A form of construction for *logic monolithic integrated circuits* using p-channel *field-effect transistors*. See *NMOS*.

peak clipper See clipping.

peaking circuit (or network) A coupling circuit in a pulse or video amplifier including a series or shunt inductor which resonates with the stray capacitance to improve the frequency response at the upper end of the passband. *Figure P.6* gives an example of a shunt-inductance peaking circuit.

peak inverse voltage (PIV) Of a *diode* in a rectifying circuit, the maximum instantaneous voltage across the input terminals when the diode is non-

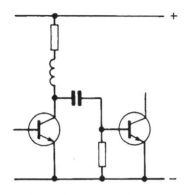

Figure P.6 A peaking circuit incorporating a shunt inductance

conductive. This voltage is made up of the alternating input to the diode plus the voltage across the load circuit and the latter can equal the peak value of the alternating input if a reservoir capacitor is included. Thus the peak inverse voltage can be twice the peak value of the alternating input to the diode, e.g. nearly 700 V if the input voltage is 240 V RMS. The diode must be capable of safely withstanding such a voltage.

peak programme meter (PPM) An instrument for measuring the amplitude of audio signals in terms of the peaks occurring within a predetermined period. The instrument responds very rapidly to increases in signal amplitude to permit accuracy in peak measurement but it takes an appreciable period (of the order of seconds) to return to zero after a peak has been indicated. This avoids the very rapid changes in reading which would otherwise occur and make peak reading difficult. The instrument incorporates a logarithmic amplifier so that the scale can have linear decibel calibrations. The instrument is used for AF measurements in sound broadcasting and recording where accurate knowledge of peak amplitudes is essential.

peak white In black-and-white television the peak excursion of the picture or video signal towards white level. In a dark scene with no highlights peak white may be considerably below *white level.*

pedestal In general a pulse with a level top which is used as a base for another pulse. In TV the difference between *black level* and *blanking level.* This is usually expressed as a percentage of the difference between *white level* and blanking level. See *Figure B.18*(a).

peek In computing, to read the contents of a *byte* of memory. See also *poke.*

Peltier effect When an electric current is maintained through the junction of two dissimilar metals the junction temperature increases or decreases depending on the direction of the current. The heat generated at the junction or lost from it is proportional to the current.

penetration Extent to which an alternating current occupies the cross section of a solid conductor. As frequency is raised alternating currents tend to confine themselves to the outer areas of the cross section (see *skin effect*). At high radio frequencies so little current is carried by the central region of the conductor that it is practicable to use tubing for RF inductors in place of solid conductors. The depth of penetration is the wall thickness of such a

231

tube which has the same outer diameter and the same RF resistance as a given solid conductor.

pentagrid tube (US) Same as *heptode*.

Pentium™ Trade name for the Intel 80586 *microprocessor*.

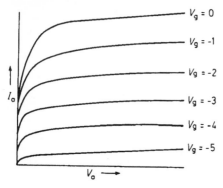

Figure P.7 Shape of I_a–V_a characteristic of a pentode

pentode An electron tube with a *cathode, control grid, screen grid, suppressor grid* and *anode*. The tube is in effect a tetrode in which the *tetrode kink* has been eliminated by the inclusion of the suppressor grid. Thus the characteristics of the tube are a good approximation to the ideal as shown in the typical example shown in *Figure P.7*. Such tubes were extensively used for amplification at RF, IF and AF and in the output stages of amplifiers. The graphical symbol for a pentode is given in *Figure P.8*.

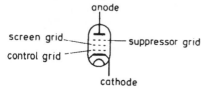

Figure P.8 Graphical symbol for an indirectly-heated pentode

percentage modulation Same as *modulation percentage*.

Percival system A system of compatible stereo sound transmission pioneered by Electric and Musical Industries in which the sound signal is transmitted as in a monophonic system and the directive information is transmitted as a narrow-band (100 Hz) signal on a subcarrier of about 22 kHz.

performance testing Testing of equipment to determine its characteristics or to determine whether it complies with the specified performance.

periodic permanent magnet (PPM) A structure of alternate permanent ring magnets and salient soft-iron pole pieces used to provide the axial magnetic field in travelling-wave and similar tubes. As shown in *Figure P.9* the annular magnets are so arranged that neighbouring salient pole pieces have opposite polarity. Such a structure avoids the need for a focusing solenoid and requires only small amounts of magnetic material.

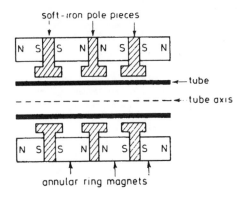

soft-iron pole pieces

N S | S N | N S | S N

← tube

← tube axis

N S | S N | N S | S N

annular ring magnets

Figure P.9 A periodic permanent magnet structure

peripheral Any equipment that is commonly connected to a central control unit. For example, to a computer user a *printer*, a *scanner* and a *modem* would be peripherals in that they must be connected to the computer for use. A programmer, however, might think of the computer's internal *disk drives* and *memory* as peripherals because they are controlled by the *central processing unit.*

permanent storage Same as *read-only memory.*

permeability Of a magnetic material, the ratio of its *absolute permeability* to the absolute permeability of a vacuum.

permeability tuning A method of adjusting the resonance frequency of a tuned circuit by varying the effective *reluctance* of the magnetic circuit of the *inductor.* Usually such tuning is effected by adjusting the position of a ferromagnetic core in the inductor.

permeance Of a magnetic circuit the ratio of the *magnetic flux density* to the *magnetomotive force* which gives rise to it. It is the reciprocal of *reluctance* and magnetic analogue of *conductance.*

permitted band Same as *allowed band.*

permittivity Of an insulator a characteristic constant equal to the ratio of the electric flux density produced in the insulator by a given electric field strength to the flux density produced by the same field strength in a vacuum. More practically it is the ratio of the *capacitance* of a capacitor with the insulator as dielectric to that of the same capacitor with air (strictly a vacuum) as dielectric. Values of permittivity for commonly-used insulators range between 2 and 10 but values as high as several thousand are possible.

persistence Emission of light from the screen of a *cathode ray tube* after removal of excitation by the electron beam. The period of persistence may be between a fraction of a second to several minutes depending on the *phosphor* used in the screen coating.

Personal Computer (PC) Strictly, any *computer* designed for use by a single user, in contrast with *mainframe* and *mini-computers* which may be used simultaneously by several users at separate *terminals.* In practice this generally means a *desktop* or *laptop microcomputer* which is self

233

contained, although capable of being connected to a *network* for access to shared resources. More specifically, the term Personal Computer (capitalised) or PC is commonly used to designate the IBM PC, introduced in 1981, and its successors and compatible machines.

Personal Computer Memory Card International Association (PCMCIA) A standard originally developed for *demountable* memory cards intended primarily for use with portable computers and other data processing equipment. The cards are about the same size as credit cards but slightly thicker and may contain *ROM*, *EPROM* or battery-backed *RAM* with storage capacities up to 80 Mbytes. They locate in a slot in the *host* device in much the same way as a *floppy disk*. They are used for the storage and transmission of software, especially in environments where the use of electromechanical storage devices such as floppy and *hard drives* is impractical because of, e.g. electromagnetic interference or extremes of temperature. The PCMCIA standard is also used for connecting other peripherals, such as modems and hard disk drives, to portable computers.

persuader An electrode in an *image orthicon* tube biased so as to direct the return scanning beam into the *electron multiplier* which surrounds the *electron gun*. The persuader is in the form of a short cylinder and is given a positive bias (with respect to the electron-gun-cathode potential) which is adjusted to give the best uniformity of illumination over the area of the reproduced image: the adjustment is usually labelled 'multiplier focus'.

perveance A factor expressing the ability of a *diode* or *electron gun* to provide an electron stream under the stimulus of a given accelerating voltage. It is equal to the *space-charge-limited* electron current divided by the three-halves power of the accelerating voltage.

p-gate thyristor A *thyristor* which is switched to the on-state by a positive signal applied between *gate* and *cathode*, the gate being a connection to the p-region nearest the cathode. It is normally a reverse-blocking thyristor, i.e. one which cannot be switched to the on-state when the anode is negative with respect to the cathode. The graphical symbol for the p-gate thyristor is given in *Figure P.10*.

Figure P.10 Graphical symbol for a p-gate thyristor

phantastron A monostable *transitron*, i.e. a pentode in which the anode is capacitively coupled to the control grid and the screen grid is directly coupled to the suppressor grid. The circuit is equivalent to a *monostable multivibrator* and, when triggered by positive signals on the control grid, generates rectangular pulses at the screen grid and linear sawtooths at the anode. The duration of the output signals can be controlled by adjustment of the standing bias on the control grid. A typical circuit diagram for a phantastron is given in *Figure P.11*.

phantom In telephony an additional independent circuit obtained from two physical balanced two-wire circuits by effectively connecting each pair of conductors in parallel. Signals in the phantom circuit travel in the same direction in the conductors of each physical circuit, as shown in *Figure P.12*, and thus do not interfere with the signals in them.

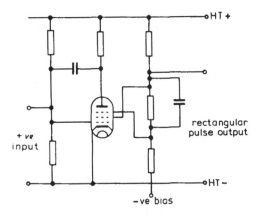

Figure P.11 A phantastron circuit

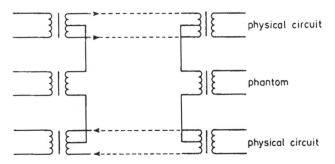

Figure P.12 Derivation of a phantom circuit from two physical circuits

phase Of a periodically-varying quantity, that part of a period (in degrees or radians) through which the quantity has advanced since some arbitrary time origin. Usually the origin is taken as the time when the quantity last passed through zero value in changing from negative to positive values.

phase-change disk and **drive** Same as *magneto-optical disk* and *drive*.

phase coefficient (or **constant**) See *propagation coefficient*.

phase comparator A circuit which compares the phases of two input signals and gives an output dependent on the phase difference. See *long-tailed* pair.

phase corrector Same as *phase equaliser*.

phase delay The time obtained by dividing the *phase shift* introduced by a circuit or equipment by the frequency. The phase angle is expressed in radians and the angular frequency in radians per second. Thus

$$\text{phase delay} = \frac{\phi}{\omega}$$

Phase delay can be regarded as the time interval between a crossing of the time axis in the sinusoidal input signal and the corresponding crossing of the time axis in the output signal. Admittedly it would not be easy to measure phase delay in this way because of the difficulty of identifying corresponding time-axis crossings at input and output. This is why delay is usually measured with a complex signal having an easily-identifiable feature such as a steep leading edge which can be timed through the equipment.

phase difference Angle which represents the extent to which one sinusoidal quantity differs in time from another of the same frequency, i.e. it is the angle between the two vectors representing the sinusoidal quantities. If the two vectors coincide the two quantities are in phase and the phase angle is zero. If the two vectors are at right angles the two quantities are in quadrature and the phase angle is 90°.

In electronics we are often concerned with the phase angle between the current in a circuit and the applied alternating voltage. If, as in an inductive circuit, the voltage reaches a given point in its cycle before the current reaches the corresponding point in its cycle, the current is described as lagging on the voltage and the phase angle may be termed the angle of lag.

In a capacitive circuit the current reaches a given point in its cycle before the voltage reaches the corresponding point in its cycle: here the current leads the voltage and the phase angle may be termed an angle of lead.

phase discriminator See *discriminator*.

phase distortion Distortion which arises when the *phase shift* introduced by a system or equipment is not directly proportional to the frequency. If the phase shift is strictly proportional to frequency, as shown in *Figure P.13*, then the *phase delay* through the system is constant and independent of frequency. All components of a complex input signal therefore arrive at the output correctly related to each other in time and there is no distortion of the output waveform. *Figure P.13* is a typical practical phase-frequency characteristic and this would cause waveform distortion, in particular an increase in the *rise time* of a pulse.

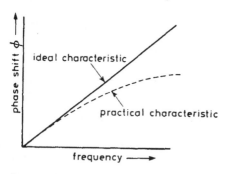

Figure P.13 Phase-frequency characteristics

phase equaliser A *network* designed to correct phase distortion occurring within a specified frequency band. If the phase equaliser is not also required to correct for amplitude-frequency distortion, a simple *all-pass network* may be used as a phase equaliser.

phase-frequency distortion Same as *phase distortion*.

236

phase inverter A circuit of which the output signal is an inverted replica of the input signal. Such stages are required where two signals of identical waveform, one inverted with respect to the other, are needed for application to a push-pull amplifier. 'Signal inverter' is a better term.

phase-locked loop (PLL) A combination of a voltage-controlled oscillator and a phase comparator so connected that the oscillator frequency (or phase) accurately tracks that of an applied frequency- or phase-modulated signal.

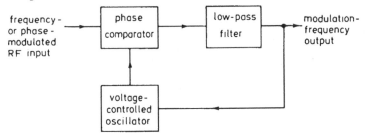

Figure P.14 Block diagram of a phase-locked loop

The arrangement is illustrated in the block diagram of *Figure P.14*. The output of the phase comparator depends on any difference in frequency (or phase) between the two input signals and contains zero-frequency and RF components. The latter are removed by the low-pass filter but, provided the *cut-off frequency* of the filter is suitable, the voltage-controlled oscillator will track any frequency or phase modulation of the input signal and the control signal for the oscillator is therefore a replica of the modulating signal.

phase modulation (PM) Method of modulation in which the phase angle of a carrier wave is made to vary in accordance with the instantaneous value of the *modulating signal*. Phase modulation is similar to frequency modulation. The difference between them is that in frequency modulation, for a constant-amplitude modulating signal, the phase shift is swept between limits inversely proportional to the modulating frequency: in phase modulation the limits are fixed. Similarly in phase modulation for a constant-amplitude modulating signal the frequency is swept between limits directly proportional to the modulating frequency: in frequency modulation the limits are fixed. In practice this means that one form of modulation can be converted to the other by including a 6 dB-per-octave filter in the modulating-signal path.

phase quadrature See *phase difference*.

phase reversal Term applied, somewhat loosely, to the inversion of a signal.

phase shift Of a sinusoidal signal a change of phase angle as a result of, for example, transmission of the signal through a *network*. Also an alteration in the difference in phase between two or more sinusoidal signals at the same frequency.

Phase Shift Key (PSK) A method of *phase modulation* used to transmit digital data over telecommunications links.

237

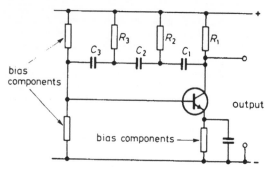

Figure P.15 Essential features of a phase-shift oscillator using a bipolar transistor

phase-shift oscillator A sinusoidal oscillator the frequency of which is determined by the phase shift in an *RC* network. In one type, illustrated in *Figure P.15*, a three-section RC network is used to give 180° phase shift at the operating frequency, signal inversion in the active device providing the positive feedback essential for oscillation. In another type of phase-shift oscillator a *Wien-bridge* network gives zero phase shift at the operating frequency, a non-signal-inverting amplifier being used to give positive feedback. This is the bias of a number of AF test oscillators.

phase splitter A circuit which provides two output signals of identical waveform, one inverted with respect to the other, for application to a pushpull amplifier. A number of circuits can be used for this purpose. See *concertina phase splitter, long-tailed pair, see-saw phase splitter.*

phase velocity Of a sinusoidal plane wave the velocity of propagation of an equiphase front in the direction of the normal. It is equal to the product of the wavelength and the frequency. If the phase velocity is independent of frequency it is equal to the envelope and *group velocity.*

phasing In general the adjustment of an oscillating system to establish a desired phase relationship with another system with the same frequency. In colour TV the process of ensuring that the time of arrival of video or synchronising signals at a particular point are within the tolerances for the system.

   In sound reproduction so connecting a (*monophonic*) audio signal to the terminals of the individual units of a multi-unit loudspeaker or of the reproducers of a stereo system that all the diaphragms move in phase.

phon The unit of loudness level. See *Fletcher-Munson curves, loudness level.*

phonograph pickup (US) Same as *pickup.*

phosphor A material capable of *luminescence* and therefore used as a coating for the screens of *cathode ray tubes*. A number of different materials are used depending on the colour and the *persistence* required.

phosphorescence Emission of light from a material after it has been irradiated by energy of a higher frequency or has been bombarded by electrons. The emission may continue for a considerable period after the stimulus has been removed. See *fluorescence.*

photicon Type of *image iconoscope.*

photo-cathode Electrode in a *photocell* or a *camera tube* which emits *photo-electrons* when irradiated by light. In general the light falling on a photocell is in the form of a spot or stripe and the tube is required to respond to variations in the intensity of the light.

In a camera tube, however, an optical image of the scene to be televised is focused on one face of the photo-cathode which is transparent so that photo-electrons are liberated from the opposite face. The number of photo-electrons emitted from any point on the photo-cathode surface depends on the illumination of that point and thus varies over the area of the photo-cathode depending on the detail in the optical image. The liberated photo-electrons are focused by an *electron lens* to form a corresponding charge image on the target of the camera tube. See *image section*.

Photo-CD A system developed by Kodak for the digital storage of photographic images on a *compact disc*. The original intention was that users purchase a dedicated Photo-CD player (which would also play audio CDs) and connect it to a domestic television receiver for viewing. Films were processed by Kodak who supplied conventional prints as well as a Photo-CD disk containing *scanned* images of the photographs. Essentially a recordable CD, this was *multi-sessional* in that more images could be added to it in a later session. Each image is stored in five resolutions allowing the user to *zoom in* on points of interest. At first the system was not popular because the hardware was expensive, access to the higher-resolution images was slow and it was sometimes inconvenient to view photographs on a domestic television receiver. Photo-CD's renaissance came with the advent of CD-ROM. With suitable software any computer having a CD-ROM drive and a high-resolution *VDU* can read a Photo-CD, display the images stored on it and copy them to graphics applications for editing or to DTP applications for incorporation into documents. Photo-CD is now an established source of high-quality *bitmaps*.

photocell A device which gives an electrical output dependent on the light falling on the input electrode. There are a number of different types of photocell depending on the nature of the light-sensitive material used in them. For example a photocell may be photo-emissive, photoconductive or photovoltaic. Cells using photo-emissive materials may be vacuum or gas-filled.

photoconductive camera tube A camera tube in which the photo-sensitive electrode is photoconductive. Such tubes are known as *vidicons* and are extensively used in colour television cameras.

photoconductive cell A photocell in which the photo-sensitive electrode is photoconductive.

photoconductivity Variation of the electrical conductivity of a material when it is irradiated by electromagnetic waves within a particular frequency range. Examples of materials which exhibit photoconductivity are selenium, antimony trisulphide and copper oxide, all of which have been used in the targets of *vidicon* camera tubes.

photo-coupler or **isolator** Same as *opto-coupler* or *isolator*.

photo-diode A pn diode the reverse current of which is dependent on the amount of light falling on the junction. The reverse current is carried by minority carriers and is greatly dependent on temperature because heat can

(a)　　　　　　(b)

Figure P.16 Graphical symbol for (a) a photo-diode and (b) a photo-transistor

liberate more covalent bonds. Light can also do this and thus a pn diode in a transparent container can be used as a light cell. The graphical symbol for a photodiode is given in *Figure P.16* (a).

photo-electric cell (PEC) Same as *photocell.*

photo-electric emission (photo-emission) The liberation of *electrons* from a material when this is irradiated by electromagnetic waves. The elements sodium, potassium, lithium, rubidium and caesium are notable for their photo-emissive effect and for each element there is a certain threshold frequency of incident light below which no emission occurs.

photo-electron An *electron* liberated from a photo-emissive surface by incident light.

photo-multiplier A photo-emissive tube in which the liberated *photo-electrons* are directed into an *electron multiplier* to give a greatly enhanced current output.

photon An elementary quantity of radiant energy which may be regarded as a fundamental particle or a train of electromagnetic waves. A photon is emitted when an *electron*, which has been transferred to an orbital of higher energy level, suddenly returns to its former orbital. See *energy level diagram.*

photo relay Same as *opto coupler.*

photo-resist An organic substance sensitive to ultra-violet light which is used in the manufacture of transistors and integrated circuits. See *planar process.*

photo-sensitivity Ability of a material to change its chemical or electrical state when irradiated by electromagnetic waves. Such materials can absorb *photons* of light and liberate *electrons* which can effect the properties of the material in three ways:

(a) the electrons may be released as photo-electric emission, i.e. as photo-electrons.

(b) the electrons, acting as charge carriers, may increase the electrical conductivity of the material.

(c) the electrons may set up an EMF proportional to the light falling on the material.

These three effects are known as the photo-emissive, photoconductive and photovoltaic effects. Thus photo-sensitivity is a generic term embracing all three effects.

photo-transistor A *bipolar transistor* in which the output current is determined primarily by the amount of light falling on the base region. Absorption of *photons* in the base region liberates charge carriers so giving rise to a collector current. In circuits using photo-transistors the base may be left disconnected or may be returned to the emitter via a high impedance. The graphical symbol for a photo-transistor is given in *Figure P.16* (b).

photo-tube A vacuum or gas-filled tube containing a *photocathode* and an *anode*. The *photo-electrons* liberated from the cathode when light falls on it are collected by the positively-charged anode to give an anode current. The sensitivity of the gas-filled tube is greater than that of the vacuum tube because of the increased anode current caused by *ionisation*. See *gas-amplification factor*.

photovoltaic effect The generation of a voltage between two dissimilar materials when their junction is irradiated by electromagnetic waves.

physical In computing, relating to the actual structure of a system or an operation or a block of data. This contrasts with its *logical* form which is presented to the user, which is not necessarily the same. For example, if a *hard disk* is *partitioned* so that it appears to the user as several separate drives, it may be described as one *physical* drive which appears as several *logical* drives.

pickup In general a *transducer* which converts signals in a non-electrical form of energy into corresponding electrical signals. Thus a TV camera tube was, at one time, termed a vision pickup tube but this term is now deprecated. The term pickup now tends to be confined to disk reproducing heads which convert the mechanical movements imparted to the stylus by the groove walls of the disk into corresponding electrical output signals.

picture element In TV the smallest area of a picture which can be reproduced. The picture is regarded as composed of a large number of elements arranged in horizontal lines or rows, the length of each element being equal to the vertical distance between neighbouring lines, i.e. the elements are regarded as squares. During transmission the elements are explored in turn by the scanning beam but the system cannot resolve detail finer than that of an element. Thus the picture signal corresponding to a row of alternate black and white elements is a sine wave, as shown in *Figure P.17*, and its frequency is equal to the upper frequency limit of the video bandwidth of the system; but see *Kell factor*.

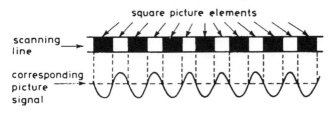

Figure P.17 Form of picture signal for a row of alternate black and white elements

In computing, the American term *pixel* is used for the smallest part of a *VDU* display or a *bitmap* that can be addressed. Pixels are most often square and, in a *VGA* screen, have a nominal size of 1/96 inch per side. Each pixel is represented in *video memory* by a number of bits, the number being the logarithm (base 2) of the number of colours in the display (or

241

bitmap). Thus a two-colour (monochrome) display requires only 1 bit per pixel (since $2^1 = 2$), a 256-colour display requires 8 bits per pixel ($2^8 = 256$) and a 16,777,216-colour (*true-colour*) display requires 24 bits per pixel ($2^{24} = 16,777,216$).

picture frequency In TV the number of times the whole picture is scanned in a second. In the British television system the picture frequency is 25 Hz and in the USA it is 30 Hz.

picture signal In TV or facsimile the signal which results from the scanning process and which, when combined with the synchronising signal, forms the video signal. See *Figure B.18* (a).

picture/sync ratio In TV the ratio of the maximum picture signal amplitude to the amplitude of the synchronising signals. In *Figure B.18* (a) it is given by: (*white level–blanking level*) : (*blanking level–sync level*). The ratio must be chosen with care for if it is too large the sync signals will be too small to hold pictures steady at the receiver and if it is too small the pictures may be too noisy to view even though synchronisation may be perfect. In the compromise adopted in the UK the picture/sync ratio is 7 : 3.

picture tube A *cathode ray tube* used to display pictures in a TV receiver or monitor. They usually have electrostatic focusing and are designed for wide-angle magnetic deflection to give a short tube suitable for mounting in a cabinet. Tubes for displaying black-and-white pictures have a single electron gun and a screen made of a continuous layer of phosphor giving an approximation to white light. For displaying colour television pictures three guns are necessary (or a single gun firing three beams) and the screen is made up of dots or stripes of phosphors giving red, green and blue light. See *delta-array picture tube*, *precision-in-line picture tube*.

Pierce oscillator A *Colpitts oscillator* in which *positive feedback* is provided by the input and output capacitances of a three-terminal *active device*, a *piezo-electric crystal* connected between input and output terminals determining the frequency of oscillation.

piezo-electric ceramic A ceramic material which, after processing, exhibits the *piezo-electric* effect.

piezo-electric crystal A crystal with piezo-electric properties. The most frequently-used piezo-electric crystals are those of quartz, tourmaline and Rochelle salt.

piezo-electric effect The development of voltages between the faces of certain crystals when they are subjected to mechanical stress and the converse mechanical deformation of the crystal when voltages are applied between the faces. These properties are exploited in crystal loudspeakers, microphones and pickups. See *bimorph*.

Slices of piezo-electric crystal suitably oriented with respect to the crystallographic axis also exhibit resonance. For example if a voltage is suddenly applied to or removed from the faces of such a slice, the crystal oscillates with diminishing amplitude at a frequency characteristic of the crystal. A voltage at this frequency applied to the crystal keeps it in continuous oscillation. The crystal behaves in fact as an LC circuit of very high Q and very high stability. Such crystal slices are used to control the frequency of oscillators where high stability is required. See *crystal oscillator*.

piezo-electric element A piezo-electric crystal or piezo-electric ceramic material cut to a required size and shape, correctly oriented with respect to the crystallographic axis of the material and provided with electrodes. Such elements are used in *microphones, loudspeakers* and *pickups*.

piezo-electric loudspeaker Same as *crystal loudspeaker*.

piezo-electric microphone Same as *crystal microphone*.

piezo-electric oscillator Same as *crystal oscillator*.

piezo-electric pickup Same as *crystal pickup*.

pilot or **pilot carrier** A signal, usually a single sinusoidal wave, transmitted over a telecommunications system to indicate the transmission characteristic or to control it. The pilot may be used, for example, for automatic control of the received level of signals transmitted over the system or for synchronisation of the oscillator used for demodulation.

pilot lamp A lamp used to indicate the condition of an associated circuit. A pilot lamp is commonly used, for example, to indicate that an item of equipment is switched on, has reached its correct operating temperature, is faulty, etc, distinctive colours being used to identify the function.

pilot-tone stereo Compatible FM system of stereo sound transmission in which the sum of the left and right signals is transmitted as in monophonic FM systems and the difference signal is transmitted as double-sideband suppressed-carrier amplitude modulation of a subcarrier. A pilot tone at half the subcarrier frequency is also transmitted at low level and is used at receivers to synchronise the demodulator which recovers the difference signal.

pinch Part of the envelope of an *electron tube* (or electric lamp) carrying the wires which support the *electrodes* and provide the electrical connections to them.

pinch effect In disk reproduction, vertical motion of the tip of the pickup stylus caused by variations in the angle between the groove walls. On recording the angle is less when the cutter is between the peaks of its swings than at the peaks themselves and thus the reproducing stylus makes two vertical movements per recorded cycle and these can result in second harmonic distortion.

pinch off In a *field-effect transistor*, cut off of the drain current. For example, the negative gate voltage which reduces the drain current to zero in an n-channel device is known as the pinch-off voltage.

pin-cushion distortion Distortion of a TV picture in which the sides of a reproduced square bulge inwards. The distortion, illustrated in *Figure P.18* is caused by non-uniformity of the field produced by the scanning coils. It is the opposite of *barrel distortion*.

Figure P.18 Reproduction of a square by a picture tube suffering from pin-cushion distortion

pin diode A semiconductor diode containing a region of *intrinsic semiconductor* between the p- and n-regions. The transit time across the intrinsic region limits the operating frequency when the diode is used as a detector to below 100 MHz and at microwave frequencies the diode behaves as a linear resistance. However, if a forward bias is applied, carriers are injected into the intrinsic region and the diode resistance falls to a low value, possibly only one thousandth of the reverse resistance. Thus the pin diode is extensively used as a modulator and switch in microwave systems.

pi-network (π-network) A *network* consisting of two shunt elements with a series element between them. The form of the network is illustrated in *Figure P.19*.

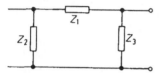

Figure P.19 A π-network

pink noise The noise signal obtained by passing a *white noise* signal through a filter with a response which falls at the rate of 3 dB per octave over the frequency range of interest.

pinout A diagram showing the arrangement of the pins of a *transistor* or an *integrated circuit* with the function of each.

pitch The position of a sound in the musical scale. For a sound with a complex waveform the pitch is determined by the fundamental frequency whether this is present as a discrete component or not. For orchestral performances the fundamental frequency for A in the treble clef is internationally agreed as 440 Hz.

pixel (US) Same as *picture element*.

planar process The formation of n-type and p-type regions in a semiconductor crystal by introducing impurities via apertures in a mask on the surface. The introduction of this process in 1960 revolutionised the manufacture of silicon transistors, making mass production possible for the first time. A very thin layer of silicon dioxide is grown on the surface of a slice of n-type monocrystalline silicon by heating it to 1100°C in a stream of oxygen. This surface is then coated with a layer of *photo-resist*. A mask defining the base areas of the transistors is placed over the surface and exposed to ultra-violet light. The photo-resist polymerises where exposed to the light and is then resistant to attack by acids and solvents. On development the unexposed areas of the photo-resist are removed, exposing the silicon dioxide. An etch removes the silicon dioxide, exposing the base area, into which a controlled amount of boron is introduced by diffusion, ion-plantation or other means. By repeating this process with another mask the emitter areas can be defined and the required amount of phosphorus is introduced. Finally a third mask is used to define the positions for connections to the base and emitter areas, the original slice acting as the collector. By this process tens of thousands of transistors can be manufactured simultaneously on a single crystal slice.

Although the above description applies only to the manufacture of bipolar silicon transistors it can obviously be used for making a wide variety of semiconductor devices, including diodes and field-effect transistors. Because the process can also be used for making interconnections between devices, monolithic integrated circuits can also be manufactured in this way. So can RAMs, ROMs and many other devices.

planar transistor A *bipolar* or *field-effect transistor* manufactured by the *planar process*.

plane of polarisation Of an *electromagnetic wave*, the plane containing the direction of the electric field and the direction of propagation.

plasma A region in an ionised gas where the number of positive *ions* equals the number of *electrons* so that there is no net charge.

plasmatron A gas discharge tube in which the conducting path between the *anode* and the main *cathode* is provided by *plasma* generated by a subsidiary cathode.

plastic effect In TV an effect of relief in reproduced images caused by exaggeration of the tonal transitions. The effect can be caused by a poor low-frequency response in the video amplifier. It can also be caused in images reproduced from an *image orthicon* tube because this type of tube tends to provide black borders around image highlights.

plate (US) Same as *anode*.

plate detector Same as *anode-bend detector*.

plated-through holes Holes drilled through the conducting tracks of a *printed-wiring board* to accept the connecting leads of components mounted on the board. The holes are plated with metal to give good electrical contact between the leads and the tracks when the components are soldered into position. See *chip carrier, dual-in-line package, surface mounting*.

platform A term sometimes used to refer to the make or type of computer. Thus an *application* is said to be available for *cross-platform* if it is available for different types of computer.

platter One of the disks present in a *hard disk drive*.

playback The process of reproducing a programme that has been *recorded*.

plotter A computer *peripheral* used for producing hard copies of drawings such as those generated in *computer aided design* software. A plotter consists of one or more pens which can move backwards and forwards (the X-axis) and lift off the paper. The paper can be wound backwards and forwards through the plotter, providing the Y axis. Any line drawing can be reproduced by the appropriate sequence of movements of the pen and the paper under computer control. A plotter is the only computer output device that can reproduce genuine curves and oblique lines; all other devices print dots on a rectangular matrix that can only simulate oblique lines and curves.

plumbicon A low-velocity TV *camera tube* with a photoconductive target containing lead. Modern plumbicon targets are of multi-layer construction and their lag and sensitivity are such that the plumbicon is now the standard tube used in colour TV cameras. See *vidicon*.

pn boundary Same as *junction*.

pnip transistor See *npin transistor*.

pn junction Same as *junction*.

pnp transistor Type of *bipolar transistor*.

point-contact diode Early form of semiconductor diode which consisted of a whisker of a metal alloy pressed against a crystal of semiconductor material. Point-contact diodes have now been superseded by *junction diodes* which are more robust and easier to manufacture.

point-contact transistor Obsolete type of transistor in which the emitter and collector connections to the base wafer were made by whiskers of a metal alloy. Manufacture of these devices was abandoned years ago in favour of *bipolar (junction) transistors* which are easier to mass produce and are more robust.

pointer In computing an *icon*, often shaped like an arrow, that the user moves around the screen by means of a *mouse, trackerball* or other *pointing device*. The pointer forms part of a *graphical user interface*. Operations are initiated by moving the pointer over fixed icons on the screen or over options in menus and clicking, that is pressing, a button on the pointing device.

point gamma In TV the slope of the curve relating the logarithm of the output of a device, equipment or system to the logarithm of the input. The input or output may be light or an electrical signal and thus there can be a point gamma for a camera tube, a picture tube or the overall system. See gamma.

pointing device In computing, the *hardware* used to move the *pointer* around the screen in a *graphical user interface*. The most commonly used pointing device is the *mouse*; others include *joysticks* and *trackerballs*.

poke In computing, to set the value contained in a *byte* of memory. See also *peek*.

polar diagram In general, of a quantity which varies with direction, the closed figure generated by the tip of the radius vector as it rotates through 360°, its length representing the magnitude of the quantity and its angle the direction.

Such diagrams are used in electronics to indicate the directivity, e.g. of *antennas* and *microphones*. An omnidirectional response such as that of a vertical radiator in the horizontal plane is represented by a circular polar diagram and a double-sided response such as that of a ribbon microphone (with ribbon vertical) in the horizontal plane is represented by a figure-of-eight polar diagram.

polarisation (1) Of a *primary cell* the liberation of gas at the electrodes during discharge. This limits the current that may be drawn from the cell. See *depolarisation*. (2) Of an electromagnetic wave the direction of the plane containing the electric vector and the direction of propagation. This is, in general, also the direction of the plane of the conductors forming the radiating antenna. Thus a vertical antenna radiates a vertically-polarised wave.

polarity Originally the distinction between the North and South poles of a magnet or the positive and negative terminals of a cell. The term is now more generally applied, e.g. to the potentials of any two terminals of a circuit and to the connections to the loudspeakers in a stereo system which can be said to be correctly polarised when the diaphragms move in phase.

pole (1) A terminal of a *network* or a *cell*. (2) In *magnetism* that region of a magnet from which the lines of force of the external magnetic field appear to diverge or converge. In a bar magnet these regions are normally near the ends and if the magnet is freely suspended at its centre it comes to rest with its axis aligned with the earth's magnetic field. The poles are therefore known as the north-seeking or more simply the north pole and the south-seeking or south pole. Magnets can, however, be made with the poles at any desired point in them. (3) For a network of pure reactances, any frequency at which the input reactance is infinite. As shown in *Figure Z.2* the reactance-frequency relationship for a *network* of pure reactances is a succession of curves, all with positive slope, which swing between minus infinity and plus infinity, passing through zero, in a manner similar to that of a tangent curve. The poles are the frequencies at which the curves approach infinity, i.e. antiresonant frequencies at which the network presents at its input terminals the equivalent of an inductance and capacitance in parallel.

pool cathode See *mercury pool cathode*.

port In a *network* a pair of terminals at which signals may be fed into the network or withdrawn from it. A two-terminal-pair network such as the simple filter shown in *Figure P.20* may thus be described alternatively as a two-port network.

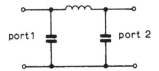

Figure P.20 An example of a two-port network

porting The process of converting an *application* for use on a computer of a different type. Porting is easier if the program is written in C which was specifically designed for use on widely differing machines.

positive feedback *Feedback* in which the signal returned to the earlier stage of the amplifier is in phase with and therefore augments the input signal at that point. Positive feedback has the effect of increasing the gain of the amplifier and was used for this purpose in early *electron tube* receivers where the feedback was called *reaction*. If, however, the amplitude of the feedback signal equals that of the normal input signal, it can take the place of the normal input and the amplifier becomes a generator. This is the basis of a large number of oscillator circuits which are distinguished from each other only in the way in which the positive feedback is obtained.

positive logic A *computer* or *data-processing system* in which the more positive of the two voltages selected as the logic levels is designated the 1-state.

positive modulation In TV transmission, *amplitude modulation* in which increased picture brightness results in increased carrier amplitude. This is the system used in the original British 405-line television system.

post-alloy diffused transistor (PADT) Same as *mesa transistor*.

post-deflector accelerator Same as *intensifier electrode*.

post emphasis (US) Same as *de-emphasis*.

post equalisation (US) Same as *de-emphasis*.

PostScript A page description language devised by Adobe. It allows pages of *text* and *graphics* to be described in a plain text file. It is widely used in *laser printers* and phototypesetting equipment. A subset, *Encapsulated PostScript*, is also sometimes used as a common language for transferring *vector graphics* images between computers of different types.

potential At a point the work done in bringing a unit electric charge from infinity to that point. It is measured in volts and is often termed the *voltage*.

potential difference (PD) The difference in electrical states at two points which causes an electric current to flow between them when they are connected by a conductive path. The PD is measured by the work done in moving a unit electric charge from one point to the other. It is measured in volts and is often termed the *voltage*.

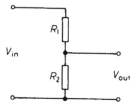

Figure P.21 A simple potential divider composed of two resistors

potential divider A circuit consisting of a number of similar elements in series connected across a voltage source, voltages being taken from the inter-element connections. Frequently the potential divider consists of two resistors as shown in *Figure P.21* and the division ratio, i.e. the ratio of v_{out} to v_{in} is given by

$$\frac{v_{out}}{v_{in}} = \frac{R_2}{R_1 + R_2}$$

See *capacitance potentiometer*.

potentiometer (1) An instrument for measuring an unknown voltage by balancing it in a bridge circuit against a known voltage. (2) A resistive two-element potential divider in which the division ratio is adjustable. Such devices are extensively used in electronic equipment for adjustable controls, e.g. to control *gain* and are constructed in the form of a resistive element carrying a sliding contact which is adjusted by a rotary or linear movement of the control knob. The graphical symbol for a potentiometer is given in *Figure P.22*.

Figure P.22 Graphical symbol for a potentiometer

power amplifier An amplifier designed to deliver substantial output power. The term is used to distinguish such amplifiers from voltage or current amplifiers which are often used to drive power amplifiers. Examples of power amplifiers are those used to drive *loudspeakers* or transmitting *antennas*.

power down The sequence of operations that must be performed when an equipment is switched off. In *computers* and *data processing equipment* all open files should be closed or valuable data may be lost.

power factor In an AC circuit the ratio of the real power to the apparent power, i.e. the ratio of the dissipation to the product of applied RMS voltage and resulting RMS current.

In a resistive circuit, voltage and current are in phase and the dissipation is equal to their product so that the power factor is unity. In general, however, there is a phase difference between voltage and current and the dissipation is then the product of the active (in-phase) current and applied voltage which is less than the product of external current and applied voltage. The ratio of the two, i.e. the power factor is in fact given by the cosine of the phase angle (cos ϕ). In a simple series circuit the power factor is given by the ratio of resistance to impedance and is thus the reciprocal of the Q factor.

power grid detection *Grid-leak detection* employing an *electron tube* with a large grid base and a high-voltage anode supply which can accept large input signals without introducing the distortion which arises from the curvature of the I_a–V_g characteristic near anode-current cut off. The time constant of the grid capacitor-grid resistor combination is made smaller than that normally used in grid-leak detection.

power pack (US) Same as *mains unit*.

power up The sequence of operations that must be performed when an equipment is switched on.

pre-amplifier An amplifier designed to raise the level from a signal source to a value suitable for driving a main amplifier. Where the signal source has a very low output, as for some high-grade microphones, the pre-amplifier may be located very close to it as a means of preserving a good *signal-to-noise ratio*. See *head amplifier*. In hi-fi equipment the pre-amplifier often incorporates facilities for disk and tape equalisation, audio mixing and tone control.

precision-in-line picture tube (PIL) Same as a *shadow-mask picture tube* in which three electron guns (or the three beams from a single gun) are arranged in a horizontal plane and the screen consists of vertical stripes of red, green and blue phosphors. As shown in *Figure P.23*, the mask has vertical slits, one for each group of three phosphor strips, so arranged that each strip is masked from two of the beams. The *convergence* for such tubes is considerably simpler than for *delta-array picture tubes*.

pre-emphasis The use of a *network* which accentuates the response at high frequencies before an audio signal is recorded or transmitted. A corresponding de-emphasis network is used at the reproducing or receiving end to ensure an overall level frequency response. The falling high-frequency response reduces the subjective effect of noise introduced during the transmission or recording process so improving the *signal-to-noise ratio*.

preferred value Series of values in which each differs from the preceding by a constant multiple and used by international agreement for the values of resistors and capacitors. The system was adopted as a means of reducing to a minimum the number of values required to cover a given range. The value of the constant multiple depends on the tolerance of the component

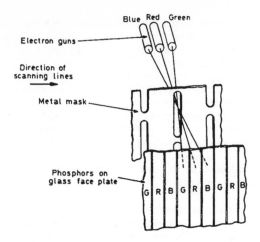

Figure P.23 Precision-in-line colour picture tube: basic arrangement of electron gun, shadow mask and screen.

value. For example the values for a 20% tolerance are

| 10 | 15 | 22 | 33 | 47 | 68 | 100 |

and for 10% tolerance the following intermediate values are added

| 12 | 18 | 27 | 39 | 56 | 82. |

presence In reproduced sound the illusion of closeness to the performer or instrument. Presence can be improved by applying a lift to frequencies in the range of 3 to 5 kHz and this is commonly done in hi-fi reproduction.

preset control A variable control intended for infrequent adjustment. An example is provided by the *trimming* capacitors in a *superheterodyne receiver* which are adjusted when the receiver is aligned after manufacture and are unlikely to require subsequent attention. Such controls are often designed for operation by a screwdriver and are not normally accessible to the receiver owner.

preshoot Form of transient distortion of a step or pulse signal in which the reproduced step is preceded by oscillation. A typical example of preshoot is given in *Figure P.24*.

time ⟶

Figure P.24 An example of preshoot

250

pressure-gradient microphone A microphone of which the electrical output is proportional to the instantaneous particle velocity of the sound wave to which it responds. In such microphones both faces of the diaphragm are exposed to the incident sound wave so that movement of the diaphragm depends on the difference of the acoustic pressure exerted on the two faces and this difference is proportional to the particle velocity.

A *ribbon microphone* in which both faces of the ribbon are exposed to the sound input is an example of a pressure-gradient microphone and one of its most useful properties is that, with the ribbon vertical, it is effectively double-sided, having a figure-of-eight polar diagram in the horizontal plane.

pressure microphone A microphone of which the electrical output is proportional to the instantaneous pressure of the sound wave to which it responds. In such microphones only one face of the diaphragm is exposed to the incident sound wave, the other face being totally enclosed.

A *moving-coil microphone* is an example of a pressure microphone and one of its most useful properties is that, if the plane of the diaphragm is horizontal, it is effectively omnidirectional in the horizontal plane, the polar diagram being circular.

Prestel The *videotex* system of British Telecom.

preventive maintenance Maintenance designed to detect ageing or failing components which can thus be replaced before they cause failure of the equipment. Typically preventive maintenance consists of a regular application of performance tests.

primary cell A *voltaic cell* in which the chemical energy of its constituents gives rise to a current in an external circuit but in which the action is not normally reversible as in a *secondary cell*. There are many different types of primary cell. See *Leclanché cell, mercury cell*.

primary electron An electron liberated from a heated cathode by thermionic emission or from a photocathode irradiated by light or from a cold cathode by an intense electric field. See *secondary electron*.

primary storage Same as *main storage*.

printed circuit Printed wiring designed to perform a function other than interconnections between components. For example a conducting path can be given a helical form so as to have sufficient inductance for UHF tuning.

printed wiring A pattern of conductors formed on an insulating base and extensively used for the interconnection of components in electronic equipment.

The insulating base is initially given a uniform coating of copper and the required conductor pattern is then printed photographically on the surface in a material which protects the copper. The unwanted copper and the protective material are then removed by immersion in baths of suitable solvents, leaving the printed wiring as a layer of copper. Connections between the printed wiring and components can be achieved by soldering the wire leads of resistors, capacitors, transistors, etc, to the copper where they project through holes drilled in the copper. There are automatic methods of making such soldered connections. An alternative method is *surface mounting*. See *chip carrier, dual-in-line, dual-in-line package, plated through holes*.

printer A device which converts suitably coded output from a computer into hard copy printed on paper or other material. All printers reproduce *text* and most also handle *graphics*. The connection to the *host* computer is often by a parallel link, e.g. *Centronics standard*, but sometimes a serial connection is used.

The printer types in most common use are:

Daisy wheel printers A spin-off of typewriter technology, these use a printhead having a detachable *daisy wheel,* so called because the radial platens bearing the reverse impressions of the characters superficially resemble the petals of a daisy. As the printhead scans the paper the daisy wheel is rotated until the required platen is opposite the print position and then the platen is pressed against the paper through an inked ribbon, usually a plastic one-trip ribbon. The system has the advantages of excellent print quality; moreover the typeface may be changed easily by swapping daisy wheels. But the machines are noisy and incapable of reproducing graphics. They are no longer widely used.

Impact dot matrix printers These employ a print head consisting of a vertical row of closely spaced pins, usually 9 or 27 in number, each controlled by a *solenoid*. As the head scans across the paper, the solenoids are momentarily energised at intervals forcing the pins to strike the paper through an ink-impregnated fabric or plastic ribbon. These printers are suitable for draft documents and internal memos, but not for high-quality document mastering. Graphics can be printed, but inaccuracies of the print head mechanism often result in vertical lines being crooked while the inking of solid areas is often inconsistent because of uneven ribbon wear.

Line printers are specialised dot matrix printers used for long print jobs. They employ a number, sometimes 33, of print heads spaced evenly across the page. Each has a single pin. The heads are moved from side to side to allow the pins to reach every position across the page. The paper is advanced one dot position at a time. Line printers can print over 400 lines of text per minute with high reliability and low operating costs. They are used mainly by medium to large businesses for printing invoices and form letters.

Ink jet and bubble jet printers A development of dot matrix printers in which the pins and inked ribbons are replaced by tiny nozzles through which minute droplets of ink are fired at the paper. The positional accuracy and consistency of inking are greatly superior to those of impact dot matrix machines and high-quality output is obtainable. Many machines offer a resolution of 360×360 dots per inch and some offer 720×720 dots per inch. Colour ink jet machines use cyan, magenta, yellow and black inks to simulate a wide range of colours, as in commercial printing.

Laser printers These use technology developed for the xerographic photocopying process. A *laser* or other light source is used to write a reverse image of the document on an electrostatically charged drum. As the light hits the drum it neutralises the charge. The drum is then dusted with powdered ink called *toner* which is attracted into those areas of the drum surface that still bear a charge. The drum is then rolled over the paper transferring the image in toner on to it. Finally the paper is momentarily heated to melt the toner so that it adheres firmly to the paper, forming a permanent image. Output is fast and print quality excellent. Resolution varies from 300 × 300 dots per inch to 1200 × 1200 dots per inch. Colour laser printers are available, but at the time of writing this technology is still in its infancy. The only disadvantage is that the entire bitmapped image of the document has to be built up at once, usually in the printer's memory. An A4 printer at 300 × 300 dots per inch requires approximately 1 Mbyte. For this reason laser printers are sometimes called *page printers*.

Other printer types currently available use wax thermal and dye sublimation technology. These are used mainly for specialised colour work. They offer the advantage that the dot structure is invisible because the dyes diffuse into each other.

printout The *hard copy document* produced by a *printer*.

print-through In magnetic tape recording the undesired transfer of a recorded signal from one part of a tape to another part when the two parts are brought into contact. This can occur when tape carrying strong signals is wound on spools. On reproduction the transferred signals give weak and usually distorted versions of the strong signals.

probe (1) A unit containing the input stage of an electronic voltmeter or oscilloscope and connected to the main part of the instrument via a multi-core cable so permitting the unit to be positioned very close to the circuit under investigation. A probe permits very short connections to be made to the signal source being measured or examined so minimising capacitive loading of the source.

(2) A conductor of resonant length projecting into a *waveguide* or *cavity resonator* to inject or abstract RF energy.

processor See *central processing unit, microprocessor*.

product detector A circuit with two inputs which gives an output proportional to the product of the inputs. Such detectors are used for reception of CW and suppressed-carrier signals. These signals constitute one of the inputs and the other is from a carrier-insertion oscillator the frequency of which is adjusted to give a suitable audio note in CW reception and is adjusted to the (missing) carrier frequency in suppressed-carrier reception. See *continuous wave, suppressed carrier modulation*.

program For a *digital computer* a set of instructions arranged in the correct sequence to achieve the desired result. The result may be the solution of a problem or the preparation of a required set of data.

programmable logic controller (PLC) A specialised *microcomputer* intended for controlling industrial equipment and processes.

programmable read-only-memory (PROM) A *non-volatile read-only memory* which can be programmed by the user. For a memory using *bipolar transistors* this can be done by applying voltages greater than the normal operating voltages. The device has fusible links in series with one of the eletrodes in each *memory cell* and the abnormal voltages destroy the links to the chosen cells, thus programming the memory. Once this is done the device is permanently programmed and there is no means of changing the program, but see *EPROM* and *EEPROM*.

programmable terminal A *terminal* with a store capable of operating under the control of a stored program.

programmable workstation (PWS) A suitably equipped microcomputer running *terminal emulation* software to allow access to a mainframe computer. An advantage of programmable workstations over *dumb terminals* is that, in a multitasking environment, an application may be run simultaneously which performs all accesses to the *host* mainframe computer automatically, providing a user-friendly interface to the mainframe.

programme meter An instrument for indicating the volume of an audio signal. Two types of programme meter in common use are the *peak programme meter* and the *volume indicator*.

programme volume In sound broadcasting the amplitude of an audio signal as indicated by a *programme meter*.

projection television System for reproducing TV pictures in which the image formed on the screen of a small picture tube is optically projected on to a larger viewing screen. In domestic projection receivers the screen is of ground glass and the optical image is focused on the rear by an optical system incorporating a concave mirror, a plane mirror and a correcting lens designed to minimise optical distortion. The system is folded, as shown in *Figure P.25*, so as to occupy minimum space and is a version of the system

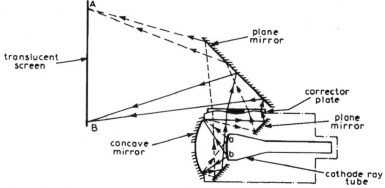

Figure P.25 Folded optical system used in projecction television receivers

originated by Schmidt for use in astronomical telescopes. For larger installations, e.g. in cinemas the optical image is projected on to the front surface of a reflecting screen.

propagation coefficient or **constant** For a wave at a specified frequency and transmitted along a *transmission line*, the natural logarithm of the vector ratio of the steady-state amplitudes of the wave at points unit distance apart.

For a *network* terminated in its iterative impedances at both ends, the natural logarithm of the complex ratio of the input current to the output current. The real part of propagation coefficient is known as the attenuation coefficient or constant and is a logarithmic measure of the ratio of the signal amplitudes. The imaginary part of the propagation coefficient is the phase coefficient and measures the phase difference between the two signals.

For symmetrical networks the real and imaginary parts of the propagation coefficient are equal to the real and imaginary parts of the *image transfer coefficient*.

protective gap Two *electrodes*, separated by a small air gap, and connected between the conductors of a *transmission line* or between a live conductor and earth to limit the voltage which can be developed. When the voltage exceeds a critical value which depends on the dimensions of the gap, a discharge occurs between the electrodes so removing the high voltage.

protocol An agreed set of rules and procedures for data transmission between two devices.

proton An elementary atomic particle with a positive charge equal to that of the electron but with much greater mass. It is one of the constituents of the atomic nucleus. See *atomic structure, electron, neutron*.

proximity effect In a conductor carrying an alternating current a concentration of the current towards the edges of the cross section caused by interaction between the conductor and the electromagnetic field set up by neighbouring conductors carrying the same current. This effect is analogous to and additional to *skin effect* and becomes more marked as frequency is raised, causing the effective resistance of the conductor to increase steeply with increase in frequency. Proximity effect is particularly marked where a number of conductors carrying the same current are in close proximity as in RF inductors and transformers.

p-type semiconductor An *extrinsic semiconductor* material in which the hole density exceeds the free-electron density. Thus the positive charge carriers outnumber the negative charge carriers. The positive charge of a hole can be neutralised by injecting an *electron* into it from a neighbouring atom in the crystal lattice. This *atom* is then left with an electron deficiency (a *hole*) which in turn can be neutralised by an electron injected from a third atom. In this way a current can be driven through the crystal under the stimulus of an applied EMF and this current can be regarded as a movement of holes (positive charge carriers) in one direction or of electrons (negative charge carriers) in the opposite direction. See *n-type semiconductor*.

public address (PA) System of sound reproduction used for diffusing speech or music to large audiences. The system includes loudspeakers and amplifiers and a source of signal such as *microphone, disk-player* or tape reproducer. Mixing facilities may also be incorporated. Such systems are widely used in public halls, factories, railway stations, airports, etc, for announcements or entertainment.

public domain software *Software* which may be copied, distributed, used or modified without any restrictions except that no one may claim copyright on it. See also *freeware, shareware*.

Puckle time base An early time base circuit in which the sawtooth voltage was generated across a capacitor charged via a *pentode* and discharged by a *multivibrator*.

pulling Same as *frequency pulling*.

pulsatance Same as *angular frequency*.

pulse A sudden departure of short duration of a voltage or current from a steady value. Often pulses approximate to rectangular form and are then

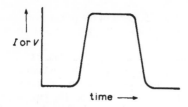

Figure P.26 A pulse waveform with appreciable rise and fall times

characterised by steep leading and trailing edges as shown in *Figure P.26*. Pulses can, however, be of sawtooth or other form. See *fall time, rise time*.

pulse amplitude modulation (PAM) Modulation in which the amplitude of the pulses in a *pulse carrier* is made to vary in accordance with the instantaneous value of the modulating signal. See *Figure P.27*.

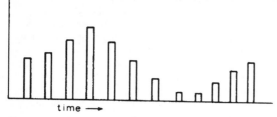

Figure P.27 Pulse amplitude modulation

pulse and bar signal In TV a test signal used for *k-rating* measurements. The signal has a duration of one line and and includes, in addition to a line sync signal, a *sine-squared pulse* and a rectangular pulse.

pulse carrier A regular train of identical pulses intended for *modulation*.

pulse code A code in which groups of pulses are used to represent information. One example is provided by the Morse code. In another example the pulses represent the quantised values of the amplitude of an *analogue signal*.

pulse code modulation (PCM) A modulation system in which the pulses constituting a *pulse carrier* are divided into groups, each group being modulated so as to represent the quantised values of the amplitude of the analogue signal to be transmitted. For example the pulses in each group can be modulated so as to represent the quantised values in binary code.

pulse counter detector A detector of frequency-modulated waves which operates by counting the number of cycles of RF in a given time interval, the variation in the totals constituting the required replica of the *modulating signal*.

pulse duration The time taken for the instantaneous value of a pulse signal to rise from and fall to a specified fraction of the peak value.

pulse duration modulation (PDM) *Modulation* in which the duration of the pulses in a pulse carrier is made to vary in accordance with the instantaneous value of the modulating signal. See *Figure P.28*.

pulse frequency See *pulse repetition frequency, pulse repetition rate*.

pulse frequency modulation (PFM) See *pulse repetition rate modulation*.

pulse interval See *pulse spacing*.

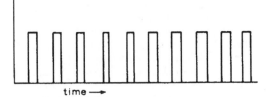

Figure P.28 Pulse duration modulation

pulse interval modulation (PIM) *Modulation* in which the spacing of the pulses in a pulse carrier is made to vary in accordance with the instantaneous value of the *modulating signal*.

pulse modulation *Modulation* of a continuous-wave carrier by pulses or modulation of a pulse carrier by another signal.

pulse position modulation (PPM) *Modulation* in which the displacement in time of the pulses in a *pulse carrier* relative to the unmodulated value is made to vary in accordance with the instantaneous value of the *modulating signal*.

pulse repetition frequency (PRF) The number of pulses per unit time. The term is used when the number of pulses per unit time is constant as in a regular train of pulses.

pulse repetition rate The average number of pulses per unit time. The term is used when the rate varies as in *pulse repetition rate modulation*.

pulse repetition rate modulation *Modulation* in which the *pulse repetition rate* of the pulses in a pulse carrier is made to vary in accordance with the instantaneous value of the *modulating signal*. See *Figure P.29*.

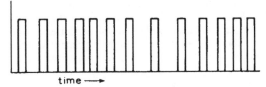

Figure P.29 Pulse repetition rate modulation

pulse shaping The process of modifying the shape of a pulse to a desired form.

pulse spacing The interval between corresponding points on two consecutive pulses.

pulse spectrum The distribution with frequency of the Fourier components of a pulse.

pulse time modulation General term for *modulation* in which the time of occurrence of some characteristic of the pulses in a *pulse carrier* is made to vary in accordance with the instantaneous value of the *modulating signal*. Thus *pulse duration modulation, pulse repetition rate modulation, pulse interval modulation* and *pulse position modulation* are all examples of pulse time modulation.

257

pulse width modulation Same as *pulse duration modulation*.

pumping (1) The process of evacuating the envelope of a vacuum tube. (2) In a parametric amplifier the supply of RF oscillation which energises the *varactor*. In a laser or maser the provision of energy in the form of light or RF which excites the atoms which then emit *radiation*.

punched card A card in which a pattern of holes is punched to represent data. A standard type of card has 80 columns, each with 12 positions at which a hole may be punched. Each column can represent a character (letter, numeral or symbol) according to the position of one, two or three holes in the 12 positions for that column. Such cards can provide the input for a *digital computer*, a punched-card reader being necessary to provide the electrical input signal for the computer.

punched tape A tape, usually paper, in which a pattern of holes is used to represent data. Each character (letter, numeral or symbol) is represented by a unique pattern of holes across the width of the tape. Commonly there are eight tracks across the width, seven used for coding the information and the eighth used for *parity checking*.

purity In colour TV the degree to which a display in one of the primary colours is free from contamination by the other primary colours.

push-pull operation A form of operation of electronic equipment in which two matched signal chains are used to carry signals in phase opposition. A phase-splitting device is used to derive the signals for the two chains and the two signals are recombined at the output. The system has the advantage of minimising even-harmonic distortion, avoids the need for polarising transformer cores and gives economy of current by permitting *class-B operation* of *active devices*.

The principles are illustrated in the circuit diagram of a typical stage of push-pull amplification shown in *Figure P.30*. The two transistors must be accurately matched in characteristics, the antiphase input signals are derived from the centre-tapped secondary winding of the input transformer and the output signals are combined in the centre-tapped primary winding of the output transformer.

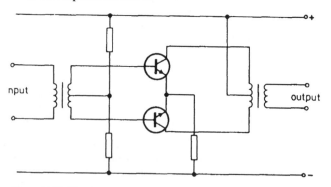

Figure P.30 Simple example of a push-pull stage of amplification

Q

***Q* factor** Of an inductor or capacitor the ratio of its *reactance* to its effective series resistance. An ideal inductor or capacitor has no series resistance and the *Q* factor is therefore infinite. Practical components inevitably have resistance and thus *Q* factors are finite: they are, in fact, a figure of merit indicating the extent to which the component approximates to a pure reactance. For inductors *Q* values of 100 to 300 are possible and for capacitors much higher values are common. For an inductor *Q* is given by $L\omega/R$ and for the capacitor $1/\omega CR$.

quadraphony In sound reproduction an extension of the stereo principle, in which two further channels are used to originate stereo sound behind the listener to give the illusion that he is surrounded by sound. Quadraphony is, in fact, sometimes called surround sound. The four channels are identified as front left, front right, back left and back right.

A number of systems for broadcasting quadraphonic sound (compatible with existing stereo and monophonic FM broadcasting) and of recording it on disk (compatible with existing stereo and monophonic disks) have been proposed and used but so far there has been no international agreement. There is, of course, no difficulty in recording the four quadraphonic channels as separate tracks on magnetic tape and such tapes are available.

quadrature The relationship between two sinusoidal signals at the same frequency when there is a 90° ($\pi/2$ radians) phase difference between them. As an example the current in a loss-less inductor or a loss-less capacitor is in quadrature with the applied EMF, the current in the inductor lagging the EMF and the current in the capacitor leading the EMF.

quadrature detector A detector used to demodulate a *quadrature modulated* signal, i.e. to abstract from a quadrature-modulated carrier replicas of the two original modulating signals. Such a detector is an example of a *synchronous detector* and is used in an NTSC or PAL television receiver to derive the two *colour-difference signals* which together constitute the *chrominance signal*.

quadrature modulation Method of transmitting two *modulating signals* independently on one carrier by splitting the carrier into two components in quadrature and using each signal to modulate one of the components. This method is used in the *NTSC* and *PAL* colour TV systems to transmit the two colour-difference signals. They are transmitted by suppressed-carrier amplitude modulation of the two quadrature components of the colour sub-carrier.

quadripole A *network* with two input terminals and two output terminals. It is sometimes termed a 4-terminal, 2-terminal-pair or 2-port network.

quality factor Same as *Q factor*.

quantising For a quantity which can have any value between certain limits the division of the range between the limits into a number of sub-ranges, any value within a sub-range being represented by an assigned value within the sub-range. The sub-ranges need not necessarily be equal. Thus when an analogue signal is quantised, the instantaneous amplitude of the resulting signals can have only a limited number of values—equal, in fact, to the number of sub-ranges originally decided. These values can be expressed, for example, in binary code and can thus be transmitted by a *pulse modulation system.*

quantising distortion (or **noise**) The distortion (or *noise*) introduced by quantising. An *analogue signal* after quantising jumps in amplitude from one quantising level to another and these jumps give rise to distortion the magnitude of which depends on the number of quantising levels employed. The greater the number of levels chosen the less is the quantising distortion but to reduce distortion to the level needed in high-quality sound reproduction requires at least 1000 levels and in practice 2^{12} (4096) are employed.

quarter-wave line (or **transformer**) A length of *transmission line* with an electrical length of a quarter of a *wavelength* at the operating frequency, used for impedance matching and for isolating at radio frequencies. Such lines are used, for example, to match the characteristic impedance Z_0 of a feeder to the driving-point impedance Z_a of an *antenna*, correct matching being achieved when the characteristic impedance of the quarter-wave section is the geometric mean of Z_0, and Z_a. If a quarter-wave line is short-circuited at one end, the other end presents an infinite impedance at the operating frequency. This isolating property enables short-circuited quarter-wave lines to be used for supporting transmission lines and antenna elements without effect on their performance.

quartz crystal Natural crystal of silicon dioxide with pronounced piezo-electric properties. Such crystals are used for controlling the frequency of oscillators where great stability is required.

quench frequency The frequency at which an *oscillation* is quenched. In a *super-regenerative receiver*, for example, the quench frequency must be ultrasonic to avoid audible interference with the received signal.

quenching The periodic suppression of an *oscillation*. The term was applied to the suppression of the primary wavetrain in the obsolete quenched-spark transmitting system and is now used to describe the process of terminating the discharge in a *Geiger-Muller tube*. It is also applied to the suppression of oscillation in a *super-regenerative receiver*.

quiescent-carrier modulation In telephony, *modulation* in which the carrier is suppressed when there is no *modulating signal* to be transmitted.

quiescent current The current taken by the *anode* (*collector* or *drain*) of an *active device* or by an equipment from the DC supply in the absence of an input signal. It is the current at the *quiescent point*.

quiescent point Of an *active device*, the position of the *operating point* when no input signal is applied. The quiescent point indicates the mean current and the mean voltage in the output circuit and so measures the dissipation within the active devices in the absence of an input signal.

quiescent push-pull operation (QPP) A push-pull circuit using two *pentodes* biased near anode-current cut off once popular in battery-operated equipment. The circuit is an example of a class B1 amplifier and its economy of HT consumption was its principal attraction.

quiet automatic gain control (QAGC) A combination of *delayed AGC* and signal suppression so designed that the audio output of the receiver is suppressed for input signals too weak to operate the AGC system. Quiet AGC has the advantage that only signals capable of operating the AGC system are received and these stand out from a quiet background. In particular the inter-station noise normally heard when the tuning control is operated is absent from a receiver with quiet AGC.

R

radar (radio direction and ranging) Equipment originally designed for measuring the range and bearing of any object by means of radio waves reflected or retransmitted from the object.

In this system a pulse of RF energy was radiated from the radar and this pulse together with the reflected pulse were displayed as blips on the screen of an oscilloscope, the distance between them indicating the range. A later development used a rotating antenna synchronised with a radial time base in the oscilloscope. By using the reflected signals to intensity modulate the oscilloscope beam it was possible to give a display in the form of a map of the surrounding area with moving objects indicated on it: a long-persistence tube was generally used for such displays so that the map persisted for the duration of the sweep. Radars of this type are particularly useful as navigational aids in ships and aircraft.

Radars are nowadays very sophisticated. In addition to the above-mentioned applications these equipments are now used for fire control of guns and for tracking the trajectory of missiles to mention only two examples.

radar beacon Same as *radio beacon*

radial-beam tube An *electron tube* in which the electron beam from a *cathode* is rotated by a magnetic field so as to contact a set of anodes arranged circumferentially around the cathode.

radiation The emission of energy in the form of electromagnetic waves or particles.

radiation counter or **radiation detector** See *Geiger-Muller tube*.

radio The use of electromagnetic waves to transmit and receive electrical signals over a distance without connecting wires. Thus radio, quite properly, embraces TV. However the term radio is often restricted to the transmission and reception of sound signals. For example receiver manufacturers commonly divide their catalogues into two parts, one describing radio (i.e. sound) receivers and the other TV receivers.

radio beacon or **radar beacon** A radio transmitter from which the signals enable a ship or aircraft to determine its position or direction of movement by use of its own receiving equipment.

radio data services (RDS) or **radio broadcast data system** (US) A system of communication in which data is transmitted in digital code by phase modulation of a subcarrier on the normal VHF FM sound broadcast transmission.

This data is ignored by conventional FM receivers but can be decoded by an RDS receiver to perform a number of functions particularly useful to users of car radios. These can include automatic tuning (to find the strongest signal for the chosen channel), instant switching to receive traffic messages on other channels and automatic switch on to a preselected programme. A visual read-out display can give the name of the received station, the time and date, and even details of music being broadcast. RDS is the sound broadcasting equivalent of *teletext*.

radio-frequency resistance Same as *effective resistance* (2).

radio microphone A *microphone* the audio output from which is used to modulate a low-power transmitter, the output of which is picked up by a nearby *receiver*. A lapel microphone is often used as a radio microphone, the miniature transmitter being carried in a pocket. This arrangement is often used where it would be inconvenient to use a cable to carry the microphone output, e.g. in TV production where an actor has a part requiring great mobility. The use of a boom microphone or a trailing microphone lead would restrict his movements but a radio microphone gives the required freedom.

raised-cosine pulse See *sine-squared pulse*. Because the square of a quantity is always positive irrespective of the sign of the quantity, a sine-squared pulse is positive for both half-cycles of the sine wave. Each sine wave thus gives rise to two sine-squared pulses and these pulses are positive, i.e. they stand above the zero axis and are thus raised. The identity

$$\sin^2\theta = (1 - \cos 2\theta)/2$$

shows that the sine-squared pulse is of cosine form and that its frequency is double that of the sine wave.

ramp generator Same as *sawtooth generator*.

random access memory (RAM) A storage device in which access to any item of data is independent of the location of the item in the store. Sometimes called direct access store. The term is used to distinguish such storage from that in which data retrieval is serial as in magnetic tape storage.

The use of a RAM is described under *digital computer*. An alternative name for the RAM is read/write memory. See also *dynamic RAM* and *static RAM*.

random noise *Noise* arising from the movements of a large number of *electrons* in a conductor (*thermal noise*) or released from a cathode (*shot noise*). Such noise has no periodicity and, if reproduced by an acoustic transducer, has the sound of a smooth hiss.

raster Display on the screen of a picture tube made up by unmodulated *scanning lines*.

raster graphic Same as *bitmap*.

rated frequency deviation The maximum value of the frequency deviation permitted in a frequency-modulated system. For FM broadcasting in Band II the rated frequency deviation is ± 75kHz and for the sound channel in 625-line TV system it is ± 50kHz.

ratio arms See *Wheatstone bridge*.

ratio detector A detector for frequency- or phase-modulated signals capable of a degree of amplitude limitation.

The circuit, a typical example of which is given in *Figure R.1*, incorporates a transformer with a tuned secondary winding, the centre tap of which is coupled to the primary winding as in the *Foster-Seeley discriminator*. However the two diodes in the ratio detector are series

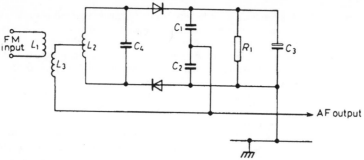

Figure R.1 Typical ratio detector circuit

aiding and feed an *RC* circuit with a low-value resistor and a long time constant. This combination imposes heavy damping on the tuned secondary circuit and acts, in fact, as a ***dynamic limiter,*** thus giving the detector a measure of protection against amplitude-modulated signals. The voltage across C_3 is independent of the frequency modulation and proportional to input-signal amplitude: it is often used to operate a tuning meter or for AGC purposes. Thus the sum of the voltage across C_1 and C_2 is constant but the individual voltages across C_1 and C_2 vary with the frequency of the input signal in the same way as in a Foster-Seeley discriminator and the audio output can be taken from across either of them. The circuit has a performance similar to that of the Foster-Seeley discriminator but has the additional advantage of a measure of amplitude limitation. It is not surprising, therefore, that this detector is widely used in FM receivers.

ray tracing A computer *application* which produces photographic-quality *bitmap* images of imaginary objects or scenes. A prerequisite is a mathematical model of the object or scene to be depicted. Details of the lighting and 'camera position' are also required. The application then builds up the image pixel by pixel by tracing a hypothetical ray of light back from each pixel to its source, taking into account the colours and other surface characteristics of the objects off which the light has been reflected. The process can be time-consuming. See also *visualisation*.

reactance (*X*) That component of the impedance of an alternating-current circuit which is due to its *inductance* and/or *capacitance*. It is the imaginary component of impedance and, for a purely-reactive circuit, the voltage and current are in quadrature. The reactance in ohms of an inductance L is given by

$$X_L = \omega L$$

and for a capacitance C

$$X_C = \frac{1}{\omega C}$$

where ω is the angular frequency.

In a purely-inductive circuit the current lags the applied voltage by 90° and for purely-capacitive circuit the current leads the voltage by 90°. For a given voltage, therefore, the currents are in antiphase and this is indicated by giving reactance a sign, inductive reactance being conventionally positive and capacitive reactance negative. The net reactance of a circuit containing inductance and capacitance in series is thus ($\omega L - 1/\omega C$).

reactance modulation *Angle modulation* brought about by the use of a *reactance* which is varied in accordance with the instantaneous amplitude of the modulating signal applied to it. A *transistor* can be made to behave as a reactance and is often used to give angle modulation, the modulating signal being applied to the base. The carrier signal is applied to the collector and, by means of an RC circuit, a signal advanced or retarded by 90° with respect to the carrier is applied to the base. The collector current is thus in quadrature with the collector voltage and the collector circuit behaves as a reactance, the magnitude of which depends on the mutual conductance of the transistor. By applying the modulating signal to the base the mutual conductance can be varied, thus varying the reactance and giving angular modulation of the carrier. In a practical reactance modulator linearity of modulation is achieved by using two transistors in push-pull with a system of *negative feedback.*

reaction *Positive feedback* used to increase the sensitivity of any form of amplifying a.m. *detector.* A typical circuit for applying such feedback is given in *Figure R.2.* The amplified modulated RF signal at the tube anode gives rise to a corresponding current in the circuit L_2C_2, the magnitude of which can be controlled by adjustment of C_2, the reaction capacitor. L_2 is coupled to L_1 and so transfers modulated RF signals to the grid circuit and, if the senses of L_1 and L_2 are correctly chosen, this increases the signal at the tube grid. If C_1 is advanced too far the circuit bursts into oscillation but if reaction is set at just below the oscillation point the gain and selectivity of the detector can be enormously increased.

reactivation Process of reviving the *thermionic emission* of the cathode of an *electron tube* or cathode ray tube. One method is to apply an

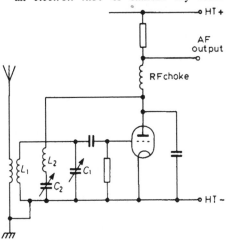

Figure R.2 A grid-leak detector with reaction applied via L_2 and C_2

abnormally-high voltage to the heater for a few minutes, a process known as flashing. An alternative approach is to increase the heater voltage permanently by say 10%. Both methods have the effect of 'boiling off' the original emitting surface of the cathode to expose a new surface.

reactive current That component of the current in an alternating-current circuit which is in quadrature with the applied alternating voltage.

read head In *computers* the electromagnetic head which reads the information from a magnetic disk or tape.

reading In *computers* and *data-processing equipment* the process of obtaining or interpreting information from a store.

read-only memory (ROM) In *computers* a *non-volatile store* the contents of which cannot be altered by instructions to the computer. Such stores are used to hold the routines needed when the computer is started up and sometimes other frequently-required programs such as those required for character generation. There are three types of semiconductor ROM:

(a) those in which the program is implanted by masking operations during manufacture and cannot be altered by the user. See *planar process*.

(b) those in which the program can be put in by the user in a once-and-for-all operation. See *PROM*.

(c) those in which the program can be put in by the user and can be altered as often as required. Erasure sometimes requires an inconvenient operation using ultra-violet light (see *PROM*) but in more recent designs can be achieved electrically (see *EEPROM*).

Also known as *fixed storage* or *permanent storage*.

real-time operation The operation of a *computer* during the time in which related physical processes occur so that the results can be used to guide the physical processes.

receiver In general a device which reproduces a signal in an intelligible form. Thus a telephone receiver reproduces an audio input signal as a sound wave. A radio receiver reproduces signals conveyed by electromagnetic waves as sound waves and a television receiver as a visual display. In hi-fi parlance a receiver is a unit containing both a *tuner* and an *amplifier*.

reciprocity theorem If an EMF introduced at one point in a linear passive *network* gives rise to a current at a second point in the network, then the same EMF transferred to the second point will give the same current at the first point.

recombination (1) In an ionised gas the loss of an *electron* by negative ions or the gain of an electron by positive ions, thus causing the gas to return to the neutral or un-ionised state. (2) In a *semiconductor* the neutralisation of holes by electrons.

reconditioned-carrier reception (US) Same as *exalted-carrier reception*.

recording Process of impressing an audio or video signal on a medium in such a way that it can be reproduced whenever required. The impression can take the form of mechanical deformation of the medium as in the lateral sound recording of records, it can be magnetic as in the audio and video recording on magnetic tape or it can be photographic on film. In all examples there is relative movement between the medium and the recording or reproducing head. To achieve such movement the medium may be in the form of a disk which is rotated beneath the heads or in the form of a tape or film which is pulled linearly past the heads.

recovery time Same as *de-ionisation time*.

rectification The process of converting electrical power in the form of alternating current into unidirectional current by use of a device with *unilateral conductivity*. A number of types of rectifiers and diodes are used for the purpose and the type chosen depends on the current and voltage required. See *full-wave rectification, half-wave rectification*.

reduced instruction set computer (RISC) A microprocessor having a limited set of instructions but capable of operating at very high speeds (in contrast with a *Complex Instruction Set Computer*). RISC processors are best suited to mathematics-intensive operations such as handling *graphics* and are used in *laser printers* and some graphical workstations. The first commercial microcomputer to use a RISC processor was the Acorn Archimedes introduced in 1987.

redundancy In information transmission that fraction of the gross information content of a message that can be eliminated without loss of essential information.

redundancy check An automatic or programmed check using extra digits inserted in the signal to help detect mistakes in the operation of *computers*.

reed relay A switch consisting of two magnetic reeds carrying electrical contacts and sealed in a glass tube containing an inert gas. The contacts can be closed by passing a current through a coil surrounding the tube or by movement of a permanent magnet near the tube. The main application of reed relays is in controlling low-level signals, e.g. in telephone exchanges.

reference level The signal amplitude chosen as a standard of reference for the measurement of signal amplitudes. Signal amplitudes are commonly expressed in decibels relative to the reference level which is taken as 0 dB. For example if the reference level for signal power is 1 mW then a power of 1 µW could be expressed as a power level of −30 dB. A reference level commonly used for signal voltage measurements in audio work is 0.775 V (which corresponds to a power of 1 mW in 600 Ω).

reference oscillator In a colour TV receiver an oscillator which is synchronised in frequency and in phase with the *colour burst* and is used to operate the synchronous detectors which decode the *chrominance signal*.

reference volume That volume which gives a reading of zero on a *volume indicator* calibrated to read zero on a steady 1-kHz tone with a voltage level of 0.775 V.

reference white level (US) Same as *white level*.

reflected binary code Same as *Gray code*.

reflection coefficient (1) Of a surface illuminated by light, the ratio of the reflected to the incident light energy. Freshly-fallen snow is a very good reflector of light, having a reflection coefficient of 0.93, whereas grass has a reflection coefficient of only 0.25.

(2) Of electromagnetic waves at the termination of a *transmission line*, the complex ratio of the reflected signal current to the incident signal current. If the line is terminated in an *impedance* equal to its characteristic impedance, signals transmitted along the line are completely absorbed at

the termination and there is no reflected wave. The reflection coefficient is therefore zero. If, however, the line has a characteristic impedance of Z_o and is terminated in an impedance of Z_r there is a reflected current and the reflection coefficient is given by $(Z_o - Z_r)/(Z_o + Z_r)$.

reflection loss (or **gain**) At the junction of a source of power and a load, and at a given frequency, the ratio of the power delivered to a load which is not matched to the generator, to the power which would be delivered to a perfectly-matched load. Generally , the ratio is expressed in *decibels* and is given by the formula $10 \log_{10} (Z_1 + Z_2)^2/4Z_1Z_2$, where Z_1 is the generator impedance and Z_2 the load impedance. See *mismatch factor*.

reflex amplification System in which an amplifier amplifies the same signal twice, e.g. before and after detection or frequency changing. The system was used in the early days of radio, one *electron tube* acting simultaneously as RF and AF amplifier.

reflex baffle Same as *vented enclosure*.

reflex detector Same as *infinite-impedance detector*.

reflex klystron A *klystron* oscillator in which the electron beam is velocity-modulated by a resonator, is reflected back so as to pass through the resonator again, so setting up oscillations the frequency of which can be controlled by the negative voltage applied to the reflector electrode.

Such klystrons can give a frequency-modulated output by applying the modulating signal to the reflector electrode (known as the repeller). Reflex klystrons are used as output stages in microwave transmitters and as local oscillators in microwave receivers. A simplified cross section of a reflex klystron is given in *Figure R.3* and the graphical symbol in *Figure R.4*.

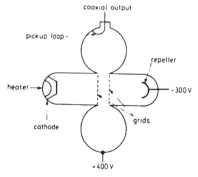

Figure R.4 Graphical symbol for a reflex klystron with indirectly-heated cathode, focusing electrode, grid, repeller and tunable cavity resonator loop-coupled to coaxial line output

Figure R.3 Simplified cross section of a reflex klystron

refraction The deflection of the path of a beam of light, radio waves or stream of electrons caused by the change in propagation velocity when the beam crosses the boundary between two different media. A familiar example of refraction is the apparent bending of the straight edge of an object when part of it is immersed in water. Refraction in this example is caused by the difference between the refractive indexes of air and water.

Radio waves suffer refraction in the upper atmosphere as they encounter regions of different electron density and as a result of such refraction waves can return to the earth's surface making long-distance radio propagation possible. Electron beams are deflected when they cross the boundary between two electric fields of different potential and this effect is exploited in *electron lenses* and *electrostatic deflection*.

refreshing In *dynamic RAMs* the automatic process of rewriting the information in each *memory cell* at regular intervals. The fundamental storage element in each cell is a capacitor which is charged to represent 1 and discharged to represent 0. The charge leaks away in time through the inevitable resistance but is regularly compared with the charge on a standard cell and is brought up to equal it if necessary.

regeneration (1) Same as *positive feedback*. (2) Of *pulses* the process of restoring them to their original amplitude, shape and timing.

register In a *central processor* a temporary *store* usually with a capacity of one word length. Processors contain a number of registers for various purposes, e.g. arithmetical functions, indexing and memory addressing.

regulation Of a power supply the variation of output voltage with variation of the current drawn from it. Regulation may also be expressed as the variation of output voltage as a result of changes in alternating voltage input or as a result of changes in temperature.

Regulation can be improved by including in series with the supply a resistance the value of which is adjusted automatically so as to give a constant output voltage in spite of variations in the current drawn from the supply or of input voltage. Many series-regulated supplies of this type exist, the series resistance taking the form of a transistor which is controlled, sometimes via a DC amplifier, by a sample of the output voltage of the unit.

Alternatively the supply may include a fixed series resistance, voltage control being achieved by an *active device* connected in parallel with the output. The current drawn by this shunt element varies automatically in such a manner that the voltage across the load circuit remains constant. Again the shunt element, normally a transistor, is fed with a sample of the output voltage.

Reinartz oscillator An oscillator in which the frequency-determining *LC* circuit is connected to the input of an *active device*, *positive feedback* being

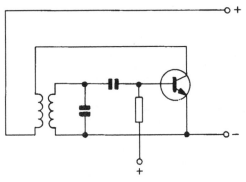

Figure R.5 A Reinartz oscillator

269

obtained via a coil connected in the output circuit of the device and coupled to the *LC* circuit. A typical circuit diagram using an npn transistor is given in *Figure R.5*.

rejection band Of a *filter*, the stop band.

rejector circuit A resonant circuit designed to present a high impedance at a particular frequency. It is connected in series with a low-impedance circuit and acts as a notch filter which attenuates signals at or near the chosen frequency. The commonest form of rejector circuit is a parallel *LC* combination. See *parallel resonance*.

relational database A *database* comprising data objects of various types which relate to each other, that is refer to each other by having common fields or clauses. Objects can be sorted on the basis of these relationships as well as types. Perhaps the best known example is IBM's DB2.

relative level See *level, reference level*.

relative permeability See *permeability*.

relative permittivity See *permittivity*.

relaxation oscillator A waveform generator the action of which is governed by one or more RC circuits, an output being delivered when the capacitors are rapidly discharged, the circuit 'relaxing' whilst the capacitors are recharged. *Blocking oscillators* and *multivibrators* are examples of relaxation oscillators.

relay In general a device which enables an input power to control a local source of power and in which there is no proportionality between controlling and controlled power. Thus *thyristors* and *thyratrons* can both be used as relays but probably the most familiar form of relay is the electromagnetic type which takes the form of a mechanical switch operated by the armature of an electromagnet. The switch may be a multi-contact type so enabling many circuits to be controlled simultaneously by energising or de-energising the electromagnet.

reluctance Of a magnetic circuit, the ratio of the magnetomotive force to the resulting magnetic flux density. It is the reciprocal of *permeance*. Reluctance is analogous to the resistance of an electrical circuit and is directly proportional to the length *l* of the magnetic circuit and inversely proportional to the cross-sectional area *A*. In fact the reluctance is given by

$$R = \frac{l}{A\mu}$$

where μ, the permeability of the magnetic material, is analogous to the conductivity of an electrical circuit.

remanence The flux density which remains in a magnetic material which has been brought to *saturation*, when the magnetising field is removed. See *Figure R.6*.

remote control Means of controlling equipment from a distance. The link between the remote-control unit and the equipment may be a cable, an ultrasonic wave or an electromagnetic wave. Many TV receivers have facilities for changing channel, adjustment of brightness, contrast, saturation and volume by remote control using an infra-red beam.

remote cut-off tube (US) Same as *variable-mu electron tube*.

repeater Equipment installed at intervals on a long communications circuit

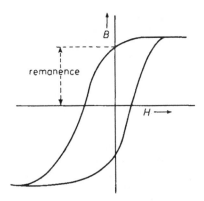

Figure R.6 A *B-H* loop illustrating remanence

to amplify and/or reshape signals to offset the effects of attenuation and distortion occurring on the circuit.

repeller See *reflex klystron*.

replay head or **reproducing head** Device for converting a sound or vision track recorded on disk, tape or film into a corresponding electrical signal.

repository In computing a specialised *database,* together with supporting *applications*, containing definitions and descriptions of the various data resources and processes in use in an organisation. Such data describing data is called *meta-data*. A repository handles not only definitions of data items, but also of the relationships between data items. Repositories evolved from *data dictionaries* and differ from them principally in their broader application; for example, they can also store diagrams of data systems and of business organisations and business processes.

Repositories are used in data administration to provide a consistent and authoritative source of definitions and to enforce standards on data types. They are widely used in software development for integrating the output of applications such as *computer-aided software engineering* tools. Increasingly they are used in business administration for such processes as business modelling and business process re-engineering. The value of a repository for such purposes lies in its comprehensive internal cross referencing, which causes any update made to one member to be reflected instantly in all other members which refer to it. Moreover, tools are usually provided which show immediately which other members would be affected by a change to one member, a process called *impact analysis*.

reservoir capacitor *Capacitor* connected in parallel with the output of a rectifier. When the rectifier conducts, the reservoir capacitor (*Figure R.7*) is charged to the peak value of the alternating input voltage to the rectifier and when the input voltage falls below the peak value, the reservoir capacitor supplies the load. In so doing the capacitor voltage falls but it is restored to the peak value the next time the rectifier conducts.

residual magnetism Magnetism which persists after the magnetising force has been removed. Residual magnetism varies enormously depending on the magnetic material. In soft iron, for example, residual magnetism is negligible whereas in steel it is so high that a single application of a magnetising field can give permanent magnetism.

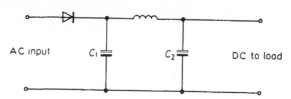

Figure R.7 A half-wave rectifying circuit with reservoir capacitor C_1 and smoothing capacitor C_2

resistance That property of a circuit which enables it to dissipate electricity as heat. It is the real component of *impedance*. For a purely-resistive circuit the voltage and current are in phase and the resistance is given by the ratio of voltage to current. The unit is the ohm; symbol Ω.

resistance-capacitance coupling In general any coupling between circuits by means of a combination of *resistance* and *capacitance*. The term is frequently applied to coupling between active devices, the resistance being the load of the first device. The capacitance couples the output of the first device to the input of the second as shown in *Figure R.8*.

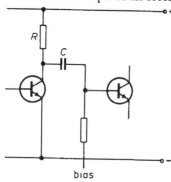

bias

Figure R.8 Resistance-capacitance coupling between cascaded transistor stages

resistance-capacitance filter A filter circuit composed only of *resistance* and *capacitance*. Simple low-pass filters composed of series resistance and shunt capacitance are often used for smoothing the output of a rectifier. Another example of an *RC* filter is provided by the *parallel-T network* of *Figure P.5* which is a *notch filter*.

resistance-capacitance oscillator An oscillator the frequency of which is determined by a network of *resistance* and *capacitance*. One example is the *phase-shift oscillator* of *Figure P.15*; another is the *Wien-bridge oscillator*.

resistivity Same as *specific resistance*.

resistor Component used in a circuit primarily because of its electrical resistance. There are many different types of resistor. Low-power resistors consist of carbon compounds imbedded in an insulating medium (the resistance value being indicated by a *colour code*) or may be thick or thin metal films on an insulating base. For high-power dissipation resistors are usually wire wound. The graphical symbol for a resistor is shown in *Figure R.9*.

—▭— *Figure R.9* Graphical symbol for a fixed resistor

resistor-transistor logic (RTL) A system in which use is made of logic elements comprising *resistors* and *transistors*.

resolution See *definition*.

resonance The condition in which an oscillating or vibrating system responds with maximum amplitude to an applied force. In an *LC* circuit resonance occurs at the frequency for which the inductive reactance is equal to the capacitive reactance. Thus

$$L\omega = \frac{1}{\omega C}$$

giving

$$\omega = \frac{1}{\sqrt{(LC)}}$$

Thus the resonance frequency is given by

$$f = \frac{1}{2\pi\sqrt{(LC)}}$$

resonant cavity Same as *cavity resonator*.

resonant circuit A circuit containing inductance and capacitance and which is therefore capable of resonance.

response curve See *frequency response*.

response time The time taken by a system to respond to an input.

results In computing, the output, usually in text form, generated by data processing operations, especially those involving a mainframe computer.

retentivity Same as *remanence*.

retrace Same as *flyback*.

retrieval In *computers* and *data-processing equipment* the process of selecting and extracting data records from a *file*.

return current (or **voltage**) In a *transmission line* the current (or voltage) wave which is set up at any impedance discontinuity and travels back to the source. The return wave combines with the forward wave to form standing or stationary waves along the transmission line.

return-current coefficient Same as *reflection coefficient* (2).

return loss The value of the reflection coefficient expressed in *decibels* or *nepers*.

reverberation time Of a room, hall or studio, the time taken for the intensity of a sound to fall by 60 dB. It is an important parameter because if speech is to be clear in a small room the reverberation time should be about 0.3 s over the audible spectrum. Similarly good acoustics in an orchestral hall require a reverberation time of about 2 s and independent of frequency.

reverse automatic gain control An AGC system in which the gain of *active devices* is reduced by use of reverse control bias. This system was extensively used with *electron tubes* which were specially designed for the purpose, having variable-mu characteristics which enabled the tubes to accept large-amplitude signals when biased back without introducing serious distortion. *Transistors* do not have such characteristics and give distortion when handling large signals in the biased-back condition. *Forward AGC* is better suited to transistors.

reverse-blocking thyristor A *thyristor* which cannot be switched to the on-state when the anode is negative with respect to the cathode. The graphical symbol for the diode type is given in *Figure R.10* and definitions of the triode types are given under *n-gate thyristor* and *p-gate thyristor*.

Figure R.10 Graphical symbol for a reverse-blocking diode thyristor

reverse-breakdown voltage Of a pn junction or a *thyristor*, that value of reverse voltage at which any increase in the voltage causes the dynamic resistance of the device to change from a high to a low value. In a pn junction this effect can result from *avalanche* or *Zener breakdown*.

reverse compatibility Of a colour TV system that property which permits a colour television receiver to reproduce a normal black-and-white television signal as a normal black-and-white picture.

reverse-conducting thyristor A *thyristor* which is inherently conductive when the anode is negative with respect to the *cathode*. Diode and triode types exist and the triode may have an n-gate or a p-gate as indicated in the graphical symbols of *Figure R.11*.

(a)   (b)   (c)

Figure R.11 Graphical symbol for reverse-conducting thyristors (a) diode, (b) triode n-gate and (c) triode p-gate

reverse recovery time Of a pn *junction* or a *reverse-blocking thyristor* the time taken for the reverse current or voltage to reach a specified value after the device has been instantaneously switched from a steady forward current by a reverse bias.

RGB colour model See *colour model*.

rheostat Old term for a *variable resistor*.

rhumbatron Same as *cavity resonator*.

ribbon cable A cable composed of insulated conductors laid side by side and moulded together to form a flat structure in which the conductors have a precise spacing (typically 0.05 in) to facilitate use of the cable with *insulation-displacement connectors*.

ribbon loudspeaker *Loudspeaker* in which the input signal is applied to the ends of a thin metal ribbon which is suspended in a strong magnetic field and acts as diaphragm, the acoustic output being increased by use of a horn. Such loudspeakers are used to radiate at frequencies above approximately 3 kHz.

ribbon microphone *Microphone* in which the electrical output is developed between the ends of a thin corrugated metal ribbon which is suspended under light tension in a strong magnetic field and acts as diaphragm. If both faces of the ribbon are exposed to the incident sound wave the microphone acts as a pressure-gradient type but if one face is screened from the sound wave then the microphone behaves as a pressure type. Ribbon microphones are widely used in radio, television and recording. See *pressure microphone, pressure-gradient microphone.*

rifle microphone   Same as *line microphone.*

ring counter   A number of *bistable circuits* connected in a closed loop and so arranged that only one can be in a particular state at any given time. On receipt of each input signal this particular state is passed on to the next bistable circuit in the ring. Thus if there are n bistable circuits, the chosen state makes one complete circuit of the loop for every n input signals and each bistable circuit gives one output signal for every n input signals.

ringing   At the output of an electronic circuit the generation of a damped oscillation following any sudden change in input signal. In a TV picture display ringing in the video amplifier gives rise to regularly-spaced vertical stripes to the right of each vertical line in the picture.

ring modulator   A *balanced modulator* consisting essentially of four diodes connected in a ring or bridge arrangement and two centre-tapped transformers. As shown in *Figure R.12* the carrier input is applied between

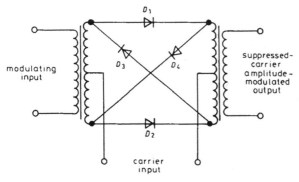

Figure R.12 Circuit diagram of a ring modulator

the centre taps on the input and output transformers and, in the absence of a modulating signal, the circuit between the transformers is completed by diodes D_1 and D_2 or alternatively diodes D_3 and D_4 depending on the polarity of the carrier. The carrier signal flows in opposite directions in the tapped transformer windings and, provided the circuit is accurately balanced, there is no carrier-frequency output.

A *modulating signal* is, however, transmitted through the circuit in an upright form during one half-cycle of carrier (when D_1 and D_2 conduct) and in an inverted form on the alternate half-cycles (when D_3 and D_4 conduct). A sinusoidal modulating signal thus emerges in the form shown in *Figure R.13*; this represents suppressed-carrier amplitude modulation. Because the carrier input acts as a switching signal its amplitude must be greater than that of the modulating signal. See *suppressed-carrier modulation.*

ring seal tube   Same as *disk-seal tube.*

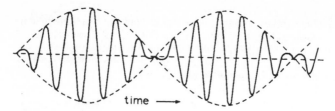

Figure R.13 Waveform of a suppressed-carrier amplitude-modulated signal for a sinusoidal modulating signal

ripple The alternating component in the output from a rectifier. The ripple frequency is equal to that of the alternating input to the rectifier if this is a half-wave type but is double the input frequency for a full-wave rectifier.

ripple filter A *low-pass filter* used to attenuate the ripple component in the output from a rectifier.

rise time A measure of the steepness of the leading edge of a *pulse*. More specifically it is the time taken for the instantaneous amplitude to change from 10% to 90% of the peak value as shown in *Figure R.14*.

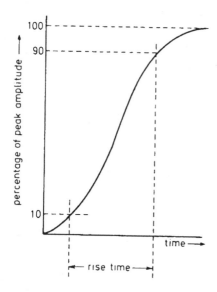

Figure R.14 Rise time of a pulse

The ability of a circuit to reproduce abrupt changes in voltage or current as required in pulses is determined by the high-frequency response of the circuit and the following simple relationship exists between rise time and upper frequency limit f_{max}:

$$\text{rise time} = \frac{1}{2f_{max}}$$

276

Thus to reproduce a pulse with a 0.1-μs rise time an amplifier requires an upper frequency limit of at least 5 MHz.

In a simple *RC* or *RL* circuit any increases in voltage or current are exponential in form and, for such a change, the rise time is simply related to the *time constant* by the approximate relationship:

rise time = 2.2 × time constant.

robot In general any machine which performs physical tasks automatically which would otherwise have been undertaken by a human operative. Robots are used widely in manufacturing industry for repetitive routine tasks such as spray painting components on a production line. They are also used in the nuclear industries and in space exploration for tasks in hazardous environments. Robots are usually remotely controlled by a computer, but some robots have built-in computers giving them limited local intelligence.

roll off Of a system or equipment the shape of the *frequency response* curve immediately outside the *passband*. For example equipment may be said to have a 12dB-per-octave roll off at the high-frequency end of the passband. High-quality audio equipment sometimes provides a choice of roll off rates, e.g. sharp or gradual at both ends of the spectrum. The shape of the roll off of an amplifier is important because it has a bearing on the degree of *negative feedback* that can be applied without instability.

root directory The top-level *directory* on a *floppy disk* or *hard disk* or other storage medium. It is created when the disk is *formatted* and must be accessed before any read or write operation relating to that storage device is possible.

Round-Travis discriminator A circuit for the detection of phase- or frequency-modulated signals containing two *LC* circuits of which the resonance frequencies are symmetrically disposed about the centre frequency and which feed diode detectors the outputs of which are

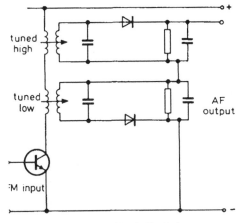

Figure R.15 A Round-Travis discriminator

connected in series opposition. The AF output is the net output of the diodes as shown in the circuit diagram of *Figure R.15*. The circuit is not in common use because others, notably the *ratio detector*, are easier to align and provide a measure of amplitude limitation.

routine A set of instructions implementing one operation in a *computer program*. The same routine may be used at many points in the program.

RS bistable A *bistable circuit* with an S (set) and R (reset) inputs. See *Figure R.16*. When the S input is at its 1-state, the output takes up its 1-state but

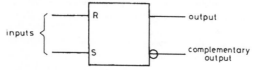

Figure R.16 Logic symbol for an RS bistable

when the S input is at its 0-state it has no effect on the output. When the R input is at its 1-state, the output takes up its 0-state but when the R input is at its 0-state it has no effect on the output. A disadvantage of the RS bistable is that if 1-state signals are applied simultaneously to the two inputs the effect on the output cannot be predicted. Bistables of this type are not used in applications where simultaneous inputs are possible. The ambiguity can be eliminated by use of a *JK bistable*.

RS 232, etc An Electronics Industries Association standard for the hardware and protocols of serial data transmission.

rumble In disk reproduction an unwanted low-frequency component usually of mechanical origin. High-quality audio equipment often includes means of attenuating such components, e.g. a bass cut facility.

S

saddle coils Coils for magnetic deflection, usually of rectangular form and shaped so as to fit the neck of a *cathode ray tube*.

safe operating area An area on the collector voltage-collector current characteristics of a power transistor within which the *operating point* should be located to avoid any possibility of overstressing the device. For safe operation of the *transistor*, the collector current and the collector voltage must not exceed certain values. Moreover the collector dissipation must be kept below a certain limit to avoid overheating. A fourth limit is set by the need to avoid second breakdown (caused by the emitter current concentrating in a particular small area of the junction). To avoid all these limits the values of I_c and V_c must be kept within a safe operating area such as that shown hatched in *Figure S.1*.

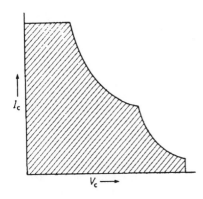

Figure S.1 Safe operating area of a transistor

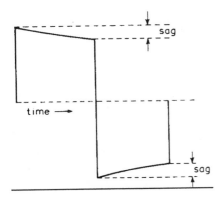

Figure S.2 Sag in the reproduction of a square wave

sag Form of *distortion* of a pulse signal in which the instantaneous amplitude falls during the period of the pulse. The extent of the fall is usually quoted as a percentage of the maximum amplitude. *Figure S.2* illustrates 10% sag. An inadequate low-frequency response is the most likely cause of sag.

sample and hold circuit A *sampling* circuit in which the instantaneous values of a sampled signal are held for a time by storing them in a capacitor. Such circuits are extensively used in *analogue-to-digital converters*.

sampling The process of obtaining a sequence of instantaneous values of an *analogue signal* at regular or irregular intervals. If the sampled signal is to represent the original at all closely, the sampling frequency must be at least twice that of the highest-frequency component in the original signal.

sanatron A circuit using two *pentodes* for producing pulses with a duration linearly related to the amplitude of a control voltage.

saturable reactor An *inductor* of which the magnetic core is operated in the region of *saturation*. If the core material is brought near to saturation by DC magnetisation, a superimposed AC can bring about large changes in inductance and hence reactance. Such devices are used in *magnetic amplifiers*.

saturation (1) In general, a condition in which further increase in input signal to a device gives no increase in output. Thus a magnetically-saturated material gives no increase in flux for an increase in magnetising force and a saturated transistor gives no increase in collector current for an increase in input signal.

(2) In colour perception the degree to which the light energy is confined to a narrow frequency band. It is thus the converse of the extent to which the colour is diluted with white light. The difference between pink and red is one of saturation.

save In data processing to store a *file* in a storage medium such as a *floppy* or *hard disk*.

sawtooth A periodic signal in which each cycle consists of a linear change followed by a rapid return to the value at the beginning of the linear change. The form of the ideal sawtooth is illustrated in *Figure S.3*. A

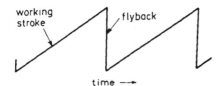

Figure S.3 A sawtooth or ramp waveform (idealised)

feature of it is that it is composed entirely of odd-order harmonics. Such signals are used extensively for feeding the deflecting systems in oscilloscopes, picture tubes and camera tubes, and also in the conversion between analogue and digital signals. The linear rise is known as the working stroke and the rapid collapse as the *flyback* or retrace.

In most generators a sawtooth voltage is developed across a capacitor by passing a constant current through it, flyback being achieved by discharging the capacitor. Such circuits may be free-running, in which case they are usually synchronised by external signals or they may be driven types, in which case external *triggering* signals are essential to drive them.

scale factor In *computers* the factor by which quantities must be multiplied to bring them within the range acceptable to the computer.

scale of ten A counting scale containing ten digits. This is the decimal scale normally used in counting. The digits in a decimal number are multiples of the power of ten. As an example the number 4062 is, if written out in full:

$$4 \times 10^3 + 0 \times 10^2 + 6 \times 10^1 + 2 \times 10^0$$

scale of two Same as *binary scale*.

scanner (1) The rotating or oscillating directional *antenna* used in *radar* systems. (2) A computer-controlled radio receiver used in communications systems that monitors each of a range of channels in rotation. (3) In computing, a *peripheral* which converts the contents of a paper document, such as a photograph or a printed page, to a *bitmap*. It contains an array of photo-sensitive cells, usually charge-coupled devices, which pass over the suitably illuminated document at a controlled speed. The output from each cell is sampled at regular intervals and assembled into the bitmap. The finished bitmap is *uploaded* to the *host* computer for viewing or editing. Text in the document may be converted to *ASCII* code using an *OCR* application.

scanning In TV the process of analysing or synthesising the light content of the elements constituting the scene. In practice this is achieved by an electron beam which moves over the target of the camera tube or the screen of the picture tube in a series of lines which embrace every element of the image.

scanning lines In TV one of the horizontal rows of elements of which the image is assumed to be composed and which is explored by the scanning beam during transmission and reception. The elements in each line are scanned in order from left to right with a rapid return to the left to start the next line below, and a rapid return to the top of the image when the bottom line has been completed. The motion of the scanning beam is, in fact, similar to that of the eye in reading. The definition of a television system is primarily determined by the number of scanning lines chosen and most modern systems use 525 or 625. Not all the lines are reproduced on receivers because a number are lost during the vertical flyback.

scanning spot In television the small area of the *target* of a *camera tube* or *screen* of a *picture tube* which is affected by the scanning beam at any instant during the scanning process.

schema The model defining the structure and organisation of a *database*.

Schmitt trigger A *bistable* device of which the input has two threshold values. The device takes up one of its two stable states when the input signal reaches a threshold value V_1 and remains in this state when the input returns through V_1 and until it reaches a second threshold V_2, at which the device switches to its alternative stable state. The input must now return through V_2 and reach V_1 again before the next change of state occurs.

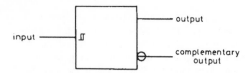

Figure S.4 Logic symbol for a Schmitt trigger

The switching operations clearly lag behind the changes in input signal and the device is sometimes described as having *hysteresis*. This is indicated in the block symbol for the Schmitt trigger shown in *Figure S.4*.

Schottky barrier diode Same as *Schottky diode*.

Schottky diode A diode using a metal/semiconductor junction. The semiconductor is usually highly-doped silicon or gallium arsenide with a thin n-type *epitaxial* layer. Conduction is entirely by *majority carriers* and there are no *minority carriers* to give reverse recovery effects. The diode thus makes an efficient mixer and detector at microwave frequencies.

Schottky diodes start conduction at a low forward voltage (between 0.1 and 0.2 V compared with 0.6 V for a silicon pn diode).

Schottky effect The increase in current from a *cathode* beyond the saturation value resulting from an increase in anode voltage. The increase is caused by the lowering of the *work function* of the cathode by the increased electric field at its surface.

Schottky noise Same as *shot noise*.

scintillation (1) Small random variations of the amplitude, phase or angle of arrival of a received radio wave caused by its passage through the ionosphere. The term is analogous to the twinkling of light from a star. (2) In an amplitude-modulated transmitter unwanted frequency modulation of the carrier. Such modulation can occur when the amplitude modulation depth is great. (3) The flash of light produced in a layer of material, notably a *phosphor*, when a charged particle strikes it at high speed.

scintillation counter Alpha particle and gamma ray counter in which the incident radiation strikes a *phosphor* to produce light which is detected by a *photocell*, the output of which is amplified to operate the counter circuit.

Scophony system System of *projection television* in which a beam of light is modulated by a *Kerr cell* and then projected on to a screen after reflection at a rotating system of mirrors (known as a mirror screw) which gives the required scanning pattern.

scrambler A device which takes an incoming signal and applies an encryption algorithm to it. The resulting scrambled signal is transmitted to the receiver which contains an *unscrambler*. This applies the reverse of the encrypting algorithm so as to reconstitute the original signal. Scramblers are used to provide security in communications systems to which the public have access.

scratch filter *Filter* giving attenuation of the higher audio frequencies and used with an AF amplifier to reduce surface noise in the reproduction of early or worn records. The noise of the reproducing stylus in the groove is negligible in modern discs but the abrasive powder formerly included in the material of records intended for reproduction by steel needles causes a hiss or scratching noise which can be reduced by use of a *low-pass filter*. If the cut-off frequency and/or the rate of fall of response are controllable it is usually possible to find a compromise setting of the controls where the hiss is not objectionable and the musical fidelity not unduly impaired.

scratch pad In *computers* and *data-processing equipment* a storage area reserved for intermediate results.

screen Of a *cathode ray tube* the layer of *phosphor* which fluoresces under the impact of the *electron beam* so as to form the required visible display.

screening The use of materials to reduce the penetration of a field into a particular region.

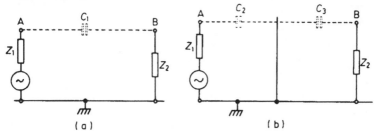

Figure S.5 (a) The capacitance between two parts A and B of a circuit and (b) interposing an earthed screen between A and B eliminates the direct capacitance between them

The principle of electric screening can be understood from *Figure S.5*(a) in which A and B represent two parts of a device or circuit. A is a generator of impedance Z_1 and B has an impedance of Z_2 to chassis. There is a capacitance C_1 between A and B determined primarily by the dimensions of A and B and their distance apart and this can convey energy from A to B. As C_1 is a small capacitance its reactance is likely to be large compared with Z_1 and Z_2, but the energy transfer increases with increase in frequency and at RF may be sufficient to cause instability if B is an early stage in an amplifier.

To prevent such unwanted feedback an earthed screen may be interposed between A and B as shown in *Figure S.5*(b). This eliminates C_1 and replaces it by two capacitances C_2 and C_3 which cannot give *feedback* provided the screen is of low-resistance material and is connected to chassis via a very low impedance. The screen need not be solid; a mesh is equally effective.

This principle was used in the *screen-grid electron tube*. The screen grid is located between anode and control grid to prevent feedback from anode to control grid. The screen must not interrupt the electron stream from cathode to anode so it is made of mesh construction and biased positively with respect to the cathode. To provide the low-impedance connection to earth the screen is decoupled by a low-reactance capacitor.

Unwanted coupling can also occur via *magnetic fields* which can induce EMFs in conductors so leading to instability. At low frequencies this can be avoided by totally enclosing the source of magnetic field in a box of high-permeability material. The stray field will then be confined to the low-reluctance path offered by the screening box. At low frequencies, such as mains and audio frequencies, magnetic screening is not often required because inductors and transformers which are likely to generate magnetic fields usually have magnetic cores which, with proper design, confine the field to the component itself so minimising stray field.

However, at higher frequencies, notably radio frequencies, unwanted coupling via magnetic fields is easily possible and the RF and IF inductors of radio receivers require screening to minimise it. The principle used here is to mount the inductors in a conducting can so that eddy currents are induced on the inner surface of the can by the magnetic field due to the current in the inductor. These currents also set up a magnetic field which opposes that due to the inductors so that, outside the can, the two fields cancel thus eliminating the possibility of instability via coupling to external components. Inside the can the two fields do not cancel. In fact the can behaves as a short-circuited turn on a transformer which reduces the effective inductance and Q factor of the inductor. To minimise this effect the can must not be placed too near the inductor: nevertheless the effect of the screening can must be allowed for in designing the inductors. There is no need, of course, for a can, acting as a magnetic screen, to be earthed but it is usual to do this so that the can acts as an electric screen also. The can should be of low-resistance material such as copper or aluminium and the thickness should be greater than the depth of penetration for the frequency used. In general the thickness required to give adequate mechanical strength is more than sufficient.

screen AC resistance See *electrode AC resistance.*

screen grid An *electrode* situated between the control grid and the anode of an *electron tube* with the object of minimising the anode-control-grid capacitance. It is effectively earthed, i.e. connected to the cathode at the operating frequency so as to screen the control grid from the anode and prevent *feedback* from output to input circuit. In practical circuits the screen grid is maintained at a positive voltage with respect to the cathode to ensure a useful value of anode current. It is decoupled at signal frequencies to the cathode by a low-reactance capacitor.

screen-grid electron tube A tube containing a *cathode*, *control grid*, *screen grid* and *anode*. The tube was developed from the triode which proved unsuitable for use as an RF amplifier because of *feedback* from output to input via the internal anode-control-grid capacitance. A screen grid was therefore inserted between these electrodes to minimise this capacitance. The screen-grid tube so developed permitted stable RF amplification but unfortunately introduced a *tetrode kink* in the I_a-V_a characteristic which prevented large undistorted anode-voltage swings. The introduction of the suppressor grid (to make the pentode) eliminated the tetrode kink.

Tetrodes can be made suitable for large output signals by critical screen-grid-anode spacing (see *beam tetrode*). The graphical symbol for a tetrode is given in *Figure S.6.*

screen mode Same as *video mode.*

scrolling Effect used in *computer applications* and in TV in which text or graphics appear to move smoothly past the screen. In word processing,

Figure S.6 Graphical symbol for an indirectly heated tetrode

desktop publishing and graphics, user-controlled scrolling is one means of navigating through the *document* until the required portion is visible on the screen. In a *graphical user interface* scrolling is usually controlled by scroll bars along the right and bottom edges of the window; their positions also give some indication of the location of the portion of the document currently visible.

search coil A small coil used for the measurement of magnetic fields. The coil is inserted in the field so as to give maximum *coupling* and is then rapidly withdrawn or the field is abruptly reduced to zero. From the EMF induced in the coil it is possible to calculate the *flux density*.

SECAM colour television system (Système En Couleurs À Mémoire). A compatible colour TV system originated in France in which the *luminance signal* is transmitted by amplitude modulation of the vision carrier and colour information is transmitted by frequency modulation of the vision subcarrier, the two chrominance signals being sent separately on alternate lines. The total bandwidth is the same as that of a black-and-white system using the same line standards. A line-period delay line and an electronic switch are needed at the receiver to ensure use of both chrominance signals on each displayed line. The system is used in France and a number of other countries. See *NTSC* and *PAL colour television systems*.

secondary cell A voltaic cell the chemical state of which, after discharge, can be restored to its original state by passing direct current through the cell in the opposite direction to the discharge current. The cell effectively stores electricity in chemical form and can be recharged as often as necessary without permanent effect on the cell. See *primary cell*.

secondary electron An *electron* liberated from a surface as a result of bombardment by fast-moving primary electrons or *ions*.

secondary emission ratio The average number of secondary *electrons* released from a surface for each primary electron striking it. Depending on the nature and treatment of the surface, secondary emission ratios up to 10 can be achieved. See *dynode, electron multiplier*.

second channel In *superheterodyne receivers* unwanted signals at the *image frequency* which are accepted together with the wanted signal by the IF amplifier. Such signals can cause interference with the wanted signals in the form of unwanted modulation or whistles. The signal-frequency circuits of receivers are therefore designed to introduce attenuation at the image frequency so as to minimise second-channel interference.

second detector In *superheterodyne receivers* the detector which demodulates the IF signal. The term arose in the early days of radio when the mixer was known as the first detector.

second generation computer See *digital computer*.

section (of a filter) See *ladder network*.

sector The smallest area on a *floppy disk* or *hard disk* that can be accessed. Whole sectors are copied into a buffer for reading and copied from a buffer for writing, even if only one byte requires changing. The size of sectors varies depending on the type and format of the disk, 1 Kilobyte being a typical size.

Seebeck effect Same as *thermo-electric effect*.

Seeley-Foster discriminator Same as *Foster-Seeley discriminator*.

see-saw phase splitter A circuit in which a push-pull output is obtained from the outputs of two active devices from an input applied to one of them, the input to the other being obtained from a tapping on a resistor connected between the two output terminals. The circuit is shown in essence in *Figure S.7* using bipolar transistors.

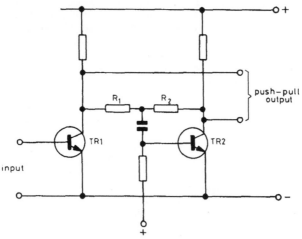

Figure S.7 Basic features of the see-saw phase-splitter circuit

If R_1 is equal to R_2 then the outputs of TR1 and TR2 cannot be of equal amplitude otherwise there would be no input for TR2; in such circumstances the output of TR2 automatically adjusts itself to less than that of TR1 so that the input for TR2 is generated at the junction of R_1 and R_2. However, if the ratio of R_1 to R_2 is made equal to $A:(A + 1)$, where A is the voltage gain of TR1 (and TR2) then equal-amplitude signals can be obtained from the two collectors.

selectance A measure of the fall in response of a resonant circuit as frequency departs from the resonance value. See selectivity.

selectivity That property of a *tuned circuit*, a tuned amplifier or a receiver which enables it to accept a wanted signal on a particular frequency and to reject signals on nearby frequencies. The phenomenon of *resonance* implies selectivity and the selectivity of an *LC* circuit is measured by its *Q factor*. The selectivity of an amplifier or receiver depends primarily on the number of tuned circuits employed and on their *Q* factors. The superior selectivity of the superheterodyne receiver compared with that of the tuned radio-frequency receiver is due to the fact that a large number of tuned circuits can be used without making tuning difficult.

Figure S.8 The self-capacitance of an inductor shown in dashed lines

self-capacitance The capacitance between parts of the same conductor, resistor or inductor. It is usually represented as being in parallel with the component as shown in *Figure S.8* in which the self-capacitance of an inductor is shown in dashed lines. The self-capacitance of an inductor sets an upper limit on the frequency to which the inductor can be tuned.

self-demagnetisation In a material in which the magnetic circuit is not closed, the reduction of the magnetic field strength by the poles produced at the ends of the magnetic circuit. The self-induced field due to the poles opposes the magnetising field and reduces the magnetisation it is possible to induce in the material.

self-inductance See *inductance*.

self-oscillating mixer In a *superheterodyne receiver* an *active device* used as a mixer which is designed to oscillate also thus obviating the need for a separate oscillator.

A typical circuit diagram of a transistor self-oscillating mixer used in a medium-wave receiver is given in *Figure S.9*. The **coupling** between L_1, L_2 and L_3 causes the transistor to oscillate. The RF input is applied to the base so that it is, in effect, connected in series with the oscillation injected into the emitter circuit. Mixing occurs by virtue of the non-linearity of the transistor characteristic and the mean emitter current must be chosen with care, being low enough to give adequate non-linearity yet high enough to sustain oscillation over the whole of the **waveband**.

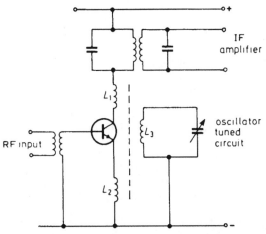

Figure S.9 Simplified circuit diagram of a self-oscillating mixer

self-oscillation The production of sustained oscillations in an electronic circuit caused by *positive feedback*.

self-quenching oscillator Same as *squegging oscillator*.

semiconductor Generally any material with an electrical conductivity between that of an *insulator* and a *conductor*. Specifically a crystalline material the conductivity of which increases on addition of certain impurities, the conductivity increasing with rising temperature over a particular temperature range as shown in *Figure S.10*. The semiconductor materials most used in the manufacture of junction diodes and transistors are germanium and silicon but others such as gallium arsenide are used also notably in light-emitting diodes. The term is also loosely used to mean semiconductor devices such as diodes and transistors. See *extrinsic semiconductor, intrinsic semiconductor, n-type semiconductor, p-type semiconductor*.

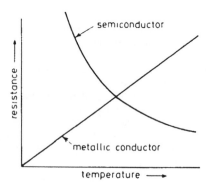

Figure S.10 Resistance-temperature characteristics for a metallic conductor and a semiconductor

semiconductor diode A two-electrode semiconductor device with unilateral conductivity.

There are two basic types described under *junction diode* and *point-contact diode*.

sensitivity In general of equipment, the smallest input signal which gives a specified output or *signal-to-noise ratio*. For example, the sensitivity of a radio receiver is the smallest RF input which gives a specified signal-to-noise ratio at the output, the modulation frequency and depth of modulation being specified. As another example the sensitivity of a television camera tube is the quotient of the output current and light input at a specified wavelength. The sensitivity of a measuring instrument is the ratio of the deflection to the input causing it.

sensor A *transducer* whose output, usually electrical, is used as the input to measuring or monitoring equipment.

separation In twin-channel stereo the degree to which the signal in one channel is free from interference from the signal in the other channel. More specifically when a signal is applied to one channel, the level of that signal which appears in the other channel, the separation being the difference in the two levels expressed in dB. In stereo pickups it is difficult to design a mechanism which can separate the two components of the stylus movement without mutual interference and it is therefore useful to know what degree of separation is possible.

sequencer In electronic music a machine, often a computer running a suitable *program*, which plays a series of *sampled* sounds in a previously defined order, thus providing a musical part or even a whole performance.

sequential scanning Scanning system in which all the lines composing the image are scanned in succession during each vertical downward sweep of the scanning agent. It is therefore non-interlaced scanning. In all practical TV systems *interlaced scanning* is used because of the economy in bandwidth that it gives.

serial storage (access) Storage in which the data are arranged in sequence in the storage medium so that the access time for a particular item depends on its position. See *parallel storage*.

serial transmission In data transmission a system in which the elements of a signal are sent in sequence over a single channel. See *parallel transmission*.

series connection A method of connecting components or circuits so that they share the same current, the voltage dividing between them depending on their impedance. See *Figure S.11*. The equivalent impedance Z_{eq} of a number of impedances Z_1, Z_2, Z_3, etc connected in series is given by

$$Z_{eq} = Z_1 + Z_2 + Z_3 + \text{etc.}$$

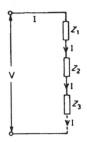

Figure S.11 A simple series circuit

series modulation Circuit for *amplitude modulation* in which the *modulator* and *modulated-amplifier* electron tubes are connected in series across the HT supply. The modulated amplifier acts as load for the modulator and the modulation frequency voltage generated across it effects modulation. The circuit (shown simplified in *Figure S.12*) has the advantage of requiring no modulation choke or transformer but a disadvantage is that the filament of the modulator, and the circuit which feeds it, are at high tension.

series-parallel connection Form of connection in which branches formed by series-connected elements are connected in parallel or in which groups

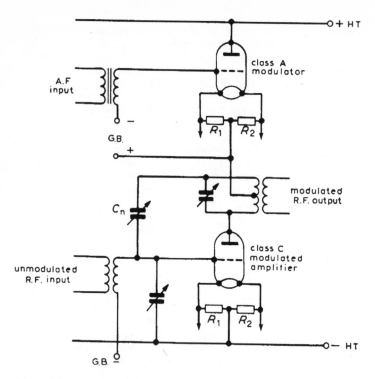

Figure S.12 Essential features of the series modulation circuit

formed by parallel-connected elements are connected in series. Two simple examples are given in *Figure S.13*.

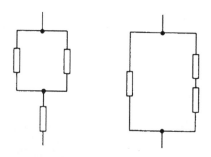

Figure S.13 Simple examples of series-parallel connection

series regulator See *regulation*.
series resonance Resonance in which the applied signal is connected to an inductor and capacitor in series. See *Figure S.14*. Resonance is indicated by

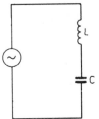

Figure S.14 Series resonant circuit

the impedance of the *LC* circuit which is a minimum at the resonance frequency given by

$$f = \frac{1}{2\pi\sqrt{(LC)}}$$

as for parallel resonance. See *acceptor circuit*.

series stabilisation See *regulation, stabilisation*.

service area Of a broadcast transmitter the area in which the field strength exceeds a specified value.

servo A system which provides automatic help with an operation or adjustment. For example, the playback head of a video tape recorder is guided along the magnetically recorded tracks by a servo which automatically corrects any deviation from the track.

set-up (US) Same as *pedestal*.

seven-segment display A method of displaying the numerals from 0 to 9 by illuminating two or more elements out of seven arranged in the form shown in *Figure S.15*. If all the elements are illuminated the display resembles 8; if elements 1, 2, 7, 5 and 4 are illuminated a 5 is displayed. This form of display using *light-emitting diodes* or *liquid-crystal* elements is extensively used in electronic equipment, e.g. video cassette recorders, calculators, digital watches.

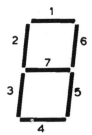

Figure S.15 Arrangement of elements in seven-segment display

shading signals In TV, components in the output of a camera tube which give rise to undesired shading effects in reproduced images. High-velocity camera tubes generate such signals and, in the early days of TV, these signals were neutralised by adding to the camera output specially-generated waveforms at line and field frequencies (known as tilt and bend waveforms).

shadow-mask picture tube A colour picture tube with three *electron guns* (or a single gun firing three beams) in which the screen is composed of dots or stripes of red, green and blue phosphors, a shadow mask near the screen ensuring that each beam always strikes the same colour phosphor. In some tubes the phosphor dots are arranged in groups of three in a triangular pattern known as a triad (see *delta-array colour tube*) and in others as vertical stripes (see *precision-in-line colour tube*).

shaft encoder A *transducer* for indicating the rotational position of a shaft.

Shannon A unit of information in a binary scale.

shape factor In receiver design a measure of the extent to which the shape of the selectivity curve approximates to the ideal rectangular shape. More specifically it is the ratio of the attenuation band to the pass-band, i.e. the ratio of the bandwidth between the 60-dB attenuation points to that between the 3-dB attenuation points (see *Figure S.16*). The ideal selectivity curve has a shape factor of unity and a communications receiver is likely to have a value between 1 and 1.5, a normal domestic receiver having a value between 2.5 and 4.0.

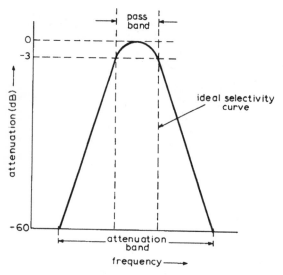

Figure S.16 Ideal and practical receiver selectivity curves

shaping network A *network* designed to modify the *waveform* of a *signal*. The simplest and most common forms of shaping network are *differentiating* and *integrating circuits*.

shareware *Software* which may be copied and distributed subject to the conditions stated in a text file which forms part of it. Usually these stipulate that the software must not be modified in any way or incorporated into any other package. The user may freely use the software for a while, but if he wishes to continue using it he is urged to send a payment to the originator.

This payment often results in a more complete or more recent version being sent to the user. See also *freeware, public domain software*.

sharp cut-off tube An *electron tube* in which the I_a-V_g characteristic is straight so that anode current is sharply cut off as negative grid bias is increased. Cf *variable-mu electron tube*.

shielding (US) Same as *screening*.

shift register In digital equipment a *store* in which an ordered set of characters can be moved to the left or right. In particular a row of bistable elements into which a binary word can be fed by serial input to the first bistable and then proceed to the next and so on, at a speed determined by the clock.

shock excitation In a resonant circuit the generation of *oscillations* at its natural frequency by the sudden application of an external signal or the sudden release of energy stored in the system.

For example an *LC* circuit can be shock excited into oscillation by the momentary application of an external EMF. Alternatively, if the capacitor is initially charged, shock-excited oscillation occurs if an inductor is suddenly connected across the capacitor.

short circuit An intentional or accidental connection between two points in a circuit by a conducting path of low resistance. If there is normally a difference in potential at the two points, a short circuit can result in an abnormally-large current flow.

short-circuit impedance Of a two-terminal-pair *network* the *impedance* measured at one pair of terminals when the other pair is short circuited.

shot noise *Noise* arising from random variations in the emission of *electrons* from a *cathode*. The number of electrons liberated from a cathode may be constant when they are measured over appreciable periods but there are momentary surfeits or deficits in the number measured over short periods and their variation with time constitutes shot noise. The magnitude of the shot-noise current I_n is given by

$$I_n{}^2 = 2eI_o\Delta f$$

where e is the charge on the electron
I_o is the current from the cathode
Δf is the bandwidth over which the noise is measured.

Substituting the value for e gives

$$I_n = 5.45 \times 10^{-4}\sqrt{(I_o\Delta f)}\,\mu A$$

This is the noise component in the total emission from the cathode, i.e. it would be the value if every emitted electron were collected by say the anode of an *electron tube*. In practice the anode current is limited by the space charge surrounding the cathode to about two thirds of this value. Thus, as an example, if I_o is 10 mA and Δf is 5.5 MHz the noise current is given by

$$I_n = 0.67 \times 5.45 \times 10^{-4}\sqrt{(10^{-2} \times 5.5 \times 10^6)}$$
$$= 0.086\,\mu A$$

and, if the anode load of the tube is $2\,\text{k}\Omega$ the noise voltage generated across it is $171.2\,\mu\text{V}$.

shunt connection The connection of a circuit element in parallel with another so diverting (shunting) some current from it. The term shunt is thus synonymous with parallel but often it is reserved for applications in which one end of the shunt element is connected to earth as in filter circuits.

shunt feed Same as *parallel feed*.

shunt regulator See *regulation*.

sideband A band of frequencies containing the new components introduced by modulation of a carrier wave. The term is also used to denote one particular component within a *sideband*: strictly such a component should be termed a *side frequency*. Amplitude modulation and angle modulation of a carrier wave result in the generation of numerous pairs of side frequencies symmetrically disposed about the carrier frequency. These components occupy two frequency bands, one above and the other below the carrier frequency as shown in *Figure S.17*. These are the upper sideband and the lower sideband respectively.

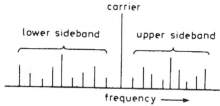

Figure S.17 Spectrum diagram showing the side frequencies constituting the two sidebands

side frequency One of the new components generated by the process of *modulation*. When a sinusoidal carrier of frequency f_1 is amplitude modulated by a signal of lower frequency f_2, two new components with frequencies $(f_1 + f_2)$ and $(f_1 - f_2)$ are generated. Thus the two side frequencies are situated symmetrically above and below the carrier frequency and are spaced from it by an interval equal to the modulating frequency as shown in *Figure S.18*. This shows the amplitude of the two side-frequency components as equal to one half that of the carrier: this is true for 100% modulation.

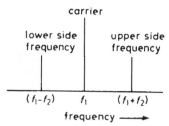

Figure S.18 Spectrum diagram illustrating the pair of side frequencies generated by 100% amplitude modulation by a sinusoidal signal

294

When the modulating signal is complex, as in sound broadcasting, each component in it gives rise to a pair of side frequencies so that two patterns of side frequencies are produced symmetrically disposed about the carrier. The spacing between the adjacent side frequencies is equal to the fundamental frequency of the *modulating signal*.

Side frequencies are also generated in *frequency modulation and phase modulation* but here a sinusoidal modulating signal produces several side-frequency pairs some of which can have an amplitude greater than that of the carrier as shown in *Figure M.10*. The side frequencies are symmetrically disposed about the carrier frequency and have a spacing equal to the modulation frequency. A complex modulating signal thus gives rise to a very large number of side frequencies theoretically extending to an infinite frequency. In practice, however, the amplitude of the outer side frequency components is small compared with that of the others and these extreme components can be filtered out without introducing appreciable distortion. Even so a frequency-modulated signal occupies a considerably greater bandwidth than the comparable amplitude-modulated signal.

signal In electronics an electrical quantity varying with time so as to convey information. In equipment, signals are usually in the form of varying voltages or currents but in the link between a radio transmitter and receiver signals have the form of modulated electromagnetic waves.

signal distance The number of digit positions in which the corresponding digits of two equal-length binary words differ. For example the signal distance between the two words 11100101 and 00100011 is 4.

signal generator Equipment for producing voltages or currents of controllable waveform, amplitude and frequency and used for the testing and alignment of electronic equipment. For receiver testing, signal generators are available giving modulated RF outputs.

signal processing or **signal conditioning** (US) The operations it is necessary to carry out on a *signal* to achieve the desired result. Typical operations are amplification, frequency changing, waveform shaping, mixing and gating.

signal-to-noise ratio In general the ratio of the power in the wanted signal to that of the noise which tends to interfere with it or mask it. It is usually expressed in dB. The precise way in which the *signal* and the *noise* are measured depends on the nature of the signal and the noise. For example RMS measurements are used for sound signals and random noise, peak-to-peak measurements for TV signals and impulsive noise.

Signal-to-noise ratio is a most important parameter of a communications system because system efficiency depends, in the final analysis, on the signal-to-noise ratio obtainable. Systems differ considerably in signal-to-noise ratio. For example amplitude limiting in FM systems permits a better signal-to-noise ratio than AM systems. In systems using *pulse code modulation* the ability to regenerate clean pulses from received pulses which are only marginally above the noise level makes possible very good signal-to-noise ratios.

silicon A tetravalent element (atomic number 14) widely used in the manufacture of semiconductor devices. In its pure state it is an insulator but by suitable doping, p-type or n-type conductivity of a value suitable for semiconductor device manufacture can be obtained. At normal temperatures leakage current is negligible in silicon devices. For this reason and also because silicon is more readily obtainable (in the form of its oxide it forms about 25% of the earth's crust) it has virtually displaced *germanium* as the basic material used for semiconductor devices.

silicon alloy transistor An *alloy transistor* in which the base wafer is of silicon. The usual alloying element was aluminium but the transistors were not very successful primarily because of the difficulties caused by the different coefficients of expansion of silicon and aluminium.

silicon chip Popular name for silicon *integrated circuit*.

silicon controlled rectifier (SCR) Same as *thyristor*.

silicon controlled switch Same as *tetrode thyristor*.

silicon disk A bank of *random access memory* configured to simulate a *floppy* or *hard disk* and used for temporary storage of programs and data.

silicon epitaxial planar transistor A *transistor* manufactured by the *planar process* and making use of *epitaxy*.

silicon on sapphire (SOS) An example of a technology in which *monolithic integrated circuit* elements are fabricated in silicon on an insulating (sapphire) *substrate*. In this way the leakages and parasitic capacitances can be minimised so that the operating speed of a store fabricated in this way using *field-effect transistors* can approach that of a bipolar store.

silicon target vidicon A *vidicon* in which the *target* is principally composed of *silicon*. Silicon has high photoconductivity but its low resistance is an embarrassment. It is used in the form of a mosaic of individually-insulated pn diodes (with boron as the impurity) which are reverse-biased in normal use.

simplex operation A method of operation in which communication between two stations can take place only in one direction at a time. To effect two-way communication the direction of the link must be reversed. This can be achieved, in a radio system for example, by simultaneously switching the antennas at both ends from receiver to transmitter or vice versa. The same carrier frequency can be used for both directions. See *duplex operation*.

simulation In general, representation of one device by another. The second device must have characteristics similar to those of the first and is used in preference to the first because it is more convenient to use or easier to understand. For example an *LCR* network may be used to simulate the impedance of an antenna.

sinad ratio In mobile communications the ratio of the wanted signal, together with any distortion terms and noise to the distortion terms and noise. It is usually expressed in dB thus:

$$\text{sinad ratio} = 20 \log_{10} \frac{s + n + d}{n + d}$$

where s = wanted signal voltage
d = distortion terms voltage
n = noise voltage.

sine-squared pulse A single unidirectional pulse equal to the square of a sine wave and used for testing in TV. Such a pulse has a limited spectrum and the pulse duration can be so chosen that it is well suited for testing TV circuits. It is used in *k-rating* determinations. See *raised cosine pulse*.

singing In a *transmission system* unwanted sustained oscillation caused by *positive feedback*. The term is mainly used in telephony. In electronics the effect is usually termed *instability*.

singing point Of a system with *positive feedback* the point at which the gain of the system is just sufficient to cause sustained oscillation. The term has a more precise meaning in telephony. See *Dictionary of Telecommunications* (Butterworths).

single-ended push-pull A circuit in which a pair of *active devices* operating in push-pull are connected in series across the supply, the output being taken from their common point. The essential features of the circuit are illustrated in *Figure S.19*. If the two *transistors* are of the same type, as shown in (a), then the input signals applied to their bases must be in push-pull and a phase-splitting stage is necessary. If, however, the transistors are complementary, as shown in (b) they can both be driven by the same input signal and a separate phase-splitting stage is unnecessary.

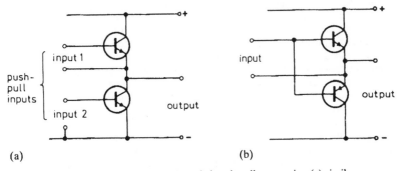

(a)   (b)

Figure S.19 Essential features of single-ended push-pull stage using (a) similar transistors and (b) complementary transistors

single in-line memory module (SIMM) An *integrated circuit* containing, usually, dynamic RAM. SIMMs are widely used in microcomputers where they make memory upgrades simple. To upgrade from 4 Mbytes to 16 Mbytes, e.g. might require the removal of four 1-Mbyte SIMMs and their replacement by four 4-Mbyte SIMMs.

single-shot circuit (US) Same as *monostable circuit*.

single-sideband transmission (SSB) A system of *amplitude modulation* in which only one *sideband* is transmitted, the carrier and other sideband being suppressed. The system has the advantage that it occupies only half the bandwidth of the corresponding double-sideband system. Reception is more difficult, however, because the missing carrier must be reintroduced at the receiver to permit detection.

sink A device which drains energy from a system. For example a heat sink is attached to electronic components with considerable dissipation to remove heat from them.

sinusoidal waveform A waveform in which the variable quantity is a sine or cosine function of time. For example a sinusoidal voltage waveform can be represented by the expression

$$v = V_0 \sin \omega t$$

where v is the voltage at a time t, V_0 is the amplitude of the signal and ω is the *angular frequency*.

skiatron Same as *dark-trace tube*.

skin effect In a *conductor* carrying an alternating current a concentration of the current towards the edges of the cross section caused by interaction between the conductor and the electromagnetic field set up by it. The effect becomes more marked as the frequency of the current is raised and causes the effective resistance of the conductor to increase sharply with increase in frequency. At high radio frequencies so little of the current is carried by central areas of the cross section that a tube is as good as a solid conductor of the same diameter. It is common practice, in fact, to use copper tubing for the tuning inductors of radio transmitters. Skin effect can also be minimised by using a conductor consisting of a number of individually-insulated strands such as Litzendraht. See *penetration*.

skirt Of the amplitude-frequency response curve of a resonant circuit, those parts of it away from the peak where the two halves of the curve diverge.

slant polarisation See *elliptical polarisation*.

slew rate (1) Of a power supplier a measure of the speed with which the output voltage compensates for changes in the current drawn. It measures the maximum rate of change of voltage across the output terminals. (2) Of a differential amplifier the rate of change of the output voltage when a step signal is applied to the input terminals.

slicing The process of suppressing those parts of a signal waveform which lie outside two amplitude boundaries, allowing only those parts which lie between the boundaries to pass on. The process is illustrated in *Figure S.20*.

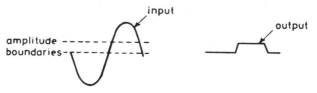

Figure S.20 Input and output waveform illustrating slicing

slope Same as *mutual conductance*.

slope detector A detector for *phase-* or *frequency-modulated signals* which depends for its operation on the slope of the skirts of the amplitude-frequency response curve of a resonant circuit.

If the difference between the resonance frequency and the centre frequency is suitably chosen the frequency variations are converted to amplitude variations (*Figure S.21*) which can be demodulated by an AM detector. Unfortunately the amplitude-frequency relationship of the skirts is not linear and the slope detector inevitably gives considerable distortion. This can, however, be minimised by using the skirts of two resonant circuits in a push-pull arrangement as in the *Round-Travis discriminator*.

298

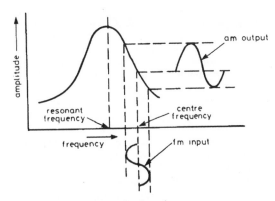

Figure S.21 Operation of a slope detector

slope resistance Of a non-linear resistance the ratio of a small change in applied voltage to the resulting change in current. It is measured by the reciprocal of the slope of the current-voltage characteristic at the operating point and in general depends on the position of that point. For example in *Figure S.22* the slope resistance is high at A, and low at B and infinite at C.

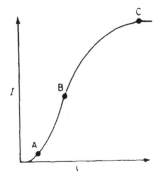

Figure S.22 Characteristic for a non-linear resistor showing variation of slope resistance with the position of the operating point

slow-wave circuit (US) Same as *slow-wave structure*.

slow-wave structure In *travelling-wave tubes* a structure used to reduce the axial velocity of electromagnetic waves to a value equal to the electron-beam velocity to permit the interaction which leads to amplification. The usual form of slow-wave structure is a helix, the ratio of the pitch to circumference being chosen to give the required axial wave velocity.

slug (1) A small cylindrical plug of permeable or conducting material introduced into a coil to adjust the *inductance*. A permeable slug increases inductance whereas a conducting slug (acting as a short-circuited turn) reduces it. If the slug and coil former are threaded, inductance can be adjusted by screwing the slug in or out, so making slug tuning possible.

(2) A short-circuited winding placed on the core of a relay to introduce a time lag in its operation.

small-scale integration (SSI) See *monolithic integrated circuit*.

Small Computer Systems Interface (SCSI) A standard for linking computers to *peripherals* such as printers and scanners. The advantage of SCSI over other standards is that a number of peripherals can be *daisy-chained* together so that they share the same *port* on the computer. This reduces the requirements for ports on the computer and can also save cabling costs.

smart A vague term used to describe almost any kind of device which, with the aid of electronics, has acquired some facility for automatic control.

smart card A device which looks like a credit card, but in fact contains electronic circuitry. When this is connected to a suitable host device, it may provide a variety of services. It may, e.g. carry confidential information about the card's owner, such as a full medical record. Alternatively it may contain details of the owner's bank account, which are automatically updated as transactions take place.

smearing In TV, blurring of the verticals in a reproduced image.

Smith chart or **Smith diagram** Same as *circle diagram*.

smoothing The attenuation of the ripple component in the output of a rectifier.

smoothing circuit Same as *ripple filter*.

sneak current An unwanted current that flows in a circuit, usually as a consequence of poor circuit design rather than of defective components.

soak testing Of a component or equipment, a prolonged period of operation under normal conditions. The test is often used as a means of detecting intermittent faults, diagnostic instruments being connected to the equipment to call attention to the faulty condition.

soft keys Same as *function keys*.

soft tube See *gas-filled tube*.

software For *computers* and *data-processing equipment* the programs, routines, data and on-line documentation required for their operation. See also *hardware*.

solar battery or **solar cell** Device for the direct conversion of solar energy to electrical energy. Such devices commonly use the photovoltaic effect of silicon diodes and a single cell gives an EMF of about 0.5 V. A series-parallel arrangement of cells is necessary to provide a battery with a useful output voltage and current. Solar batteries are used to power telephone repeaters in isolated situations and to power space vehicles.

solder An alloy of tin and lead used for making permanent low-resistance joints between metal surfaces. The solder must 'wet' the surface for a satisfactory joint and a flux is used as a cleaning agent.

solenoid An electromagnetic device consisting of an inductor with a hollow core in which a bolt of ferromagnetic material, the *armature*, is free to move. The magnetic field set up by an electric current in the inductor attracts the armature deeper into the core of the inductor. When the current stops flowing and the magnetic field disappears, the armature may return by spring action or may remain in its present position until another inductor in line with the first is energised and draws it away. Solenoids form part of electromagnetic *relays* and are widely used in the electrical and electronic control of machinery and industrial processes.

solid state Adjective used to describe a device, the operation of which depends on the structure and properties of a solid, particularly a *semiconductor*.

sound Subjectively, the sensation produced when the ear responds to vibrations within a certain frequency range. Objectively, the vibrations themselves which are propagated as longitudinal waves in gases, liquids and solids. See *audio frequency*.

soundblaster The popular name applied to the sound synthesis cards fitted in some IBM *PC-compatible* computers to provide the sound effects used in games and *multimedia* presentations.

sound cell A piezo-electric element used in *microphones* and consisting of two crystal *bimorphs* (bender types) mounted back-to-back with a small air space between them. The construction of a sound cell is illustrated in *Figure S.23*. A number of such elements may be connected in parallel to increase microphone sensitivity.

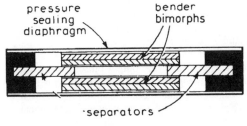

Figure S.23 Construction of a sound cell

source Of a *field-effect transistor* the connection to the channel from which majority carriers enter the channel. It corresponds with the *emitter* of a *bipolar transistor* and the cathode of an *electron tube*.

source AC resistance See *electrode AC resistance*.

source code A *computer program* in the form from which it is *compiled* or *assembled* into machine language (the *object code*). See *computer language*.

source follower The *field-effect-transistor* equivalent of the *emitter follower* and the *cathode follower*.

space charge In general the net electric charge due to a collection of *ions* or *electrons* in space. In an *electron tube* the accumulation of electrons in the region between the cathode and the control grid. The electrons leave the heated cathode with an initial velocity which is soon reduced to zero by the negative charge on the grid and the positive charge on the cathode caused by the loss of electrons. Some electrons return to the cathode but others are being constantly liberated so that a cloud of them gathers in the grid-cathode space. A similar space charge occurs in a pentode between the screen grid and the suppressor grid. See *virtual cathode*.

spark Conduction of an electric current of high density and of short duration through an ionised gas. See *arc*.

specific inductive capacity (SIC) Same as *permittivity*.

specific resistance Of a material, the *resistance* between opposite faces of a cube of the material of 1-cm side. The unit is the ohm-cm but for most good conductors the specific resistance expressed in this unit is very small and it is more convenient to use the micro-ohm-cm. For example the specific resistance of copper is 1.66 μΩ-cm.

spectrum (1) A continuous range of frequencies for which the waves have a specified common characteristic. For example the audio spectrum occupies the range between 20 Hz and 20 kHz approximately, the common characteristic being that the human ear can respond to waves of such frequencies. See *electromagnetic waves*. (2) Of a complex signal, the distribution of the components with frequency. For example a rectangular wave has a spectrum consisting of uniformly-spaced components, the amplitudes of which decrease with increasing frequency.

spectrum analyser An instrument which resolves complex waveforms into their components and displays the result on the screen of a cathode ray tube, the horizontal axis representing *frequency* and the vertical axis the *amplitude*.

speech coil Of a *moving-coil loudspeaker* or *microphone*, the coil which is attached to the diaphragm and to which the audio-frequency input is applied or from which the audio-frequency output is taken.

speed-up capacitor Capacitor bridging the coupling resistor in a multivibrator circuit to improve the rise time of the output pulses. In a *bistable* circuit as shown in *Figure S.24* the edges of the output waveforms are not as steep as they could be because the inter-transistor couplings consist of the resistors R_3 and R_4 which, in conjunction with the input capacitances of the transistors, introduce a high-frequency loss. This can be offset by bridging R_3 and R_4 by capacitors C_1 and C_2 which, if of a suitable capacitance, can make the coupling circuit aperiodic.

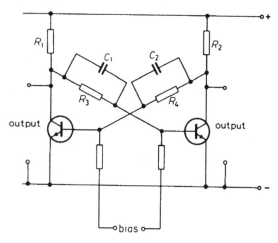

Figure S.24 Speed-up capacitors C_1 and C_2 in a bistable multivibrator circuit

spherical aberration In an optical lens or mirror, lack of definition in the image caused by rays emanating from a point in the object failing to meet at a point in the image area. The defect is caused by use of spherical surfaces in the lens or mirror: it could be eliminated by use of parabolic surfaces.

spherical wave A wave of which the front is a spherical surface. A wave originating from a point source has such a front, the source being the centre of the spherical surface.

spike A transient of short-duration which occurs during the reproduction of a *pulse.*

spiral scanning A form of *scanning* in which the spot on the screen of a *cathode ray tube* moves around the centre of the screen at constant angular velocity gradually moving inwards or outwards in a spiral path, with a rapid return to the edge or centre at the end of the scanning stroke.

split-beam oscilloscope Same as *double-beam oscilloscope.*

spot In *cathode ray tubes* the small area of *luminescence* on the screen where the *electron beams* strike it. Efforts are made in the design of cathode ray tubes and the associated circuitry to keep the spot area as small as possible in order to improve the definition of the display.

spot wobble A technique used in black-and-white TV receivers to make the line structure of the displayed image less obvious by superimposing a vertical oscillation of very small amplitude on the scanning spot by use of an oscillator working at a frequency well above the video band.

spread Of characteristics the limits between which the properties of a manufactured component may vary from the published nominal value.

sprite Same as *bitmap.*

spurious emission or radiation From a radio transmitter any radiation at a frequency other than the intended frequency, e.g. at a harmonic of the carrier frequency.

spurious oscillation See *parasitic oscillation.*

spurious response Any response from a device or equipment other than the wanted response. For example second-channel response in a *superheterodyne receiver* is a spurious response.

square-law detector A detector for which the output signal is directly proportional to the square of the input signal over the useful range of the device. The *anode-bend detector* can be regarded as an example because the I_a-V_g characteristic of an *electron tube* near anode-current cut off is an approximation to a square-law curve.

square wave A wave which regularly alternates between two fixed values of amplitude, spending equal time at each, the times of transition between the values being negligible compared with the period of wave. The waveform of an ideal square wave is shown in *Figure S.25*. It contains a wealth of odd

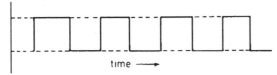

time ⟶

Figure S.25 Idealised square waveform

harmonics and is extensively used for testing and for timing purposes in electronics.

squegging The generation of an oscillation, amplitude modulated by a relaxation oscillation. Squegging can occur in an oscillator in which there is very tight *coupling* between input and output circuits, resulting in a very rapid build-up of oscillation amplitude which is sufficient to cut off the active device. Oscillation therefore ceases until the active device is able to conduct again, whereupon the process repeats itself. A *blocking oscillator* is an example of a squegging oscillator in which the periods of conduction are sufficient for only half a cycle of oscillation.

squelch Same as *quiet automatic gain control.*

s-signal In stereophonic sound broadcasting one half the *difference signal.*

stabilisation (1) Of a power supply the process of ensuring that the output voltage is always at the intended value. Stabilisation implies regulation and some information on the circuits used for the purpose are given under *regulation.*

(2) Of the working point of an *active device* the process of ensuring that the mean current through the device is always at the intended value. A number of circuit techniques are available for stabilising the working point. An example is shown in the transistor circuit of *Figure S.26.* R_1 and R_2 form a potential divider across the supply and impress a particular value of voltage on the base. By emitter-follower action this voltage appears at the emitter (less 0.7 V approximately in silicon transistors) so fixing the emitter voltage. R_3 is now chosen so that the desired mean emitter current is obtained. Thus if the base voltage is 3.0 V, the emitter voltage is 2.3 V and if R_3 is made, say 1 kΩ, the emitter current is stabilised at 2.3 mA.

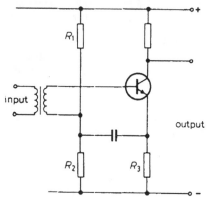

Figure S.26 One circuit for stabilising the mean emitter current of a transistor

stabilising mesh The mesh situated near the target of a low-velocity TV camera tube and biased positively with respect to it to prevent instability of the target potential by effectively limiting the maximum potential which it can develop. For small light inputs to the camera, the mesh collects the secondary electrons emitted from the target with the result that the target develops a positive charge image. If, however, any area of the target tends to become more positive than the mesh potential then the electric field between mesh and target returns the secondary electrons to the target until the target potential equals the mesh potential. Thus no area of the target can develop a potential exceeding that of the mesh. See *camera tube*, *image orthicon*, *orthicon*.

stability In general, of equipment, the extent to which a specified feature of its performance can be maintained in spite of variations in supply voltage, temperature or other factors which can affect performance. For example the stability of an oscillator is measured by the deviations in frequency from the nominal value.

stagger tuning The process of obtaining a level response over a desired bandwidth by tuning the resonant circuits to particular frequencies within the passband. This technique is used, for example, in the IF amplifiers of TV receivers and in the multi-cavity *klystrons* used in TV transmitters.

standard cell A primary cell the EMF of which remains constant with time and is used as a reference standard. To maintain the constancy of the EMF very little current must be taken from the cell which is not therefore used as a source of power. The best-known example of a standard cell is the Weston cell.

standard radiator Same as **full radiator**.

standby mode The state in which an equipment is not fully operational, but can be brought into operation instantly if required.

standing current Same as **quiescent current**.

standing wave On a *transmission line* a stationary pattern of voltage or current produced by interaction between a sinusoidal electromagnetic forwards-travelling wave and the wave travelling in the opposite direction after reflection at an impedance discontinuity. The interaction gives rise to alternate *nodes* and *antinodes* of voltage and current along the line, the distance between neighbouring voltage nodes (or antinodes) being equal to the wavelength. Acoustic standing waves can also occur, e.g. in an enclosure such as a room, by interference between a wave before and after reflection at a wall. Such standing waves are responsible for the acoustics of a hall.

star-delta transformation A method of simplifying network problems by replacing a star (or delta) network by a delta (or star) network which at a particular frequency has the same properties. The transformation is the same as that between T- and π-networks and for equivalence the following relationships apply (see *Figure S.27*):

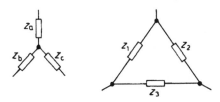

Figure S.27 Star and delta networks

$$Z_1 = Z_a + Z_b + \frac{Z_a Z_b}{Z_c}$$

$$Z_2 = Z_b + Z_c + \frac{Z_b Z_c}{Z_a}$$

$$Z_3 = Z_c + Z_a + \frac{Z_c Z_a}{Z_b}$$

$$Z_a = \frac{Z_1 Z_3}{Z_1 + Z_2 + Z_3}$$

$$Z_b = \frac{Z_1 Z_2}{Z_1 + Z_2 + Z_3}$$

$$Z_c = \frac{Z_2 Z_3}{Z_1 + Z_2 + Z_3}$$

star network A *network* with three or more branches, one terminal of each being connected to a common point. *Figure S.28* shows a star network. When there are three branches the network may alternatively be termed a *T-network* or *Y-network*.

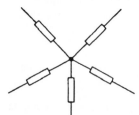

Figure S.28 A star network

starting anode or **starting electrode** In a gas-filled tube an *electrode* used to initiate the *ionisation* which causes the main discharge.

static characteristics Of an *active device*, characteristics representing the relationship between the current at one electrode and the voltage at the same or other electrode, with all other voltages constant and with no load connected to the device. By superimposing a *load line* on the static characteristics it is possible to deduce the *dynamic characteristics* representing working conditions.

static error An error which is independent of a time-dependent variable. C.f. *dynamic error*.

staticise In computers to store serial or time-dependent data.

static random-access memory (SRAM) A *volatile store* in the form of a matrix of cells, each consisting essentially of a *bistable* of *bipolar* or *insulated-gate field-effect transistors*. Within each cell additional transistors provide inputs and outputs for reading and writing purposes, giving a total of 4 or 6 transistors per cell. Such a store retains inserted data provided the power supply is not interrupted. There is no need for *refreshing* as in a *DRAM*.

static sensitive device A transistor or integrated circuit which may be damaged or even destroyed by the static charges generated by handling it if special discharging procedures are not adopted. CMOS devices are often of this type.

static store Same as *static RAM*.

stationary wave Same as *standing wave*.

steady state Of a system or equipment the condition when all transient phenomena have ceased and the voltages and currents are steady or varying sinusoidally between constant peak values.

steering diodes In a bistable circuit, *diodes* which direct incoming trigger signals alternately to the two elements of the circuit so that it is successively switched between its two possible states. Thus if the triggering signals are regular the steering diodes ensure that the bistable circuit generates square waves at half the trigger frequency.

step function A function which has a value of zero until a particular time at which it instantaneously assumes and maintains a constant finite value.

step recovery diode A semiconductor diode notable for the speed with which it ceases conduction when reverse-biased. It consists of a *substrate* of n-type silicon carrying a thin region of lightly-doped n-type silicon and an upper region of p-type silicon. When the diode is forward-biased, charge is stored in the centre region. When the diode is reverse-biased this charge falls and, when it is fully depleted, current through the diode cuts off very suddenly. If such a diode is regularly pulsed into conduction, the current contains a train of regularly-spaced steep edges which can be used to excite a resonant circuit into oscillation, so enabling an input at one frequency to give an output at a different frequency. See *frequency multiplier*.

step response Of a system or equipment, its response to a signal in the form of a step function.

stereocasting (US) Stereophonic broadcasting.

stereophonic system A system of sound broadcasting or recording and reproduction which gives listeners an impression of directionality and spatial distribution of sound which greatly enhances the realism of reproduction. The system may use a number of spaced *microphones*, an equal number of channels and the same number of spaced *loudspeakers* but, in broadcasting and recording, only two channels are normally employed and the system is compatible so that satisfactory monophonic reproduction of the stereo signal is possible.

sticking An effect in a TV camera tube which has been subjected to a stationary optical image for a long period whereby the tube continues to give a picture signal of the image after the image has been removed. In *image orthicon* tubes the effect occurs immediately after switching on because the target resistance is too high and the charge image persists longer than in normal operation.

stopband See *filter*.

stopper See *parasitic stopper*.

storage battery or **storage cell** Same as *secondary cell*.

storage camera tube A camera tube in which a charge image corresponding to the optical image input grows in magnitude on the target until neutralised by the scanning beam. This occurs in all the camera tubes such as the *iconoscope, image iconoscope, orthicon, image orthicon* and *vidicon* which have been successfully used in television services. Storage occurs by virtue of the capacitance of the target and it brings about a considerable increase in the sensitivity of the tube. In fact it was not until the advantage of the storage principle was appreciated that tubes sensitive enough for use in television services became possible.

storage element Of a computer *store*, the smallest element of storage, e.g. a *memory cell*.

storage tube An electron tube into which information can be fed for later reading. The stored signal can be displayed as a visible image on the screen or can be retrieved as an electrical signal.

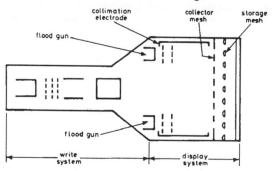

Figure S.29 Basic features of storage tube

A simplified diagram of a storage tube is given in *Figure S.29*. The phosphor target is backed (on the viewed face) by a transparent conductive layer. The writing gun is maintained negative with respect to the target to attract electrons which are focused to a fine spot on the target, causing it to luminesce. By applying potentials to the deflecting plates the beam can be deflected so as to write on the screen.

Part of the writing beam is intercepted by the storage mesh and strikes it with sufficient velocity to release secondary electrons. For each beam electron striking the mesh, several secondaries are released (i.e. the *secondary emission ratio* is greater than unity). The secondaries are attracted to the nearby collector mesh, causing the storage mesh to go positive in the areas struck by the beam. Thus the display is written on the storage mesh as a positive *charge image*. The cathodes of the flood guns are only slightly negative with respect to the target potential but electrons are attracted by the positive areas of the storage mesh and pass through it to enhance the display on the phosphor.

Once a charge image has been established on the storage mesh the trace continues to be displayed on the screen for a considerable period—under favourable conditions for up to an hour. In fact if the tube is switched off, a charge image can last for a week.

To erase the display a positive pulse can be applied to the target. This attracts electrons from the flood guns which are designed to flood the entire area of the screen with electrons so 'writing' the entire screen area. As the target potential falls at the end of the pulse the screen is erased.

By restricting the intensity of the beam from the writing gun it is possible to write on the phosphor screen without building up a charge image on the storage mesh—a technique known as 'writing through'.

Such tubes have very high resolution and give a flicker-free display.

store Of a *computer*, the device into which information can be fed for retention until required. There are a number of different types of storage device, some using magnetic disks or tape, some using capacitors or bistable electronic elements. See *bistable circuit*, *magnetic core*, *magnetic drum*, *memory cell*, *RAM*, *ROM*.

'straight' receiver Same as *tuned-radio-frequency receiver*.

stray capacitance In electronic circuits the *capacitance* arising from all sources other than from capacitors. Thus stray capacitance can arise from within active and passive components and from the proximity of components, conductors and chassis. Stray capacitance is undesirable because it limits gain at high frequencies by shunting loads but it is sometimes put to good effect as part or whole of the tuning capacitance of a resonant circuit. See *distributed constants*.

streaking In TV an effect in which areas of incorrect tonal rendering appear to the right of vertical lines in the image. In black-and-white television the streaks may be black or white following black or white lines and can occupy any fraction up to the whole of the picture width. The effect is caused by anomalies in the amplitude and/or phase response in the video chain at frequencies around the line frequency.

string In computing, a sequence of *bytes*, often representing *text*.

strip (US) Same as *scanning line*.

strobe To select a particular fraction of the period of a recurrent event. This is often necessary in detailed examinations of a regular *waveform* using an *oscilloscope*. A strobe pulse which coincides in time with the wanted part of the waveform is used to gate the waveform and then to display it.

Structured Query Language (SQL) A language devised by IBM for creating, maintaining and viewing data in a *database*. It is now becoming an industry standard and may provide a means for different computer types running different databases to extract data from each other.

stub An open-circuited or short-circuited length of *transmission line*, usually less than a quarter wave, connected across a transmission line to facilitate matching of the line to a load.

subaudio A frequency below the normal audio range. The lower audio limit cannot be accurately defined but a subaudio frequency can be taken as one below 15 Hz.

subcarrier A carrier, usually with modulation, which is used to modulate another (main) carrier. Subcarriers are used extensively in telephony. They are also used in stereo sound broadcasting (to carry the difference signal) and in colour television (to carry the colour information). The subcarriers are ignored when compatible monophonic or black-and-white reproduction is required.

subdirectory Any *directory* other than the *root directory*.

subharmonic Of a complex wave a sinusoidal component with a frequency that is an exact submultiple of that of the fundamental component. For example the second subharmonic has a frequency equal to one half that of the fundamental component.

subjective grading Of the performance of an equipment or system a method of grading in which a panel of observers record their assessments according to a scale. For example interference to a sound or TV programme can be classified as negligible, slightly distracting, distracting, very distracting or completely distracting.

submodulator Of a *transmitter*, the penultimate stage of the *modulator*.

subroutine In computers part of a program generally used more than once in the program and dedicated to the carrying out of particular operations. A typical example is the display of text characters.

subschema The user's view of data in a *database*.

subset A piece of *software* which is a shortened and therefore incomplete version of some other software. Subsets of computer languages, for instance, may contain only the principal keywords.

substrate In general, the supporting material on which or in which a *monolithic integrated ciruit* is fabricated. In many semiconductor devices the substrate is a layer of monocrystalline silicon. In a *field-effect transistor* the substrate supports the channel, the source and the drain. In some types the substrate forms a pn junction with the channel: a connection to the substrate can then be used as a second gate terminal. In other types of FET, however, the substrate is insulated from all parts of the transistor.

summing amplifier An *operational amplifier* with a *negative feedback* network so designed that the output is equal to the sum of the input signals. Such amplifiers are used in *analogue computers*.

sum signal In stereo broadcasting the signal formed by adding the signals in the left and right channels. In the pilot-tone system of transmission the sum signal is used to frequency modulate the main carrier and it is this signal which is reproduced by a *monophonic* receiver.

super-alpha pair Same as *Darlington pair*.

superconductivity The virtual disappearance of electrical resistance in certain metals at temperatures approaching absolute zero. The effect is pronounced in copper and zinc around 5°K and currents induced in a ring of these metals by movement of a magnetic field continue to flow for some hours afterwards.

super Emitron Same as *image iconoscope*.

superheterodyne receiver A radio receiver in which the modulation of all incoming signals is transferred to a new fixed carrier frequency (the intermediate frequency), most of the gain and selectivity of the receiver being achieved in the IF amplifier.

Frequency changing is achieved by mixing the incoming signals, possibly after amplification, with the output of an oscillator, the tuning of which is ganged with that of the signal-frequency circuits so that the mixer output remains at the intermediate frequency for all settings of the tuning control. The IF amplifier is followed by a detector or discriminator and subsequent modulation-frequency amplifiers as required.

The block diagram of a superheterodyne receiver is shown in *Figure S.30* and this applies to AM and FM sound receivers and to the basic structure of

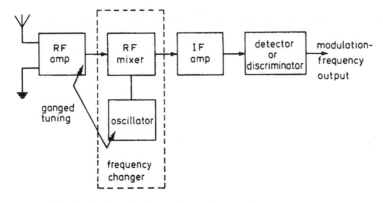

Figure S.30 Block diagram of a superheterodyne receiver

TV receivers. Practically all modern receivers utilise the superheterodyne principle. The term is a telescoped form of supersonic heterodyne, a reference to the fact that the IF is above the audible range.

superposition theorem The current flowing at a point in a linear passive *network*, as a result of the simultaneous application of a number of voltages distributed in any manner throughout the network, is the sum of the component currents at that point which would be caused by the individual voltages acting separately. Similarly the potential difference existing between any two points in the network as a result of the several applied voltages is the sum of the component potential differences between the points which would be caused by the individual voltages acting separately.

super-regenerative receiver An AM receiver incorporating a detector which is taken in and out of oscillation at an ultrasonic rate, so achieving higher gain than is normally possible without instability. A *grid-leak detector* with *reaction* is capable of high gain but the gain is limited by the onset of oscillation. The super-regenerative principle avoids this limitation by allowing oscillation to build up and then quenching it at an ultrasonic frequency. It is possible to achieve RF and ultrasonic oscillation in a single active device but better results are possible with a separate ultrasonic oscillator.

suppressed-carrier modulation A system of *amplitude modulation* in which both sidebands are transmitted, the carrier being suppressed. Because the carrier component remains constant, whether there is amplitude modulation or not, it is clear that the carrier can be suppressed, all the information being contained in the sidebands. It is necessary, however, to re-insert the carrier at the receiving end to permit detection.

suppressor grid An *electrode* situated between the screen grid and anode of an *electron tube* to prevent secondary emission from the anode from reaching the screen grid. The suppressor grid is normally connected to the cathode to form a decelerating field near the anode which returns secondary electrons to the anode. In this way the *tetrode kink* can be eliminated so improving the ability of the tube to handle large output-voltage swings.

311

surface acoustic wave device (SAW) A device which makes use of radio waves in the form of surface deformations on piezo-electric materials. Such waves can be excited by suitable *transducers* and they have wavelengths about 10^5 times smaller than those in free space. Thus the wavelength of a 35-MHz wave is less than a millimetre and it is possible by suitably shaping the piezo-electric surface to produce an extremely-small filter by techniques similar to those used in microwave filters. *Bandpass filters* of this type are used in the IF amplifiers in TV receivers and there is likely to be considerable development in this type of structure in the future.

surface-barrier transistor A *bipolar transistor* in which the *emitter* and *collector* regions are formed by electrodeposition of a trivalent or pentavalent element on opposite faces of a thin base wafer.

surface mounting A method of mounting specially-designed components and integrated circuits on a *printed wiring* board by soldering the contacts of the components directly to the tracks, so eliminating the need for holes in the board. The method permits greater component packing density, gives higher speeds of operation and much improved thermal performance.

surface noise In record reproduction the noise produced by the stylus passing through the groove. There is very little surface noise from modern records but earlier types were of shellac incorporating an abrasive material and the minute irregularities in the groove walls caused a pronounced hiss on reproduction. See *scratch filter*.

surfing Jargon for browsing on the *Internet*, which some computer users find an absorbing pastime.

surge impedance Same as *characteristic impedance*.

surround sound Same as *quadraphony*.

susceptance (B) That component of the *admittance* of an alternating-current circuit which is due to its *inductance* and/or *capacitance*. It is the imaginary component of admittance and, for a circuit containing only susceptance, the voltage and current are in quadrature. Susceptance is the reciprocal of *reactance*. For an inductance L the susceptance (in mhos) is given by

$$B_L = 1/\omega L$$

and for a capacitance C

$$B_C = -\omega C$$

where ω is the angular frequency. The signs follow the same convention as for reactance.

susceptibility Of a magnetic material the ratio of the intensity of magnetisation to the magnetising force. For an *isotropic* material the susceptibility is equal (in rationalised units) to the relative permeability less one.

sweep The regular movement of the spot across the screen in a *cathode ray tube* or the change in frequency across the band in a wobbulator output.

sweep oscillator (US) Same as *wobbulator*.

swept-frequency generator Same as *wobbulator*.

swing The difference between the extreme values of current, voltage or frequency of a signal.

swinging choke A series inductor used for smoothing and located immediately following the output of a rectifier. To improve regulation the choke is designed so that its effective inductance decreases with increase in the current through it.

switch A device for breaking (opening) or making (closing) an electric circuit. Switches can be mechanical, e.g. those controlling electric lights, or they can be electronic (consisting of transistor or thyristor circuits) such as those used in electronic telephone exchanges.

switch-mode power supplier A DC power supplier with an AC input and in which the smoothed output from the rectifier is chopped into pulses which are then rectified and smoothed to provide the output, the mark/space ratio of the pulses being automatically adjusted to regulate the output voltage. The chopping device may be self-oscillating or may be driven from a pulse generator, such as an astable multivibrator, within the power supplier. Such power suppliers can be made highly efficient and much more compact than conventional types. One reason for this is that by using a high pulse frequency only small smoothing components are needed. (In power suppliers for television receivers the pulse generator is often synchronised at line frequency.) Another reason is that the active device acting as chopper operates as a switch and (except for the instants of switching) has zero current in it or zero voltage across it; dissipation in the device is thus minimal. See *regulation*.

syllabic companding *Companding* in which the changes of gain are fast enough to follow the syllables in speech but not the individual cycles of the speech waves.

symmetrical deflection Of a *cathode ray tube, electrostatic deflection* in which the deflecting voltages applied to the plates are symmetrical about a reference potential (usually the final-anode potential). Such a system gives less geometric distortion of the image than asymmetric deflection.

symmetrical operation Same as *push-pull operation*.

synchrodyne A system of *heterodyne* reception in which the oscillator operates at the carrier frequency of the received signal and is synchronised by the received signal. It is an example of heterodyne reception in which the intermediate frequency is zero, the output from the frequency changer being at modulation frequency.

synchronising signal In TV the *signals* which ensure that the horizontal and vertical scanning circuits in picture-display equipment are synchronised with the picture-generating equipment. In colour TV the synchronising signal includes the colour burst which synchronises the chrominance decoder. The synchronising signal is added to the picture signal to form the video signal.

synchronous detector A detector sensitive to phase and amplitude variations in the modulated signal applied to it. Such detectors are used for independently recovering the two components of the chrominance signal from the quadrature-modulated subcarrier in the *NTSC* and *PAL colour TV systems*. For successful operation the detector must be synchronised with the reference signal (colour burst) contained in the TV colour signal.

synchronous logic In *data-processing* logic in which all operations start on receipt of a signal from a clock. C.f. *asynchronous logic*.

synchronous tuning Tuning in which all the resonant circuits are adjusted to the same frequency.

sync level In TV the level reached by the peaks of the synchronising signals. See *Figure B.18* (a).

synthesiser An electronic musical instrument which produces a wide repertoire of sounds using digital techniques to construct the required output waveform.

system A complex of conductors, components and equipments interconnected so as to perform a particular function. For example, a radio receiver can be said to have an *a.g.c.* system and a collection of antennas can form a *diversity* system.

T

tachogenerator A generator mechanically coupled to a rotating shaft to give an electrical output proportional to the rotational speed of the shaft. This output may be used to indicate the shaft speed or may form an input to a speed-regulator circuit.

Tagged Image File Format (TIFF) A *bitmap file* standard widely used for storing images and transferring them, especially between computers of different types. A very versatile format, it can handle images having up to 32 bits per pixel and various compression algorithms. It is, perhaps, the closest approach to an industry-standard bitmap file format.

tandem connection Same as *cascade*.

tank circuit In general an oscillatory circuit. In particular a parallel *LC* circuit forming the load of a tuned amplifier (e.g. in a transmitter) or forming the frequency-determining element of an oscillator.

tantalum bead capacitor A type of polarised capacitor offering high capacitance with low losses and leakage in a compact package.

tape See *magnetic recording*.

tape streamer A device for recording data at high speed on tape in a data cartridge. The most common use for tape streamers is for backing up *hard disk drives*, but they are also used for distributing large collections of software.

target The electrode in an *electron tube* on which the *electron beam* is focused. Thus the target of a cathode ray tube or a picture tube is the luminescent screen. In a simple camera tube such as an *iconoscope*, *orthicon* or *vidicon* the target is photosensitive and an optical image of the scene to be televised is focused on it. As a result the target develops a charge image which is discharged by the scanning beam to generate the tube output.

In camera tubes with an image section the target is of secondary-electron emitting material and the charge image is generated by bombardment of the target by photo electrons from the *photo-cathode*.

target language (US) Same as *object code*.

T bistable A *bistable circuit* with an input such that every time the input takes on its 1-state the output of the bistable goes to the complementary state. When the T input takes up its 0-state it has no effect on the output.

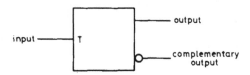

Figure T.1 Logic symbol for a T bistable

315

tearing In TV, erratic breaking up of part of a reproduced image by lateral displacement of some of the scanning lines caused by faulty line synchronisation.

tecnetron An early form of *field-effect transistor* pioneered by the French.

telecommunication The process of sending information from one point to another by electromagnetic means. The information can be of any kind, e.g. sound or picture signals, and the link between the two points may also be of any kind, e.g. by wire, radio, optical means or a combination of any of them.

telegraphy A system of telecommunication for the transmission of information consisting of graphic symbols, usually letters and figures, by use of a signal code. The system is used primarily where permanent records are required of the messages sent. The link may be of any kind, e.g. line or radio, and the code is often that used by *teleprinters*.

telemetry A system of telecommunication for the transmission of information on measurements, e.g. the water level in a reservoir is represented by an electrical signal suitable for transmission over the link which may be a line or radio or a mixture or both.

telephony A system of telecommunication for the transmission of speech or other audio signals. The link may be by line or radio or a mixture of both and is normally such as to provide two-way speech communication.

telepresence A computer technique similar to *virtual reality* in that the user dons a body suit or headset and gloves to provide an input. In telepresence, however, there is no virtual world. The computer controls a *robot* which mimics the user's actions. Telepresence has applications in industry, exploration and military operations where some task must be performed in a hazardous environment. The robot enters the hazardous area and takes the risks, while the human operator remains safe.

teleprinter A machine capable of converting a written message containing letters and figures into a coded form suitable for transmission over a telegraph link to a similar machine which recreates the original message. The teleprinter has a keyboard similar to that of a typewriter and messages typed on it appear on a paper roll at the transmitting and receiving machines so that there are identical records of the message at both ends of the link.

teletext In TV a system of transmitting information in the form of letters, figures and simple diagrams, by signals incorporated in the television waveform. The information can be displayed on the screen of a suitably-equipped television receiver in place of or superimposed on the normal programme.

In this service, termed *Ceefax* by the BBC and *Oracle* by the IBA, the teletext information is transmitted as a pulse-coded signal occupying one of the normally-empty forward scanning-line periods in the field blanking interval. Several hundred different pages of teletext information can be transmitted using this system giving, for example, details of news items, weather reports, sports results etc. A receiver equipped with a teletext decoder enables the user to select any desired page.

teletypewriter (US) Same as *teleprinter*.

television The process of creating instantaneously at a distance a black-and-white or coloured image of a real or recorded scene and the movement occurring in it by means of an electrical system of communication. The term recorded scene includes pictures from cinema film, videotape, videodisc or transparency.

television camera Equipment containing the optical and electronic components required to generate a *picture signal* from the optical image of a scene. In a black-and-white camera the principal components are an optical lens system and a camera tube. The camera tube requires facilities for deflecting and focusing the electron beam and the camera may include a head amplifier for the camera tube output. A colour television camera incorporates at least three camera tubes and an optical device for splitting the optical image into its red, green and blue components. In addition some colour cameras have a fourth camera tube to generate the luminance signal.

Most television cameras include a viewfinder which may be optical or may be a picture tube fed from the camera output.

television camera tube See *camera tube*.

television system The particular combination of television signal waveform, method of transmitting vision and sound, channel width, channel spacing etc. chosen. There are a number of different systems and they have been classified by the CCIR. The system used in the United Kingdom is system I.

television waveform See *video signal*.

temperature coefficient In general a factor expressing the degree to which a quantity is affected by changes in temperature. More specifically it is equal to the relative change in the quantity per degree Celsius. Some electronic components are marked with their temperature coefficient. For example a capacitor marked N750 has a temperature coefficient of capacitance of -750 parts in 10^6.

temporary storage In computers *storage* reserved for intermediate results.

terminal A point where a signal is applied to or withdrawn from a *network* circuit or equipment. It is also known as an external node. See also *computer terminal*.

termination The external *impedance* connected to the two terminals of a *transmission line*. If this impedance is equal to the characteristic impedance of the line, the termination is described as matched and the wave is not reflected at the termination. For any other value of termination there is a reflected wave.

testing The carrying out of operations on a component, equipment or system to determine if it is functioning correctly or, for a new component etc, to determine if it is likely to function correctly when put into service. There are a number of different types of test. See *accelerated life test, acceptance test, life test, maintenance test*.

tetrode An electron tube with two grids between cathode and anode, that nearest the cathode being the control grid and the other carrying a fixed positive bias. There are two main types of tetrode:

(a) The screen-grid tetrode in which the positively-biased grid is designed as an electrostatic screen between control grid and anode virtually eliminating the grid-anode capacitance and so enabling the tube to be used as a stable RF amplifier. The I_a–V_a characteristic contains the *tetrode kink* but if the tube is used only as a small-signal amplifier this is not a serious disadvantage.

(b) The kinkless tetrode in which the positively-biased grid is used, not to eliminate anode-grid capacitance but to eliminate the tetrode kink in the I_a–V_a characteristic. Thus the tube is capable of large output-voltage swings and is used in power output stages, e.g. in AF amplifiers. See *beam tetrode, critical-space tetrode, screen-grid electron tube*.

tetrode kink See *dynatron*.

tetrode thyristor A *thyristor* with two gate terminals, one connected to the inner n-region (anode gate) and the other to the inner p-region (cathode gate) and which can be switched to the on-state and to the off-state by signals applied to one of these gates. It is sometimes known as a silicon controlled switch. The graphical symbol is given in *Figure T.2*.

Figure T.2 Graphical symbol for a tetrode thyristor (silicon controlled switch)

tetrode transistor See *insulated-gate field-effect transistor*.

text Data consisting of printable characters such as alphanumerics and punctuation marks. Such data uses a code such as *ASCII* or *EBCDIC*.

text file A *file* representing a string of printable characters such as alphanumeric characters and punctuation marks, perhaps in *ASCII* format.

thermal breakdown See *thermal runaway*.

thermal noise *Noise* arising from the thermal agitation of the *electrons* in a *conductor*. Although in a conductor with no applied voltage, the number of electrons passing a given point in one direction equals the number passing it in the opposite direction when the measurement is made over an appreciable period, there are momentary surfeits or deficits of electrons and their variation with time constitutes thermal noise. The magnitude of the noise voltage V_n is given by

$$V_n^2 = 4kTR\Delta f$$

where k = Boltzmann's constant,
    T = absolute temperature,
    R = resistance,
and Δf = bandwidth over which the noise is measured.
Substituting the value for k and putting T = 290°K for normal room temperature we have

$$V_n = 1.25 \times 10^{-4}\sqrt{(R\Delta f)}\,\mu V.$$

For example the voltage generated across a 2-kΩ resistor over a bandwidth of 5.5 MHz is 131 µV.

thermal relay A *relay* which depends for its operation on the heating effect of an electric current.

thermal resistance The difference in the temperature at two points divided by the power flow between them when thermal equilibrium has been established. It is thus the temperature difference per unit dissipation and can be measured in degrees C per watt. The concept is useful in the design of *heat sinks* which must have a minimum thermal resistance to be able to dissipate the power transmitted through them satisfactorily. The thermal resistance depends chiefly on the mass and the surface area of the heat sink.

thermal runaway Cumulative effect which occurs in *semiconductors* in which the heat generated in the device by the current through it increases the current, so raising the temperature further. This results in a rapid rise in device temperature which can damage or even destroy the device. Where there is danger of thermal runaway, transistor circuits are designed to stabilise the mean current, a process known as DC stabilisation or *stabilisation* of the operating point.

thermionic emission The emission of *electrons* from a heated conductor. At normal temperatures the free electrons in a conductor have a certain mobility but their energy is not normally sufficient to enable them to escape from the surface of the conductor. If, however, the conductor is heated the electron energy is increased and some can escape into the surrounding space. The more the conductor is heated the greater is the number of electrons which escape. Thus the electrons liberated from the cathode of an *electron tube* are released as a result of thermionic emission.

thermionic tube An *electron tube* containing an electrode, normally the cathode, which must be heated to release the electrons or ions necessary to give conduction through the tube. See *gas-filled tube*.

thermistor A *resistor* with a large negative temperature coefficient of resistance. The temperature-sensitive materials used in thermistors are oxides of manganese or nickel formed into rods and fired at high temperature. The resistance of a thermistor decreases with increase in temperature and thus with increase in the current through it. Its behaviour, therefore, is the opposite of that of most metallic conductors. Small thermistors are heated by passing the current directly through the temperature-sensitive element but there are also indirectly-heated types where the current is passed through a winding electrically-insulated from the element.

Thermistors are used for surge suppression, e.g. a directly-heated type may be connected in series with the tube heaters in a radio receiver so as to limit the surge of current when the circuit is switched on and the heaters are cold and therefore of low resistance. Thermistors are also used for temperature measurement.

thermocouple Junction of two different metals which, when heated, generates an EMF which depends on the temperature of the junction. Thermocouples are used for measuring radio-frequency currents (by their heating effect) and for measuring high temperatures.

thermo-electric effect The generation of an EMF at the junction of two dissimilar metals when the junction is heated. The effect is sometimes described as the Seeback effect. See also *Peltier effect*.

thermostat A switch which operates automatically at a predetermined temperature and which is used, in conjunction with a heating or cooling device, to maintain a constant temperature in an enclosure. The switch is often operated by a bimetallic strip. Such an arrangement is used at radio transmitters to maintain a constant temperature for the piezo-electric crystal which determines the carrier frequency.

Thevenin's theorem If an impedance Z' is connected between any two points of a linear network the current which flows in it is given by $V/(Z + Z')$, where V is the voltage and Z is the impedance measured between the points before Z' was connected.

thick-film circuit A circuit in which resistors and their interconnections are formed on a *substrate* by applying special inks via a stencil screen or metal mask and firing them in a furnace. Resistance accuracy of ±15% can be achieved. Where necessary the resistors can be adjusted by abrasive trimming or by a laser beam to an accuracy of 0.1%. Discrete components (resistors, capacitors, inductors, transformers, semiconductor diodes and transistors) can be mounted on a thick-film circuit.

thin-film circuit A circuit in which components and their interconnections are formed on a *substrate* by evaporation or sputtering of resistive, conductive or dielectric material. The components can be resistors, capacitors or inductors and the substrate can be glass, alumina or silicon. Discrete components (resistors, capacitors, inductors, transformers, semiconductor diodes or transistors) can be mounted on a thin-film circuit.

third-generation computer See *digital computer*.

three-state output In logic circuitry a device output which in addition to the usual states representing the two logic levels has a third possible state (usually open-circuited) which has no logic significance. It is represented in diagrams as shown in *Figure T.3*. The importance of the open-circuit condition is that it enables an output to be effectively disconnected from its external circuit. Thus a number of outputs may be connected in parallel, e.g. on a bus without interference provided no more than one is enabled at a time.

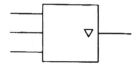

Figure T.3 Representation of three-state output

threshold The minimum amplitude at which a signal will activate a device, e.g. in logic circuits, the minimum input voltage that will be recognised as a logic-1 rather than a logic-0.

threshold frequency See *photoelectric-emission*.

throat Of a horn-loaded *loudspeaker* that end of the horn where the cross section is a minimum.

thumbnail A reduced-size reproduction of a computer *graphic*. Some applications create illustrated *catalogues* of directories containing graphics files with a thumbnail of each for ease of identification and selection.

thyratron A hot-cathode gas-filled *triode* or *tetrode* in which the *control grid* initiates anode current but does not control it. Provided that the anode voltage is high enough a grid voltage exceeding a critical value initiates anode current which is then limited only by the HT supply voltage and the resistance of the external anode circuit. To cut off the anode current both anode and grid voltages must be below certain values.

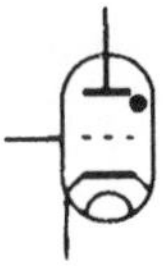

Figure T.4 Graphical symbol for a triode thyratron

Thyratrons were extensively used as electronic switches in control circuits, e.g. for controlling the power fed to a load from the mains. The thyratron was connected in series with the load and acted as a half-wave rectifier, the angle of conduction of which could be controlled by the bias applied to the control grid. *Thyristors* have now replaced thyratrons in most applications. The graphical symbol for a thyratron is given in *Figure T.4*.

thyristor A bistable *semiconductor* device which can be switched to the on-state (and in some types to the off-state also) by a signal applied to the gate terminal.

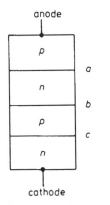

Figure T.5 Basic construction of a 2-terminal thyristor

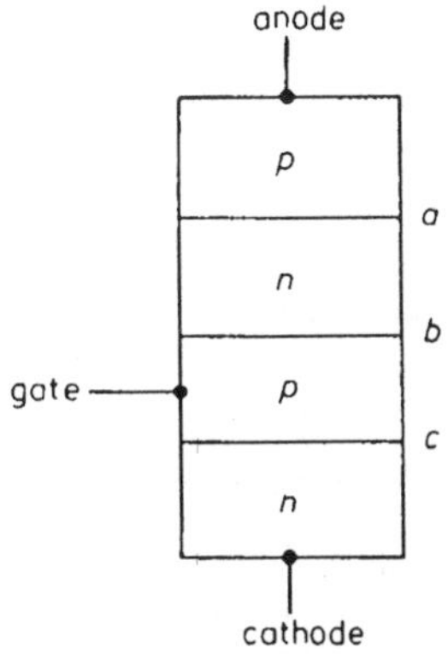

Figure T.6 Basic construction of a 3-terminal thyristor

Thyristors are of four-layer pnpn construction and the outer layers are known as the anode and cathode as indicated in *Figure T.5*. If the anode is made positive with respect to the cathode, pn junctions a and c become forward-biased and therefore of low resistance but junction b is reverse-biased thus limiting the current through the device to the leakage current of this junction. If, however, the anode voltage is increased sufficiently avalanche breakdown occurs in the centre junction and a large current flows through the device, limited only by the voltage and resistance in the external circuit: this is the on-state. The anode voltage can now be reduced to a low value but the on-state will persist until the current through the device falls below a critical value when the device becomes non-conductive again (the off-state). If the anode is made negative with respect to the cathode, junction b becomes conductive but junctions a and c are reverse-biased and prevent significant current flow: for such voltages the device remains in the off-state. Such a device is known as a *reverse-blocking*

321

diode thyristor. Two such devices connected in reversed parallel form a bidirectional diode thyristor. A variant of the diode thyristor which remains in the on-state for negative anode voltages is known as a reverse-conducting diode thyristor.

By applying a signal to one of the inner layers the pnpn structure can be triggered into the on-state even at low anode voltages and again will remain in the on-state until the current has fallen below the critical value. The inner region is termed the gate as indicated in *Figure T.6* and the three-terminal device is termed a reverse-blocking triode thyristor or, more simply, a thyristor.

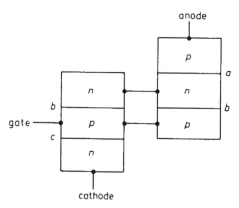

Figure T.7 Two-transistor analogue of the thyristor

The mechanism of the triggering process is best understood by reference to the two-transistor analogue of the thyristor. As shown in *Figure T.7* a thyristor can be regarded as a combination of an npn and a pnp transistor with the collector of each connected to the base of the other (see *multivibrator*). Suppose such a device has a small positive anode voltage but is in the off-state. As explained, junctions a and c will be forward-biased and junction b reverse-biased: these are the normal conditions for the two constituent transistors to operate as amplifiers. When a positive voltage is applied between gate and cathode it increases the forward bias of the npn transistor with the result that current crosses junction c and this, by normal transistor action, gives rise to a collector current in the inner n-region which is also the base region of the pnp transistor. By normal amplifying action in the pnp transistor this gives rise to a collector current in the inner p-region which is also the base region of the npn transistor. In this way positive feedback occurs and, if the combined current gain of the two transistors exceeds unity the feedback is regenerative causing a very rapid build-up of current which puts the device into the on-state. The gate voltage can now be removed but the thyristor will remain conductive until the current in it falls below the critical value—behaviour typical of a bistable multivibrator. From the symmetry of the device structure it is clear that the thyristor could alternatively be triggered by a negative signal

322

applied to the inner n-region. Thus there are two types of reverse-blocking triode thyristor one with a p-gate (cathode gate) and the other with an n-gate (anode gate).

There are corresponding reverse-conducting triode thyristors and a type which can be triggered into the off-state known as a turn-off thyristor. A very useful thyristor can be constructed of two triode thyristors in reverse parallel with a common gate connection: this is known as a triac and can be triggered into conduction by a positive or negative signal applied to the gate. If external connections are made to both inner layers of the pnpn structure the thyristor so obtained can be triggered into the on-state and into the off-state by appropriate signals applied to one of the two gates. This is a tetrode thyristor alternatively known as a silicon controlled switch which is used, amongst other applications, as a field oscillator in television receivers.

The chief application for thyristors is as rectifiers in mains-operated power-control systems where the conduction angle is controlled by signals applied to the gate. For this reason thyristors are often known as silicon controlled rectifiers.

The general graphical symbol for a thyristor is given in *Figure T.8*. The symbols for individual types of thyristor are given in the particular definitions.

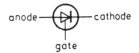

Figure T.8 General graphical symbol for a thyristor

tilt Same as *sag*.

time base In cathode ray oscilloscopes that deflection of the electron beam which is defined with respect to time. Usually this is the horizontal deflection, the waveform to be displayed providing vertical deflection.

time-base generator Equipment for generating the *signals* providing the *time base* for an oscilloscope. For *electrostatic deflection* the time base generator must provide a sawtooth voltage and, for flexibility in the applications of the oscilloscope, a wide range of sawtooth frequencies is required.

time constant Of a quantity varying exponentially with time, the time taken for the quantity to vary by 63% (i.e. $1 - 1/e$) of the full extent of the change. When a capacitor C charges from a DC supply via a resistor R the voltage across the capacitor reaches 63% of its final value after RC seconds, where R is in ohms and C in Farads. When the capacitor is discharged its voltage reaches 37% of its initial value in RC seconds as shown in *Figure T.9*. RC is thus the time constant of the circuit. If an inductor L is connected across a DC supply via a resistor R the current grows exponentially with time and reaches 63% of its final value (E/R) in L/R seconds where L is in Henries and R in ohms.

time division multiplex (TDM) A multiplex system in which each *signal* is allowed use of the communications path for a short time interval. Usually there are a number of signals and they use the path in turn.

323

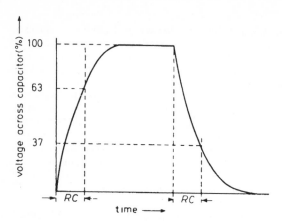

Figure T.9 The time constant for an RC combination when charging and discharging

time slot A recurrent interval of time which can be uniquely defined. For example a particular forward line-scanning period during the field blanking interval is the time slot occupied by the *teletext* signal.

timing The process of ensuring that all the *signals* necessary to make up a complex waveform or to carry out an operation arrive in the correct sequence and at the correct intervals.

T metal-oxide semiconductor transistor (TMOS) An *insulated-gate field-effect transistor* so constructed that the current flow is for the most part perpendicular to the plane of the layers as in *planar* bipolar transistors; the source consisting of a large number of sites each associated with a short channel, the parallel connection of which yields low switching times and low source-drain resistance in the on condition. Large source and drain areas permit the transistor to give substantial output power.

T-network A *network* consisting of two series elements with a shunt element connected to their *junction*. *Figure T.10* illustrates a T-network: it may be regarded as an example of a star network with three branches.

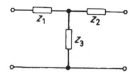

Figure T.10 A T-network

toggle (US) Same as *bistable circuit*.

tone (1) In audio a sinusoidal signal of constant frequency used for test purposes or to identify circuits. (2) In popular speech the term is used to describe the quality of a musical sound. For example a cello may be described as having a mellow tone. (3) In photography and TV the degree of light or shade of an image or an element of an image.

tone control A circuit for adjusting the *frequency response* of an audio amplifier. The chief ways in which the response may be modified are indicated in *Figure A.12* and high-quality equipment is usually capable of all such responses but in simple receivers with a single tone control this usually introduces top cut only.

tone wedge In TV a test image consisting of a series of areas of which the tone varies in steps between black and white. Most TV test cards incorporate a tone wedge which can be used for the adjustment of receivers.

top cut *Attenuation distortion* in which the upper frequencies of the audio range are attenuated.

top lift Accentuation in which the upper frequencies of the audio range are emphasised.

totem-pole output Same as *single-ended push-pull*.

touch sensor or **switch** A switch which operates when a finger comes into direct connection with two closely-spaced contacts or when the finger provides a capacitive link between the contacts. The direct-connection type of switch is fed from a DC supply and the presence of the finger causes a small direct voltage to be injected into a *monostable* or *bistable* circuit which performs the required switching action. The capacitive type of switch is fed from an AC supply and the presence of the finger on an insulating cover over the contacts causes a small alternating voltage to be injected into the following circuit which brings about the required switching action. Touch switches are extensively employed, e.g. in controlling lifts and in channel switching on television receivers.

trace Of a *cathode ray tube* the pattern generated on the screen by the movement of the *electron beam*.

track A channel containing stored data in a storage medium such as a tape or *floppy disk* or *hard disk*.

trackball or **trackerball** In computing a *pointing device* used as an alternative to a *mouse*. The trackerball is a fixed unit containing a captive ball whose uppermost portion is exposed. The user manipulates the ball with his thumb or the palm of his hand to guide the pointer around the screen. Trackerballs are also used in closed-circuit television systems, especially for security surveillance, to pan and tilt remotely mounted surveillance cameras.

tracking (1) In a *superheterodyne receiver* maintenance of the correct difference between signal frequency and oscillator frequency as the tuning is altered. When similar sections of a ganged tuning capacitor are used for signal-frequency and oscillator circuits, *trimming* and *padding capacitors* are included in the oscillator circuit to help maintain this frequency difference. In fact even with such aids the difference frequency can be made correct (i.e. equal to the *intermediate frequency*) only at three settings of the capacitor. These are therefore positioned one near each end of the tuning range and the third at the centre: this is known as three-point tracking.

(2) In record reproducing equipment the accuracy with which the plane of vibration of the reproducing stylus is maintained at right angles to the direction of the groove at the stylus point as the reproducing head moves across the disk. By offsetting the head on the arm (or curving the arm) and

by arranging for the stylus to overshoot the disk centre to a critical extent the tracking error can be minimised.

(3) The formation of carbon conducting paths across the surface of insulating materials by electric stress. Certain insulating materials, notably those made of fibrous material such as paper and paxolin, are very subject to tracking.

transceiver A device, usually portable, which can be used as a *transmitter* or a *receiver*. Much of the circuitry of a transceiver is common to the transmitter and the receiver.

transconductance (US) Same as *mutual conductance*.

transducer A device which responds to an input signal in one form of energy and gives an output signal bearing a relationship to the input signal but in a different form of energy. The forms of energy are usually acoustic, mechanical and electrical. For example a *microphone* is a transducer which converts acoustic into electrical energy. This is an interesting example because there are, in fact, two energy conversions in a microphone. The original acoustic energy (oscillations of air) is first converted into mechanical energy (vibrations of a diaphragm) which is then converted into electrical energy (changes in voltage) at the output terminals.

transductor A ferromagnetic device in which the degree of saturation of the core (and hence the effective inductance) can be controlled by the current flowing in one of the windings. It is sometimes known as a *saturable reactor*.

transfer characteristic (1) Of an *active device*, the curve obtained by plotting the output current against the input voltage or current. For an *electron tube* or *field-effect transistor* the input is taken as a voltage and for a *bipolar transistor* as a current. (2) Of a *camera tube*, the curve obtained by plotting the output current against the light input. (3) Of a magnetic recording system, the curve obtained by plotting the magnetic flux density against the magnetising force.

transfer coefficient (or **constant**) Same as *image transfer coefficient*.

transfer impedance Between any two pairs of *terminals* in a *network*, the complex ratio of the voltage applied to one pair to the resultant current at the other pair, all other terminals being terminated in a specified manner, e.g. if there are other generators these should be replaced by their internal impedances. The term is defined in the same manner as *mutual impedance* but the terminating conditions differ.

transformer A device consisting of two or more inductively-coupled windings. Transformers are widely used in electronics and the following are typical applications:

(*a*) For changing the voltage of an alternating supply.

Where equipment requires a supply voltage which differs from that of the mains, a transformer can be used to give a suitable alternating voltage and at the low mains frequency a laminated ferromagnetic core can be used.

(*b*) For isolation.

Where it is undesirable to have physical connection to the mains, a 1:1 transformer can be used to supply the power. Such transformers are often used with musical instruments such as electric guitars.

(c) For impedance matching.
When the *impedance* of a load is necessarily different from the optimum value required by the generator, a matching transformer can be included between the two impedances to achieve maximum power transfer. If the impedance ratio is n:1 the transformer turns ratio should be $\sqrt{n}$:1.

(d) For coupling.
Step-up transformers are used between stages in electron-tube amplifiers to give voltage gain and step-down transformers between *bipolar-transistor* stages to give current gain. At low frequencies such transformers use laminated ferromagnetic cores but at radio frequencies ferrite cores or even air cores can be used. In RF transformers one or both of the windings may be tuned (and the degree of coupling adjusted) to give the required passband. In oscillators a transformer may be used to provide coupling between input and output circuits of an active device so providing the positive feedback essential for oscillation.

(e) To convert between balanced and unbalanced operation.
Transformers with centre-tapped windings are used extensively in circuits where balanced operation is required e.g. in *push-pull amplifiers*, *ring modulators*.

Where there is no objection to a physical connection between primary and secondary circuits, a transformer need have only one winding with tapping points, the primary or secondary circuit being connected between suitable tapping points. This is known as an autotransformer and line output transformers in television receivers often have a number of tapping points to feed the scanning coils and the various auxiliary circuits which rely on the line output stage for their power.

The graphical symbols for (a) an air-cored transformer and (b) an autotransformer with a ferromagnetic core are shown in *Figure T.11*.

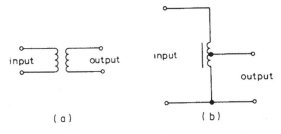

(a)   (b)

Figure T.11 The graphical symbol for (a) air-cored transformer and (b) an auto-transformer with a ferromagnetic core

transient In an acoustic or electrical system, the *signal* which persists for a brief period following a sudden disturbance to the steady-state conditions. Transients have an irregular and non-repetitive waveform. The term is applied particularly to the damped oscillations which follow shock excitation of a resonant system.

327

transient response The ability of a circuit or device to reproduce transients without *distortion*.

transistor A semiconductor device of which the output current can be controlled by the signal(s) applied to one or more input terminals. See *bipolar transistor, field-effect transistor.*

transistor-transistor logic (TTL) A system in which use is made of logic elements comprising chiefly bipolar *transistors*. In one TTL circuit the output stage is *single-ended push-pull* in which one transistor is conductive at a time. Speed of operation is high but is limited by the fact that the bipolar transistors are driven into saturation and, because of *carrier storage*, switch off more slowly than they switch on. This limitation has been overcome in a recent modification to the TT system in which *Schottky diodes* are connected between the collector and the base of the transistors to prevent them from saturating. This has resulted in a significant improvement in speed of operation.

transit angle The transit time expressed as an angle. Transit angle is thus the product of the transit time and the angular frequency.

transition In a waveform an instantaneous change from one value of amplitude to another. An ideal pulse thus contains two transitions. The term 'edge' is sometimes used instead, e.g. the trailing edge of a pulse.

transition frequency (1) Of a bipolar transistor the frequency (f_T) at which the modulus of the common-emitter current amplification factor has fallen to unity. It thus measures the highest frequency at which the transistor can be used as an amplifier; it is, in fact, the gain-bandwidth product for the transistor. (2) See *turnover frequency.*

transitron A *pentode* in which use is made of the virtual cathode set up between the *screen grid* and the negatively-charged *suppressor grid*. The virtual cathode, suppressor grid and anode can be treated as electrodes of a triode, the real cathode, control grid and screen grid being the electrodes of a second triode. Thus the pentode can be used as the equivalent of some two-triode circuits notably multivibrators. The equivalent of a *monostable multivibrator* is known as a *phantastron*.

The negative-resistance characteristic between suppressor grid and screen grid which gives the circuit its name can be used as the basis of an oscillator.

transitron oscillator An oscillator which utilises the negative-resistance kink in the screen-grid characteristic of a pentode with a negatively-biased suppressor grid. See *dynatron oscillator, transitron.*

transit time In general the time taken for a charge carrier to cross a given gap. In *electron tubes* the time taken by *electrons* to travel from the *cathode* to other electrodes is of great importance and, in fact, limits the upper frequency at which the tubes can be used successfully. Thus tubes for high-frequency applications are constructed with very small inter-electrode spacings to minimise transit time. See *disk-seal tube*. On the other hand microwave tubes such as *travelling-wave tubes* make use of transit time in their operation.

transmission line A system of conductors connecting one point to another along which electric or electromagnetic energy can be sent. Thus telephone lines and power-distribution lines are examples of transmission lines but in electronics the term usually implies a line used for the transmission of RF energy, e.g. from a radio transmitter to the antenna.

An essential feature of such a transmission line is that it should guide energy from the sending end to the receiving end without loss by radiation. One form of construction commonly used consists of two similar conductors mounted close together but maintained at a constant separation. These two conductors form the two sides of a balanced circuit and any radiation from one of them is neutralised by that from the other. Such twin-wire lines are used for carrying high RF power, e.g. at transmitters.

For low-power use the coaxial form of construction is often employed, one conductor being in the form of a cylinder which surrounds the other at its centre and thus acts as a screen. Cables so constructed are extensively used to couple FM and television receivers to their antennas.

At frequencies above approximately 1000 MHz transmission lines are usually in the form of *waveguides* which may be regarded as coaxial lines without the centre conductor, the energy being launched into the guide or abstracted from it by probes or loops projecting into the guide.

See *characteristic impedance*, *coaxial cable*.

transmitter In general a device which sends a signal. A telephone transmitter converts sound waves to a.f. signals for transmission over telephone circuits. A radio transmitter sends electromagnetic waves conveying sound or television programmes or other forms of data.

transparent In computing and telecommunications a term used to describe a vital process which occurs automatically and without affecting the user; ideally the user should be unaware that it is happening. An example is the handing over of a mobile user from one base station to another in *cellular radio*.

transponder A combination of radio receiver and transmitter designed to radiate an acknowledging signal automatically when a correct interrogating signal is received.

transputer A combination of microcomputer-type devices designed to achieve higher operating speeds than a conventional microprocessor.

In a normal microprocessor the various operations needed to obtain the desired end result are carried out sequentially as the program indicates. However, in the transputer a higher operating speed is achieved by sharing the operations between a number of microprocessors which operate simultaneously. This parallel mode of working is not easy to achieve technically. Moreover conventional computer languages cannot be used and a new language OCCAM has been designed for use with the transputer.

transverse magnetisation In *magnetic recording*, magnetisation of the tape across its width, i.e. at an angle to the direction of tape movement. This method is used in video tape recorders as a means of achieving the high tape-to-head speed necessary to record the high video frequencies whilst still keeping a low speed of tape movement. The high transverse scanning speed is achieved by using a number of heads around the circumference of a drum which rotates so sweeping the heads across the tape.

trapezium distortion Same as *keystone distortion*.

trapezium modulation A system of *amplitude modulation* used at sound broadcast transmitters to permit a higher average modulation depth and so give improved intelligibility in areas of low field strength. To achieve this result the AF signal is clipped at an adjustable amplitude level and passed through a low-pass filter to suppress the harmonics introduced by the clipper. Thus peak signals become trapezoidal in waveform.

trap valve amplifier Amplifier with a number of outputs with a high degree of isolation between them used for distributing a broadcast programme to a number of destinations.

travelling-wave tube (TWT) An *electron tube* used at VHF and UHF in which the input wave travels along a *slow-wave structure* so that its axial velocity equals that of an electron beam fired along the axis, the resulting velocity modulation and consequent bunching providing the required amplification.

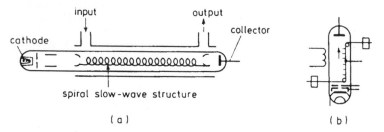

(a)   (b)

Figure T.12 (a) Simplified construction of a travelling-wave tube and (b) graphical symbol for an O-type travelling-wave tube with indirectly-heated cathode, intensity-modulating electrode, focusing electrode, slow-wave structure with connection, collector, focusing coil and window couplers to input and output waveguides

As shown in *Figure T.12*(a) the tube contains an electron gun at one end and a collector at the other, the intervening space being occupied by a helix or other form of slow-wave structure. The input signal is fed into one end of the slow-wave structure and the amplified output is abstracted from the other. The slow-wave structure is designed to reduce the velocity of electromagnetic waves which thread the structure so that the axial velocity is equal to that of the electron beam. Interaction between the wave and electron beam produces amplification. For a helix the factor by which the wave velocity is reduced is the ratio of the circumference to the pitch. Travelling-wave tubes are extensively used as microwave amplifiers and can be designed to have very wide bandwidths. The graphical symbol for one type of travelling-wave is shown in *Figure T.12*(b).

triac See *bidirectional thyristor, thyristor.*

trigatron Same as *trigger tube.*

trigger circuit A two-state circuit which can switch from one state to the other automatically or by external signals. See *astable circuit, bistable circuit, multivibrator.*

triggering The initiation of a change of state in a device by means of an external signal. The significant feature of triggering action is that the change of state continues automatically and independently of the external signal once it has been started. See *bistable circuit, monostable circuit.*

trigger tube A *gas-filled tube* containing a spark gap in which the main

Figure T.13 Graphical symbol for a trigatron or trigger tube

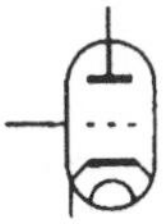

Figure T.14 Graphical symbol for an indirectly-heated triode

discharge can be initiated or terminated by a signal applied to an auxiliary electrode. If the external signal causes conduction in the tube, the state of conduction persists after the triggering signal has ceased. The graphical symbol for a trigger tube is given in *Figure T.13*.

trimming In general the process of fine adjustment of a variable resistor, capacitor or inductor. In particular a trimming capacitor is a small preset component connected in parallel with each section of a ganged tuning capacitor and which is adjusted to compensate for small differences in capacitance between the sections or to achieve three-point *tracking* in *superheterodyne receivers*.

trinitron A precision-in-line television picture tube pioneered by the Japanese.

triode An *electron tube* with three electrodes namely a *cathode*, control grid and *anode*. This was the first amplifying tube and was invented by Lee de Forest in 1905 when he inserted a grid in a diode. By suitably biasing the grid with respect to the cathode it is possible to control the density of the electron stream from the cathode which reaches the anode. Thus the grid voltage controls the anode current and by including a load in the anode circuit, the anode voltage becomes an amplified copy of the signal applied to the control grid. For linear amplification the grid must be suitably biased.

Such tubes have been widely used in electronic equipment since their invention, their use being confined to audio and subaudio frequencies. The internal anode-to-control grid capacitance makes triodes unsuitable for RF amplification. The graphical symbol for an indirectly-heated triode is given in *Figure T.14*. For most applications triodes have now been superseded by *transistors*.

triple detection See *double superheterodyne reception*.

tri-state output Same as *three-state output*.

trochotron A multi-electrode tube used for *counting*. It has a single *cathode* surrounded by a number of target electrodes and a magnetic field is employed to ensure that the electron beam moves around the cathode to strike the targets in sequence.

tropicalising The practice of *encapsulating* whole equipments or circuit boards in materials designed to survive harsh environments, such as high humidity, temperature extremes or the presence of corrosive chemicals.

true colour Colour stored digitally as 24 bits per *pixel*, that is 8 bits to represent the levels of each of the red, green and blue primaries. This gives a *palette* of 2^{24} = 16,777,216 colours, more than the human eye can distinguish. See *colour model*.

trunk Same as *bus*.

331

truth table For a logic function a table showing all the possible combinations of input signals and, for each, the output signal. As a simple example the following is the truth table for an AND function showing that a logic-1 output is obtained only when both inputs stand at the 1-state.

input 1	input 2	output
0	0	0
0	1	0
1	0	0
1	1	1

If the logic convention is reversed so that logic-1 becomes logic-0 and vice versa the truth table takes the form shown below

input 1	input 2	output
1	1	1
1	0	1
0	1	1
0	0	0

which shows that a logic-1 output is obtained when either input is at logic 1. This is the truth table for an OR gate and it shows that whether a given gate acts as an AND gate or an OR gate depends on the logic convention adopted.

tube See *electron tube*.

tuned circuit A resonant circuit of which the resonance frequency has been adjusted to the desired value.

tuned radio-frequency receiver (TRF) A radio receiver incorporating one or more stages of radio-frequency amplification before the detector and one or more audio-frequency stages after it. A block diagram of a TRF receiver is given in *Figure T.15*. Such receivers were used in the 1920s but the

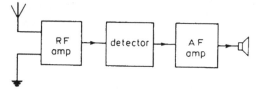

Figure T.15 Block diagram of a tuned radio-frequency receiver

number of tuned circuits it was possible to include in them was limited by the difficulty of ganging the tuning controls. The selectivity was therefore poor and TRF receivers were superseded by *superheterodyne* types in the early 1930s.

tuner (1) Short for *tuning capacitor*. (2) A radio or television receiver containing little or no amplification after the detector and no internal means of monitoring reception. It is normally used in conjunction with an external amplifier, monitor or tape recorder, e.g. in a hi-fi system.

tuning (1) Of a *tuned circuit* the process of adjusting the resonance frequency to a desired value. For an LC circuit this can be done by

332

adjusting the capacitance, the inductance or both. (2) Of a *receiver* the process of adjusting it to respond to a wanted signal.

tuning capacitor A variable capacitor forming part of a tuned circuit and used to adjust the resonant frequency.

tuning indicator Of a *receiver* a device giving visual indication of correct tuning. A number of devices are in common use. The correct tuning point may be indicated by a simple meter movement, by the area of *luminescence* on the screen of a small *cathode ray tube* or, for TV receivers, by the width of a line of *luminescence* on the screen.

tunnel diode A *pn diode* which makes use of the *tunnel effect*. The forward characteristic of a tunnel diode exhibits a region of negative resistance as shown in *Figure T.16* (a) which can be exploited in an oscillator, an electronic switch or for low-noise amplification at frequencies up to 1000 MHz. The graphical symbol for a tunnel diode is shown in *Figure T.16* (b).

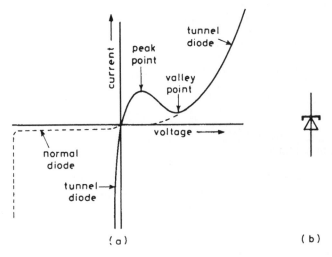

Figure T.16 (a) Characteristics for a tunnel diode (solid) and a normal diode (dashed) and (b) graphical symbol for a tunnel diode

tunnel effect Penetration of the potential barrier at the junction between highly-doped p- and n-regions by *electrons* which have insufficient energy to surmount the barrier. This effect is impossible to explain in terms of classical physics but is in agreement with quantum mechanics. Because of the tunnel effect the forward characteristics of a tunnel diode differs from that of a normal pn diode. For example current begins to flow across the junction at a very low value of forward bias and it increases with increase in bias up to a maximum (known as the peak point) at about 0.1 V for a germanium diode and then falls reaching a minimum (known as the valley point) at about 0.3 V where the characteristic joins that for a normal pn diode as shown in *Figure T.16* (a). Thus the tunnel effect gives the

333

characteristic a region of negative slope between the peak and valley points. The reverse characteristic for a *tunnel diode* also differs from that of a normal pn diode in that breakdown occurs almost at zero bias as for a *backward diode*.

turn-off thyristor A *thyristor* which can be switched from the on-state to the off-state by a signal applied to the gate. n-gate and p-gate types exist and their graphical symbols are given in *Figure T.17*.

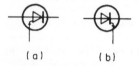

(a)   (b)

Figure T.17 Graphical symbol for turn-off thyristor (a) with n-gate and (b) with p-gate

turnover frequency (US) In disk recording the frequency at which the recording characteristic changes from a *constant-amplitude* to *constant-velocity* type.

tweeter A small loudspeaker unit designed to operate at audio frequencies above approximately 2 kHz. Tweeters are often piezo-electric types and are used in combination with other units and crossover networks in high-quality wide-range loudspeakers.

twin-interlaced scanning See *interlaced scanning*.

twin-T network Same as *parallel-T network*.

twisted pair A cable consisting of two individually insulated wires twisted tightly together. The twisting brings the two conductors in close proximity so that they respond equally to any interfering fields. The induced voltages therefore cancel, minimising the effects of the interference.

two-quadrant multiplier A multiplier the operation of which is restricted to one sign only of the input signals.

two-terminal-pair network A *network* of which the characteristics are specified at four particular terminals, two acting as input and the other two as output. The network may have other terminals. Alternatively known as a two-port network.

U

ultra-linear amplification An audio-frequency push-pull output stage using *tetrodes* or *pentodes* in which the screen grids are fed from tappings on the primary winding of the output transformer. This arrangement feeds a fraction of the output voltage to the screen grids thus giving *negative feedback* and by suitable choice of the position of the tapping points odd harmonic distortion can be reduced to a minimum. Even harmonics are, of course, eliminated by the use of push-pull and thus an output stage of this type is capable of excellent linearity.

ultrasonics The study of acoustic waves in material media, the frequency being above that of the audible range. Such waves can be generated by *magnetostriction* and *piezo-electric* devices and have many practical applications. For example they are used to assist machining operations on brittle materials, to assist soldering, for echo-sounding purposes and in remote control of television receivers. The term is preferred to 'supersonic' which is now reserved for speeds greater than that of sound: supersonic persists, however, in the word superheterodyne.

ultraudion Early form of triode detector using *positive feedback*.

ultra-violet radiation Electromagnetic waves with a wavelength between approximately 10 and 380 nanometres, i.e. immediately adjacent to the visible spectrum. Although invisible to the human eye, such radiation can affect photographic plates and can produce ionisation.

unbalanced circuit See *asymmetrical circuit*.

uncommitted logic array (ULA) An *array* of logic elements fabricated without interconnections on a substrate. Such arrays can be mass produced by the *planar* process and can be used for a number of different purposes depending on the interconnections between the elements. Customers specify the interconnections they require and these are made in a subsequent metallising operation.

undercoupled circuits Two circuits resonant at the same frequency and between which the degree of *coupling* is intentionally less than the critical value. The effect of such coupling is to give a response with a single peak at the common resonance frequency and with an amplitude proportional to the degree of coupling but less than that for critical coupling.

undershoot Form of transient distortion of a step or pulse signal in which the response makes a temporary excursion before the main transition and in the opposite direction. See *Figure U.1*.

unijunction transistor A *transistor* consisting essentially of a filament of say n-type semiconductor with a p-junction near its centre, connections being brought out from the ends of the filament and from the p-region. The device is well suited for use in simple pulse-generating circuits. The graphical symbol for a unijunction transistor is given in *Figure U.2*.

unilateral conductivity Property of certain devices, notably *electron tubes* and semiconductor pn junctions, of conducting electricity easily in one

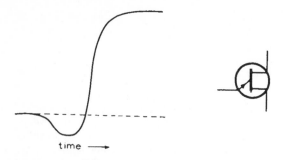

Figure U.1 Reproduction of the leading
edge of a pulse showing undershoot

Figure U.2 Graphical symbol for a unijunction transistor

direction and not at all or only slightly in the opposite direction. Such devices have obvious applications as *rectifiers*, *detectors* and *switches*.

unilateral impedance A mutual impedance through which power can be transmitted in one direction only. *Active devices* such as *electron tubes* and *transistors* are examples of unilateral impedances because they enable the current in one circuit to be controlled by the current in another circuit but the receiving-circuit current does not affect the sending-circuit current. Alternatively known as control impedances.

unilateralisation Neutralisation circuit used with *bipolar transistors* and in which the *internal feedback* is counteracted by an external series *RC* circuit, the constants of which are chosen to neutralise the internal collector-base resistance and capacitance. Such neutralisation, if properly adjusted, makes the input and output circuits of the transistor independent of each other so making stable RF amplification possible.

unipolar transistor A transistor which operates by virtue of only one type of charge carrier. An n-channel *field-effect transistor* is an example in which the charge carriers are *electrons*.

unit step See *step function*.

unitunnel diode A *tunnel diode* in which the current at the valley point is approximately equal to that at the peak point. See *Figure T.16*(a).

universal asynchronous receiver/transmitter (UART) In *computers* and *data-processing equipment* an interface unit between the central processor and a peripheral device.

universal receiver A receiver which operates from AC or DC mains. Such receivers were common when many parts of the country still had DC mains. The receivers had series-connected electron-tube heaters operating via a dropper resistor from the mains supply and HT was obtained via a half-wave rectifier which acted as a series resistor on DC mains. Also termed AC-DC receiver.

univoltage lens An *electrostatic lens* consisting of three apertured electrodes, the outer two with a fixed common bias relative to electron-gun cathode potential, focus being affected by variation of the voltage of the inner electrode. Such a lens has a very short focal length and is often used in *electron microscopes*.

UNIX A *multitasking operating system* available on a wide variety of computers. It allows multiple users to use the same computer and multiple applications may be run simultaneously.

unscrambler See *scrambler*.

unstable state Of a circuit a state which cannot persist, the circuit automatically reverting to an alternative state. When a circuit is placed in an unstable state, changes occur within it which automatically terminate the state after a certain time. Usually the change is the discharge of a capacitor through a resistor which, when the discharge has reached a certain point, precipitates the return to the alternative state. Thus the duration of the unstable state depends on the time constant of an *RC* combination and can be controlled by suitable choice of value for R or C. *Monostable circuits* have one unstable state, *astable circuits* have two.

upload To transmit data to another computer.

upward modulation Same as *positive modulation*.

user-friendly Simple and convenient to operate, needing no specialised skills or knowledge.

utility A piece of *software* that performs a useful function but which would normally be used in conjunction with other *applications*. It is not sufficiently comprehensive to be regarded as an *application* in its own right. For example, a program to *load* various *bitmap* formats would be regarded as a utility. It might be used in conjunction with graphics or desktop publishing software needing to display the bitmap.

V

vacuum tube A tube which has been exhausted to a high degree of vacuum so that the amount of gas present in the envelope is too small to have any effect on the electrical characteristics. The tube may be sealed or continuously pumped. Currents in a vacuum tube are carried by *electrons* not *ions*.

valence The force which binds atoms together to form a molecule.

valence band In an *energy-level diagram* the range of energies corresponding to states which can be occupied by the valence electrons binding the atoms together.

valence electrons *Electrons* occupying the outermost orbits of an atom and which are shared when atoms combine to form molecules. See *covalent bonds*.

value (1) The numerical level of the primary property of a passive component, for instance, the number of Ohms of resistance offered by a resistor. (2) In computing, the contents of a *variable*. (3) In the HSV *colour model* the overall level of the light emitted by the colour.

valve Term originally applied to a device which allowed current to pass through it in one direction only (i.e. a *diode*) but later extended to include vacuum devices with grids which could amplify and oscillate, and gas-filled devices such as *thyratrons*. The term has been replaced by *electron tube* which is used throughout this dictionary.

vapour cooling Of a high-power *electron tube* a method of removing heat which utilises the latent heat of vaporisation of water. The external anode has thick copper walls containing vertical passages and is immersed in distilled water which boils in the passages, the latent heat required to turn the water into steam being taken from the tube.

varactor A reverse-biased *junction diode* the *capacitance* of which can be controlled by adjustment of the bias voltage. Such diodes are used in *parametric amplifiers*, for automatic frequency control and for tuning in radio and TV receivers.

variable In computing, a block of data held in a reserved area of memory and identified by an unique name. The data held may be numeric or may consist of a text *string* and may be accessed by *programs*; instructions to output the variable yield its current contents, not its name. See also *array*.

variable-capacitance diode See *varactor*.

variable-carrier modulation Same as *floating-carrier modulation*.

variable-mu tube An *electron tube* in which the *control grid* is wound with a pitch which varies along its length so permitting the mutual conductance to be controlled by choice of grid-bias value. A typical I_a–V_g characteristic for a variable-mu tube is shown in *Figure V.1*: for a small value of grid bias as at A the slope (and gain) is high whereas for a point such as B the slope (and gain) is low. An advantage of these tubes is that they can accept a larger input signal at B than at A for comparable distortion levels. Such

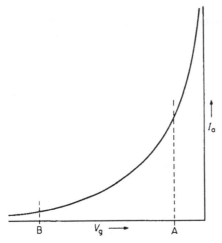

Figure V.1 I_a–V_g characteristic of a variable-mu electron tube showing high mutual conductance at A and low at B

tubes were used in situations where it is required to control gain by choice of bias value as in *automatic gain control* systems.

variable-reluctance pickup A *pickup* in which the movements of the reproducing stylus vary the reluctance of a magnetic circuit and generate corresponding EMFs in a stationary coil. The moving-iron pickup is an example of a variable-reluctance pickup.

varicap diode Same as *varactor diode*.

variocoupler A mechanical arrangement permitting control of the degree of magnetic coupling between two coils. For example one coil may be fixed and the other brought up to it to increase the mutual inductance between them.

variometer A variable inductor consisting of two coils, one of which can be rotated within the other so as to vary the mutual inductance between them. The coils are connected in series and the mutual inductance can add or subtract from the series inductance depending on the position and sense of the rotatable coil.

varistor A resistor of semiconductor material with a non-linear current-voltage characteristic.

vector graphics A system of computer graphics in which lines and shapes are stored as co-ordinate-based data. See *graphics*.

velocity factor Of a *transmission line* the ratio of the velocity of an electromagnetic wave along the line to its velocity in free space. Depending on the construction of the line, values of velocity factor lie between 0.6 and 0.97.

velocity modulation The process of periodically altering the velocity of an electron stream by subjecting it to a high-frequency electric field which alternately accelerates and decelerates the beam. If the period of the variation is comparable with the *transit time* of the electrons in the space

339

concerned, the electrons subsequently gather into bunches. *Bunching* makes possible microwave amplification and oscillation in *klystrons* and *travelling-wave tubes*. See *Applegate diagram*.

velocity microphone (US) Same as *pressure-gradient microphone*.

vented enclosure A cabinet housing a loudspeaker unit, closed except for a forward-facing vent, cabinet and vent forming a *Helmholtz resonator* tuned, by choice of dimensions, to a low audio frequency. Radiation through the vent reinforces that from the diaphragm at low frequencies so extending the effective frequency response of the loudspeaker.

vertex Same as *node* (1).

vertical amplifier In oscilloscopes the circuits which amplify the signals responsible for vertical deflection of the beam. Also known as a Y amplifier.

vertical blanking Same as *field blanking*.

vertical-groove metal-oxide-semiconductor transistor (VMOS) An *insulated-gate field-effect transistor* so constructed that the current flow is nearly perpendicular to the plane of the layers as in planar *bipolar transistors*, a groove-shaped gate giving very short channel lengths. Such transistors thus have very high switching speeds and low drain-source resistance in the on-condition. Large source and drain areas permit the transistors to give substantial output power. A simplified diagram of the structure of a VMOS transistor is given in *Figure V.2*.

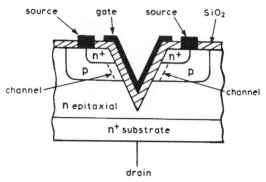

Figure V.2 Simplified construction of VMOS transistor

vertical hold See *hold control*.

vertical polarisation Property of an electromagnetic wave in which the plane of polarisation of the electric field is vertical.

vertical timebase In a *cathode ray tube* the circuits generating the signals which give vertical deflection of the beam. In TV this is usually termed the field timebase.

very-large-scale integration (VLSI) See *monolithic integrated circuit*.

vestigial sideband transmission (VSB) A system of *amplitude modulation* in which one sideband is transmitted in full, only part of the other sideband being transmitted (usually that corresponding to the lower modulating frequencies).

The system is used for transmitting the vision component of a TV signal and has the advantage that it occupies less bandwidth than if both sidebands were transmitted in full.

vibrator converter An equipment in which a tuned reed, maintained in vibration by a low-voltage DC supply, is used to generate DC at a higher voltage. The equipment is, in effect, a *vibrator inverter* in which the alternating output voltage is rectified to give the required DC supply often by contacts on the vibrating reed itself so avoiding the need for a separate rectifier.

vibrator inverter A device in which a tuned reed, maintained in vibration by a low-voltage DC supply, is used to generate alternating current usually at higher voltage. The interrupted current from the low-voltage supply is applied to a transformer which generates the required alternating voltage. Such equipments are used, for example, for powering mobile equipment normally designed for operation from mains supplies.

video cassette recorder (VCR) A *video tape recorder* which uses cassette tapes. In practice domestic VCRs also incorporate a UHF or VHF television *tuner*, allowing television broadcasts to be recorded off air without the need for a separate television receiver. A VCR usually incorporates a *dedicated microcomputer* which can be programmed to record future television broadcasts automatically.

video digitiser An equipment, often contained in an *expansion card*, which converts video frames to *bitmaps*. It is also sometimes called a *frame grabber*.

video disk A disk containing video data, usually in a digital format. The first video disks used a recording technique similar to that in *Compact Disc*, but were 12 inches in diameter and could accommodate a 1½-hour feature film. With advances in compression and storage technology, such as *MPEG*, it is now possible to store a feature film on a 5 inch compact disk.

Video Graphics Array (VGA) A VDU display standard introduced by IBM on its PS/2 computer in 1987 and since adopted as an industry standard. The highest resolution in the original VGA standard was 640 × 480 pixels with 16 colours.

video memory The RAM which stores the image on a VDU. The video memory requirement depends on the *video mode* since it is proportional to the number of pixels in the display and to the logarithm (base 2) of the number of colours to be displayed. The original VGA standard (640 × 480 × 16 colours) required 640 × 480 × 4 = 1,228,800 bits = 150 kilobytes. A computer fitted with 1 Mbyte of video memory can support a display of 640 × 480 pixels in true colour (24 bits per pixel) which requires 900 kilobytes or a display of 800 × 600 pixels at 16 bits per pixel (65,536 colours) which requires 937.5 kilobytes.

video mode or **screen mode** The specification of a *VDU* display. The data include: the number of *pixels* across the screen, the number of pixels down the screen, the nominal width and height of a single pixel, the maximum number of colours that may appear on the screen without a change of *palette* and the refresh rate, that is the number of frames per second. Most computers allow a choice of screen modes to suit various purposes.

video on demand A proposed system in which subscribers may demand any

341

of a wide range of video, audio or computer material from a remote service provider. The requested material is then transmitted in compressed format over a landline, assumed to be a fibre-optic cable, to the subscriber's terminal. At the time of writing pilot schemes are in operation.

video signal In TV the signal obtained by combining the picture signal with the synchronising signal. The way in which the signals are combined is illustrated in *Figure B.18* (a). See *picture signal, sync signal*.

video still camera An electronic camera which takes individual frames and stores them as *bitmaps*. These are subsequently *uploaded* to a computer for processing and printing as computer graphics.

video tape recorder (VTR) An equipment which records video and audio, usually in analogue format, on tape. See also *video cassette recorder*.

videotex A system whereby *signals* sent over telephones lines can be displayed as letters, figures or diagrams on the screen of a TV receiver. The system uses the same symbols and techniques as *teletext* and British Telecom is now using such a system under the name Prestel to make available to telephone users some hundreds of thousands of pages of information, e.g. on stock exchange prices, railway timetables, sports results.

videotron Same as *monoscope*.

vidicon A low-velocity TV *camera tube* with a photoconductive target. The construction of a vidicon tube is illustrated in *Figure V.3*. The target consists of a transparent signal plate on which is deposited a layer of photoconductive material. The scanned face of the target is stabilised at electrongun cathode potential as in any low-velocity tube and the signal plate is given a small positive bias. Current thus flows longitudinally through the target thickness and the magnitude of this current at any point on the target depends on the target resistance which in turn depends on the amount of light in the optical image at that point on the target. These longitudinal currents set up a charge image on the scanned face of the target and this is

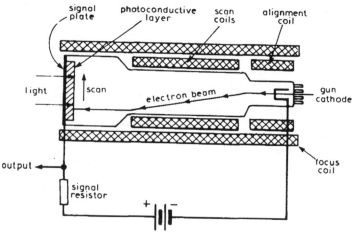

Figure V.3 Vidicon tube

neutralised by the scanning beam once per picture period. The target has a significant capacitance between front and rear faces and this is charged by the longitudinal currents in the interval between successive scans by the beam so that the charge image grows during the picture period: in other words the tube has the charge storage which is essential to achieve adequate sensitivity.

Early vidicon tubes suffered from lag, i.e. there was an appreciable delay between a change of light input and the corresponding change in target resistance but intensive work on photo-conductive materials has resulted in satisfactory sensitivity and lag. Modern tubes (plumbicons) have multi-layer targets of lead oxide.

The vidicon is a particularly simple and compact tube and it is not surprising that it is the standard tube for inclusion in colour TV cameras.

viewdata Same as *videotex.*

virtual cathode A region between the electrodes of an *electron tube* which contains an accumulation of *electrons* and can be used as a source of electrons for nearby electrodes. Such regions occur when the electron stream through the tube encounters a retarding field, e.g. between the screen grid and suppressor grid of a pentode. See *phantastron*, *space charge*, *transitron*.

virtual memory Memory which appears to be available to a computer, although it does not exist as physical RAM. Usually virtual memory is a file on a *hard disk* accessed via a buffer. Using virtual memory a computer can process files that are larger than the physical RAM fitted in the machine.

virtual reality (VR) An *interactive* computer simulation of a three-dimensional environment, such as a building, a town, a ship or a region of outer space, stored as a mathematical model. Using *vector graphics* techniques the computer repeatedly generates in *real time* the instantaneous view of the environment as seen by a hypothetical observer moving around within it. In the most sophisticated form of VR users wear special headsets containing stereoscopic LCD displays, a binaural sound system and sensors which detect movements of the head. Movement and interactive functions are controlled through further sensors in gloves, or even a whole body suit, worn by the user. These give him the illusion of being inside the virtual world, able to explore it and to interact with it in various lifelike ways. Simpler versions of VR use a conventional monitor and pointing device. VR has many serious applications in such fields as medicine, architecture, town and country planning and training (such as flight simulators). Its leisure uses, however, have attracted much publicity, especially the fantasy games in which several players interact with each other within the virtual environment. See also *telepresence.*

virus A portion of *computer program* devised by a malevolent programmer. Usually viruses are embedded in applications software and they replicate themselves, by embedding themselves in other applications, when the host application is run or even when its icon appears in a directory display. They spread from one system to another when the applications harbouring them are copied and transmitted to other systems. Some viruses are comparatively harmless, perhaps displaying a humorous message on a certain day, but others cause serious damage, overwriting or corrupting valuable data and programs.

vision carrier In TV transmission the carrier wave which is modulated by the *video signal.*

vision frequency In TV transmission the frequency of the *vision carrier.*

vision pickup tube Same as *camera tube.*

vision signal In TV transmission the carrier wave modulated by the *video signal.*

visual display unit (VDU) In *computers* or *data-processing equipment* a *cathode ray tube* used for displaying characters or graphical information under program control. Often a keyboard is associated with visual display unit for displaying the input data.

visualisation The use of computers to create high-quality images of objects that do not exist except as mathematical models. In industry such techniques are used to create photographic-quality images of proposed products (such as automobiles) from engineering drawings. Sometimes *ray tracing* is used, but sophisticated *vector graphics* applications also yield acceptable images, depending on the nature of the subject.

voice coil (US) Same as *speech coil.*

voice recognition Computer technology that recognises and is able to respond to spoken words picked up using a microphone and converted to digital format. It has many applications in security systems and in providing facilities for the handicapped who cannot use a keyboard or pointing device.

volatile Liable to disappear. Data stored in normal RAM is volatile in that it will disappear if the equipment is switched off or its power supply is interrupted. Data stored in battery-backed RAM or on magnetic media such as a floppy or hard disk is non-volatile and is not lost at power-off.

voltage The value of an electromotive force or potential difference expressed in volts.

voltage amplifier A circuit incorporating one or more *active devices* and designed to amplify voltage waveforms. The term distinguishes such amplifiers from those designed to amplify current waveforms or to deliver power to a load. It is the ratio of source impedance to input impedance and of output impedance to load impedance which determines whether a device or amplifier is best regarded as a current amplifier, power amplifier or voltage amplifier. If both ratios are small compared with unity the signals transferred from source to input and output to load are best regarded as voltage waveforms. The associated signal current waveform is of minor interest in a voltage amplifier and is usually of small amplitude. Because *electron tubes* and *field-effect transistors* have a high input impedance and a lower output impedance they are usually regarded as voltage amplifiers.

voltage-controlled oscillator (VCO) An oscillator of which the frequency can be controlled by adjustment of an input voltage. Such an oscillator forms part of *phase-locked-loop* systems. The oscillator may be an *LC* type generating a sinusoidal waveform, the frequency being controlled by a varactor connected across the oscillatory circuit. Alternatively it may be an astable *multivibrator* the duration of the unstable state (and hence the free-running frequency) being controlled by the base bias voltage.

voltage-dependent resistor (VDR) A non-linear resistor the resistance of which decreases markedly with increase in applied voltage. The current-voltage relationship for a voltage-dependent resistor is of the form $I = kV^n$ where k is a constant and n lies between 3 and 7. Such resistors are usually of silicon carbide pressed with a ceramic binder into disks or rods and fired at about 1200°C.

voltage divider Same as *potential divider*.

voltage dropper A *resistor* connected in series with a power supply to drop the voltage applied to the load to a desired value. An example occurs in radio and TV receivers where the electron-tube heaters are connected in series. Where the total heater voltage is less than the mains voltage a resistor is inserted in the circuit to act as a voltage dropper. From Ohm's law the required resistor value is given by (volts to be dropped)/current.

voltage feedback A system in which the negative feedback signal is directly proportional to the voltage across the load. Frequently the feedback signal is derived from a potential divider connected across the load as shown in *Figure V.4*, the potential divider having a total resistance large compared

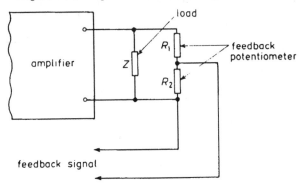

Figure V.4 Basic principle of voltage feedback

with the load value. The effect of voltage negative feedback is to improve linearity, decrease gain and to decrease the effective output resistance of the *active device* driving the load, i.e. it tends to make this device a constant-voltage source.

voltage gain Of an *active device* or amplifier the ratio of the output voltage to the input voltage. Where the voltage gain depends on the value of the load resistance it is necessary to specify the load value when the voltage gain is quoted. The gain may be expressed as a ratio or in decibels as $20 \log_{10} V_{out}/V_{in}$.

voltage gradient In a *conductor* the difference in potential per unit length. In an electric field the difference in potential per unit distance along the normal to the lines of force at the point in question.

voltage level See *level*.

voltage-multiplier rectifier A combination of *rectifiers* and *capacitors* which produces an output voltage approximately equal to an exact multiple

345

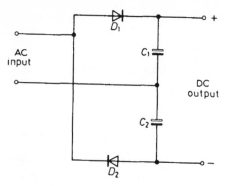

Figure V.5 A voltage-doubler rectifier

of the peak value of the alternating input voltage. As an example *Figure V.5* gives the circuit diagram for a voltage doubler. When D_1 conducts C_1 is charged to the peak value of the alternating input. On the next half cycle D_2 conducts and C_2 is also charged to the peak value of the input. C_1 and C_2 are connected in series and their voltages add so yielding an output voltage equal to twice the peak value of the alternating input.

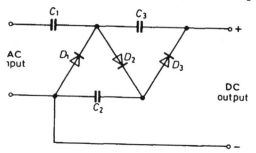

Figure V.6 A voltage-multiplier rectifier

By using a circuit of the type shown in *Figure V.6* it is possible to obtain an output of n times the peak value of the input voltage where n is the number of rectifiers (and the number of capacitors) employed. The particular example illustrated is, of course, a voltage tripler.

voltage-reference diode A semiconductor device which when carrying a current within a specified range develops across its terminals a particular voltage of specified accuracy which can be used for reference purposes.

The device consists of two *Zener diodes* connected in series back-to-back as indicated in the graphical symbol shown in *Figure V.7*. If the applied voltage has the polarity shown here and is increased sufficiently, diode D_1 is reverse biased and breaks down at its Zener voltage, D_2 being forward biased and so acting as a low-value series resistor. D_1 has a positive temperature coefficient and D_2 is designed to have an equal

346

Figure V.7 Graphical symbol for a voltage-reference diode

negative coefficient so that the device as a whole has a breakdown voltage independent of temperature provided the bias current through the device is maintained constant.

voltage-reference tube A cold-cathode gas-filled tube which when carrying a current within a specified range develops across its terminals a particular voltage of specified accuracy which can be used for reference purposes.

voltage-regulator circuit See *regulation*.

voltage-regulator diode A *Zener diode* used for voltage regulation. A simple circuit is shown in *Figure V.8*. The supply voltage must exceed the *Zener breakdown* voltage so that the voltage across the diode is at breakdown value. The series resistor R_s is so chosen that even when the load takes its maximum current there is still some current through the diode so maintaining the voltage across it at the Zener value.

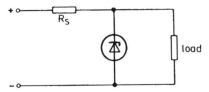

Figure V.8 Simple voltage-regulator circuit using a Zener diode

voltage-stabiliser circuit See *stabilisation*.

voltaic cell A cell in which chemical action takes place between two electrodes of different materials immersed in an electrolyte and gives rise to an EMF between the electrodes.

voltmeter An instrument for indicating or recording the magnitude of a voltage in volts. For high voltages the measurement may be in kilovolts (kV) and for small voltages in millivolts (mV) or microvolts (μV). The resistance of a voltmeter must be high so that, when connected across a circuit, it takes negligible current and so has little effect on the voltage being measured.

volume (1) See *programme volume*. (2) In some computers *volume* refers to a storage medium. This might be a *floppy disk* or a whole *hard disk* or a *partitioned* area on a *hard disk*.

volume compression See *compression*.

volume expansion See *expansion*.

volume indicator or **volume meter** See *programme meter, VU meter*.

volume unit A transmission unit for measuring programme volume. For a sinusoidal signal one volume unit equals one decibel. The VU meter is calibrated in volume units.

von Neumann's principle John von Neumann can be regarded as the father of the modern computer because it was he who in 1944 suggested that the program should be stored electronically within the computer itself so that it could be executed at a speed limited only by that of the logic circuits of the computer. Internal storage also meant that the program could be modified as a result of processing. Hitherto the program had been stored externally on cards or tape and was fed into the computer at low speed. 'Stored program control' is thus the essence of von Neumann's principle.

VU meter A *voltmeter* for measuring programme volume. It consists of a moving-coil instrument of a specified type fed from a bridge rectifier which is, in turn, fed from the programme line via a series resistor. Unlike the *peak programme meter* the VU meter needs no source of power for its operation other than the signal input. The VU meter measures the total energy of the input signal and not the peak value and its scale is markedly non-linear. It is, however, less expensive than the peak programme meter and is extensively used for measuring programme volume particularly in audio tape recording.

W

wafer (1) A thin layer of insulating material carrying switch contacts. Complex switches often consist of several wafers operated by a common control. (2) A thin slice of semiconductor material which, after treatment, becomes a multiplicity of *planar transistors* or of *integrated circuits*.

wafer loudspeaker (1) A *moving-coil loudspeaker* with very little depth designed for use in portable receivers or cars. The small depth is achieved by mounting the magnet within the conical diaphragm. (2) A piezo-electric *tweeter* with very little depth.

walkie-talkie A colloquial term for a portable *transceiver*.

water cooling Of a high-power *electron tube* a method of removing heat by pumping water through a jacket or tubing in direct contact with the external anode or collector, the water giving up the acquired heat in an air-cooled heat exchanger outside the equipment.

Watt The unit of electrical power, symbol W or VA. Practical power values in electronics range from many kilowatts (kW or kVA) through milliwatts (mW) to microwatts (μW).

wave A disturbance propagated through a medium or through space. The disturbance can be, for example, the longitudinal mechanical displacement of a material medium constituting a sound wave or the variations in the transverse magnetic and electric fields constituting a radio wave. The graphical representation of such a disturbance is also termed a wave.

wave analyser Same as *harmonic analyser*.

waveband A frequency range in the electromagnetic spectrum allocated to a specific radio service. For example the range 525 kHz to 1.6 MHz approximately, known as the medium waveband, is devoted to amplitude-modulated broadcasting.

waveform The shape obtained by plotting the instantaneous amplitude of a varying quantity against time in rectangular coordinates. The waveform of a signal can also be displayed on an oscilloscope by applying the signal to the Y plates and a time base to the X plates.

waveguide In general a system of boundaries between materials or media of different refractive index capable of guiding electromagnetic waves. Thus a length of solid dielectric can act as a waveguide, the wave being propagated along it by successive reflections at the walls. The most familiar type of waveguide is, however, the hollow conducting tube, usually of rectangular cross section. There are various modes of propagation within waveguides and the wave may be launched in it or abstracted from it by a probe or a loop projecting into the waveguide. For efficient propagation the cross-sectional dimensions must not be less than half a wavelength and thus to keep the cross section to a reasonable size the frequency of the elctromagnetic waves must be high: it is usually in the microwave region.

wavelength (λ) Of a periodic wave the distance between points of equal phase in consecutive cycles measured at any instant in the direction of propagation of the wave. It is equal to the distance travelled by the wave in one period of oscillation. For electromagnetic waves the wavelength λ is related to the frequency f by the expression

$$\lambda = \frac{v}{f}$$

where v is the velocity of propagation. Thus a frequency of 1 MHz corresponds to a wavelength of 300 m, the velocity of propagation being 3×10^8 m/s.

wavemeter Instrument for measuring the wavelength of an electromagnetic wave either directly or, more usually, indirectly by measuring the frequency. See *absorption frequency meter*.

wavetrap Same as *rejector circuit*.

weighting The modification of objective measurements for example to obtain better correlation with subjective assessments. As an example audio noise measurements may be made via a frequency-discriminating (weighting) network which allows for the way in which the annoyance value of noise varies with frequency over the audio band.

Weston cadmium cell A standard *primary cell* with electrodes of mercury (positive) and cadmium-mercury amalgam (negative), electrolyte of cadmium sulphate and depolariser of mercurous sulphate. The EMF is 1.108 V at 20°C and the variation of EMF with temperature is known to a high degree of accuracy. See *standard cell*.

What You See Is What You Get (WYSIWYG) A term sometimes used to describe computer graphics and DTP applications in which the VDU display accurately (within its capabilities) represents the printed output.

Wheatstone's bridge A *network* of four *resistors* in bridge formation used for the measurement of unknown resistances. The circuit is shown in *Figure W.1*. If the ratio of R_1 to R_2 is the same as that of R_3 to R_4 the PD across R_2 is equal to that across R_4 and the galvanometer gives zero deflection. At balance the following relationship applies

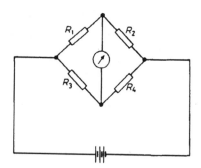

Figure W.1 Wheatstone's bridge circuit

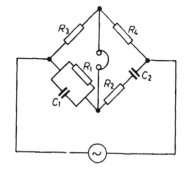

Figure W.2 Wien bridge circuit

$$\frac{R_1}{R_2} = \frac{R_3}{R_4}$$

Thus, if R_1, R_2 and R_3 are known, R_4 can be calculated. R_1 and R_2 are known as the ratio arms.

white compression or **white crushing** In TV, reduction of the gain at signal levels corresponding to white compared with the gain at black and mid-grey levels. The effect of white compression is to reduce the visibility of detail in highlight areas of the reproduced image.

white level In black-and-white TV the maximum permitted level of the picture signal (or, in colour TV, of the *luminance* component).

white noise A random noise signal with a level frequency response over a wide frequency range embracing the audible spectrum. Such a noise is used in audio-frequency testing, e.g. of loudspeakers where the noise, which sounds like escaping steam, acquires a coloration if the loudspeaker has resonances. The term is analogous to white light which has components spread over the whole of the visible spectrum.

white peak (US) Same as *peak white*.

wide area network (WAN) A *network* of computers on various premises and requiring the use of telecommunications lines to facilitate the exchange of data. Typically WANs connect *local area networks* rather than individual computers, the connection between the LAN and the telecommunications line being by a device called a *bridge* (or *router*).

width control (1) In TV receivers the control which determines the amplitude of horizontal deflection and hence the width of the displayed picture. The control is often a variable inductor connected in series with the line deflection coils. (2) In stereo sound reproduction a control which determines the apparent width of the sound source. It is often a variable resistor bridging the two channels and, in the zero-resistance position, parallels them so that the sound appears to originate from a point source located midway between the two loudspeakers.

Wien bridge An AC bridge circuit in which one arm consists of resistance and capacitance in series, an adjoining arm consists of resistance and capacitance in parallel, the remaining arms being purely resistive. The bridge, shown in *Figure W.2*, is used for measuring capacitance in terms of resistance and frequency. At balance the following relationships apply:

$$\frac{C_1}{C_2} = \frac{R_4}{R_3} - \frac{R_2}{R_1} \quad C_1 C_2 = \frac{1}{\omega^2 R_1 R_2}$$

which give the following expressions for C_1 and C_2

$$C_1^2 = \frac{R_1 R_4 - R_2 R_3}{\omega^2 R_1^2 R_2 R_3}$$

$$C_2^2 = \frac{R_3}{\omega^2 R_2 (R_1 R_4 - R_2 R_3)}$$

There is a corresponding Wien inductance bridge.

Wien-bridge oscillator A phase-shift oscillator of which the frequency is determined by the resistance and capacitance values of a Wien-bridge circuit. The network shown in *Figure W.3* consists of two equal resistors

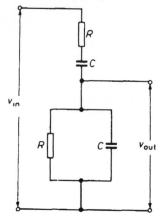

Figure W.3 Frequency-determining network used in Wien-bridge oscillator

and two equal capacitors in a Wien-bridge arrangement. This network gives zero phase shift at the frequency for which

$$\omega = \frac{1}{2\pi RC}$$

and at this frequency

$$\frac{v_{out}}{v_{in}} = \frac{1}{3}$$

If therefore an amplifier with zero phase shift and a voltage gain of 3 is connected between the output and input terminals of the network oscillation will result. This is the basis ·of a number of AF test oscillators. The amplifier gain is stabilised at 3 by negative feedback which can also serve to keep the output level constant. The frequency can be varied by using a two-gang variable capacitor for C, R being adjusted in steps to give different frequency ranges.

Winchester disk One sort of *hard disk*. See *hard disk*.

window An area in a *VDU display* dedicated to one application. In a multitasking computer system, several windows may be present simultaneously, which may belong to more than one application. Only one window at a time may be active (it is said to have the *input focus*) and this window will be at the front of the display, hiding areas of other windows which it overlaps.

Windows™ The Microsoft Windows operating environment which provides a *user-friendly multitasking* interface with MS-*DOS* and compatible operating systems.

wire broadcasting The distribution of sound and/or TV programmes to a number of receivers over a wired distribution system using audio frequencies or modulated carrier frequencies.

wired AND Same as *distributed connection*.

wired OR Same as *distributed connection*.

wobbulator An RF *signal generator* the instantaneous frequency of which can be varied continuously in a known periodic manner over a known fixed or adjustable range without change in amplitude. It is used in conjunction with an oscilloscope to display the shape of the passband of a receiver to facilitate alignment of the tuned circuits.

woofer A loudspeaker unit designed to radiate audio frequencies below approximately 200 Hz. Woofers are usually moving-coil units and are used in conjunction with other units and crossover networks in high-quality wide-range loudspeakers.

word See *binary word*.

word processor A system for writing, editing, printing and storage of documents. Most word processors are general-purpose *microcomputers* or *minicomputers* executing word-processing *programs*. Text typed at the keyboard appears on the screen of a VDU and is also stored in memory. *Software* routines allow such facilities as the insertion and deletion of matter, the movement of blocks of text, searches for the occurrence of a particular string of characters (with the option of replacing it by an alternative string) and word counts. A *hard copy* of the document may be produced on a printer, formatting routines permitting such features as the centring of titles and justification (straightening) of the right-hand margin.

work function The energy required to emit a single *electron* from a *cathode* into a field-free space. Its value depends on the material of the cathode.

working point Same as *operating point*.

working storage Same as *temporary storage*.

workstation A vague term used in various ways to indicate the data processing or communications equipment needed by one employee at his or her desk. In computing the term is sometimes reserved for specialised high-speed hardware for complex graphics work such as CAD, but the term programmable workstation (PWS) may be applied to any microcomputer being used as an *intelligent terminal* to access a mainframe computer. In telecommunications the term is sometimes used to refer to a computer and a telephone on the same desk and sharing the same access to the telecommunications network.

worldwide web A network of linked *documents* residing on the *Internet*.

WORM (Write Once Read Many times) Any *storage* system in which data can be written (recorded), but not erased. Consequently the stored data cannot be amended, although an amended version may be stored subsequently elsewhere on the media. The best known WORM system is *CD-recordable* which stores data on *Compact Disc*. This is used for backing up *hard drives* and archiving records of various kinds.

wow Distortion in sound reproduced from disk, film or tape and caused by undesired rhythmic speed variations on recording or reproduction. It is usually caused by eccentric or unbalanced driving of the film, tape or turntable. The variations occur at a rate of less than 20 Hz and cause unpleasant variations in pitch which are particularly noticeable in the reproduction of the piano and of sustained notes.

wrapped connection A method of securing low-resistance contact between a wire and a post by wrapping the wire around a sharp-cornered post under controlled tension using a hand- or power-operated tool. Wrapped connections have been used for twenty-five years and have proved a satisfactory alternative to the soldered connection.

writing In *computers* and *data-processing equipment* the process of recording information in a storage device or on a data medium. See *reading*.

writing speed Of a *cathode ray tube* the maximum speed at which the spot can travel on the screen whilst still maintaining a required degree of resolution. The more rapidly the spot moves the fainter is the trace generated. Thus there is an upper limit to the speed which can usefully be employed.

X

X amplifier Same as *horizontal amplifier*.

X-cut crystal A slice of quartz crystal cut at right angles to an X axis and parallel to the optic axis. A quartz crystal has the shape of a hexagonal prism and the electric axes (along which the piezo-electric effect is a maximum) pass through the centre and opposite corners of the hexagon as shown in *Figure X.1*. The line through O perpendicular to the plane of the

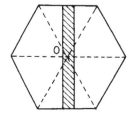

Figure X.1 Cross section of quartz crystal showing the electric axes by dashed lines: the shaded area represents an X cut

paper is the optic axis of the crystal. The shaded area on the diagram represents an X cut. Such slices are used to control the frequency of oscillators, e.g. those used to generate the carrier frequency of radio transmitters, but a disadvantage is that the natural resonance frequency of the slice (determined by its thickness) is temperature dependent so that, for high frequency stability, the temperature of the slice must be controlled within narrow limits.

X plates See *deflector plates*.

X rays Electromagnetic radiation of wavelength shorter than approximately 10^{-6} cm. The rays can be generated by directing a stream of high-velocity electrons on to a target of a suitable metal. X rays can affect photographic plates and can ionise gases but probably most applications make use of the fact that X rays can penetrate many substances which are opaque to visible light.

Y

Y amplifier Same as *vertical amplifier*.

Y-cut crystal A slice of quartz crystal cut at right angles to a *Y* axis and parallel to the optic axis. A quartz crystal has the shape of a hexagonal prism and the *Y* axes (mechanical axes) pass through the centre and the midpoints of opposite sides of the hexagon as shown in *Figure Y.1*. The line through O perpendicular to the plane of the paper is the optic axis of the crystal. The shaded area on the diagram represents a *Y* cut. Such slices

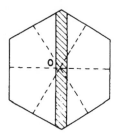

Figure Y.1 Cross section of a quartz crystal showing the mechanical axes by dashed lines: the shaded area represents a *Y* cut

are used to control the frequency of oscillators, e.g. those used to generate the carrier frequency of radio transmitters but a disadvantage is that the natural resonance frequency of the slice (determined by its thickness) is temperature dependent so that, for high frequency stability, the temperature of the slice must be controlled within narrow limits.

yoke A piece of magnetic material used to complete a magnetic circuit. The yoke is not itself surrounded by windings but it often acts as a mechanical support for windings as in the deflection coil assembly of a TV *picture tube*.

y parameters Of a *transistor*, a method of expressing the electrical characteristics in which the input and output currents are given in terms of the input and output voltages. The two fundamental equations are:

$$i_{in} = y_i v_{in} + y_r v_{out}$$

$$i_{out} = y_f v_{in} + y_o v_{out}$$

where y_i is the input admittance (for short-circuited output terminals); y_o is the output admittance (for short-circuited input terminals); y_t is the reverse transfer admittance and y_f is the forward transfer admittance.

Y plates See *deflector plates*.

Y signal Same as *luminance signal*.

356

Z

Zener breakdown In a reverse-biased pn junction a rapid increase in current which occurs at a particular reverse voltage as a result of the Zener effect. Zener breakdown occurs at a lower reverse bias than *avalanche breakdown*.

Zener diode A pn diode which exhibits *Zener breakdown* or *avalanche breakdown* at a specified value of reverse bias and which is therefore used for voltage reference purposes or for voltage stabilisation. The graphical symbol for a Zener diode is given in *Figure Z.1*.

Figure Z.1 Graphical symbol for a Zener diode

Zener effect Conduction through a reverse-biased pn junction caused by spontaneous generation of hole-electron pairs within the inner electron shells of atoms in the junction region. The hole-electron pairs are liberated, not as a result of ionisation by collision (as in avalanche breakdown) but by the intense electric field established across the junction by the reverse bias.

zener voltage The reverse voltage at which *Zener breakdown* occurs.

Zenith-G.E. pilot-tone system See *pilot-tone stereo*.

zero For a *network* of pure reactances, any frequency at which the input reactance is zero. As shown in *Figure Z.2* the reactance-frequency relationship for a network of pure reactances is a succession of curves, all with a positive slope, which swing between minus infinity and plus infinity, passing through zero, in a manner similar to that of a tangent curve. The zeros are the frequencies at which the curves pass through zero, i.e. those at

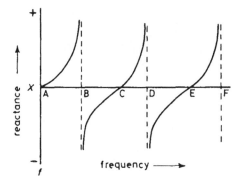

Figure Z.2 Typical reactance-frequency relationship for a network of pure reactances. A, C and E are zeros and B, D and F are poles

357

which the network presents at its input terminals the equivalent of an inductance and capacitance of equal reactance in series. See *pole*.

zero frequency (ZF) Synonym for direct current.

Zero Insertion Force (ZIF) socket An *integrated circuit* socket especially designed to allow integrated circuits to be inserted and removed quickly and easily and with minimal risk of damage to the pins. A clamping mechanism ensures that the i.c. is held securely in place, its pins making proper electrical connection with their respective sockets. A lever must be operated to release the clamp and allow the i.c. to be changed. ZIF sockets are used in equipment in which i.c.s are changed frequently, e.g. i.c. test equipment and EPROM writing equipment.

zero level See *reference level*.

zone refining Method of purifying semiconductor material which depends on the fact that impurities are more soluble in the molten material than in the solid material. An ingot of the material is moved slowly through the coils of an RF induction heater so that, in effect, a molten zone travels the length of the ingot. By repeatedly passing the ingot through the coils in the same direction in an inert atmosphere the impurities can be concentrated at one end of the ingot and the other end can be made pure enough for use in semiconductor device manufacture.

zooming In computing, especially *graphics* and *desktop publishing applications*, changing the magnification factor of the *VDU* display. Zooming in gives an enlarged display of a small area, revealing fine detail in graphics or making small print more legible. Zooming out allows the whole of a large page to be visible on screen at once; although detail is lost, the overall composition of the page can be assessed. Most applications offer a useful range of zoom settings.

z parameters Of a *transistor*, a method of expressing the electrical characteristics in which the input and output voltages are given in terms of the input and output currents. The two fundamental equations are:

$$v_{in} = z_i i_{in} + z_r i_{out}$$

$$v_{out} = z_f i_{in} + z_o i_{out}$$

in which z_i is the input impedance, z_o the output impedance, z_f the forward transfer impedance and z_r the reverse transfer impedance. These parameters are not greatly favoured largely because of the difficulty of measuring them and the *hybrid parameters* or *y parameters* are preferred.

APPENDIX: ELECTRONICS ABBREVIATIONS AND ACRONYMS

© G. W. A. Dummer

The abbreviations and acronyms in this book have been selected from many current sources—books, technical articles, papers and theses—and also from knowledgeable colleagues. Obviously it is not possible to list every acronym from specialised categories, but those in this book have been selected on the basis of their application to present-day electronics.

In the following list acronyms are usually shown in capitals and abbreviations in lower case, except where established usage directs otherwise. Full stops are not shown. Please note that individual writers, publishers and organisations may follow different conventions regarding the use of capitals and full stops in acronyms and abbreviations. For the purposes of alphabetical order Greek letters are regarded as variants of their Latin equivalents.

AC	alternating current
AAL	ATM adaptation layer
AALA	American Association for Laboratory Accreditation
ABR	auxiliary bus resonator
ABR	available bit rate
ACAST	Advisory Committee on the Application of Science and Technology to Development
ACC	automatic chrominance control
acc	accumulator
ACIA	asynchronous communication interface adapter
ACK	acknowledge
ACR	attenuator to crosstalk ratio
ACU	automatic clearing unit
A-D	analogue to digital
ADA	computer language (named after Ada Augusta)
ADC	analogue-to-digital converter
ADCCP	advanced data communication control procedure
ADDER	automatic digital data error recorder
ADF	automatic direction finder
ADP	automatic data processing
ADPCM	adaptive differential pulse code modulation

ADPCU	adaptive differential pulse conditioning unit
ADS	address data strobe
ADSL	asychronous digital subscriber line
ae	aerial
AEA	American Electronics Association
AES (USA)	Audio Engineering Society
AES	auger electron spectroscopy
AES	auto-electronic separator
AEW	airborne early warning
AF	audio frequency
AFC	automatic frequency control
AFM	atomic force microscopy
AFIPS	American Federation of Information Processing Societies
A/G	air to ground
AGC	automatic gain control
AHDL	analog hardware descriptive language
AHS	after-hours sync
AI	artificial intelligence
AIA (USA)	Automated Imaging Association
AIM	avalanche-induced migration
AIRCOMNET	Air Force Command and Administrative Network
ALC	automatic level control
ALD	automatic location device
ALGOL	ALG-orithmic computer language
ALT	radio altimeter
ALU	arithmetic/logic unit
AM	amplitude modulation
AMLCD	active-matrix liquid crystal display
AMPS	advanced mobile phone system
AND	automatic network dialling
ANI	automatic number identification
ANSI	American National Standards Institute
ant	antenna
AOI	AND/OR invert
AOI	automated optical inspection
APA	all points addressable
APC	aperiodic choke
APCM	adaptive pulse code modulation
APD	avalanche photo-diode
API	application programming interface
APL	a programming language developed by Iverson
APPC	advanced program-to-program communications
APPS	automatic parts programming system
APU	arithmetic processor unit

AQL	acceptable quality level
ARB	all routes busy
ARC	adaptive response control
ARINC	Aeronautical Radio, Inc.
ARL	acceptable reliability level
ARP	address resolution protocol
ARPA	Advanced Research Project Agency (Department of Defense)
ARQ	automatic repeat request
ARPANET	Advanced Research Project Agency Network
ARRL	American Radio Relay League
ARU	acoustical resistance unit
ASA	American Standards Association
ASA	Army Security Agency
ASCII	American Standard Code for Information Interchange
ASD	adaptive signal discriminator
ASET	Association of Super-Advanced Electronic Technologies
a-Si	amorphous silicon
ASIC	application specific integrated circuits
ASM	assembler
ASNT	American Society for Non-destructive Testing
ASR	automatic send/receive
ASSOC	application-specific optical chip
ASPR	automatic satellite position reporting system
ASQC	American Society of Quality Control
ASTE	American Society of Test Engineers
ASTM	American Society for Testing and Materials
AT	advanced technology
ATAB	area-array tape automated bonding
ATAPI	advanced technology attachment packet interface
ATC	address translation cache
ATDM	asynchronous time-division multiplexing
ATE	automatic test equipment
ATLAS	abbreviated test language for avionic systems
ATM	asynchronous transfer mode
ATPG	automatic test program generator
ATS	automatic test system
AT & T	American Telephone and Telegraph Company
ATU	aerial (or antenna) tuning unit
ATX	automatic telex exchange
aux	Auxiliary
AV	audio-visual
AVC	automatic volume control
AWACS	Airborne Warning and Control System

AWG	American Wire Gauge
BABS	beacon approach, beacon systems
BABT	British Approvals Board Telecommunications
BALUN	balanced-to-unbalanced transformer
BARRITT	barrier controlled injection and transit time (diode)
BASIC	beginners all-purpose symbolic instruction code
BBC	British Broadcasting Corporation
BBD	bucket-brigade device
BBS	bulletin board system
BCCD	buried-channel charge-coupled device
BCD	binary-coded decimal
BCD	bipolar C-MOS D-MOS
BCH	an FEC parity coding scheme
BCI	bit count integrity
BCS	binary synchronous communication
BCS	British Computer Society
BDOS	basic disk operation system
BDPSK	binary differential phase shift keying
BER	bit error rate
BFO	beat frequency oscillator
BGA	ball grid array
BICMOS	bipolar complementary metal oxide semiconductor
BIFET	bipolar field effect transistor
BIOS	basic input/output system
BIPM	International Bureau of Weights and Measures
BIST	built-in self-test
BIT	built-in test
bit	binary digit
BITE	built-in test equipment
BJT	bipolar junction transistor
BLAISE	British Library Automated Information Service
B-LD	bi-stable laser diode
BMEWS	Ballistic Missile Early Warning System
BMFT	Bundes Ministerium für Forschung und Technologie
BNC	type of co-axial connector
BORAM	block-oriented random-access memory
BP	band pass
BPS	bits per second
BPSK	binary phase-shift keying
BS	backing store
b/s	bits per second
BSB	British Satellite Broadcasting Ltd
BSC	bisynchronous or (binary synchronous) communications

BSDL	boundary scan descriptive language
BSI	British Standards Institute
BTAB	bumped tape automated bonding
BW	bandwidth
byte	8 bits
C	type of co-axial connector
CAAP	co-processor architecture and protocols
CAD	computer-aided design
CAE	computer assisted engineering
CAI	computer aided instruction
CAL	computer assisted learning
CAM	computer aided manufacturing
CAM	content-addressable memory
CAMAC	computer automated measurement and control
CAMEL	Chalmers Advanced Microelectronics Lab., Sweden
CAP	carrier amplitude phase
CAS	column address select
CAS	cordless access service
CASE	computer-aided software engineering
CAT	computer aided testing
CATT	controlled avalanche transit time (diode)
CATV	community antenna television
CAW	common aerial working
CB	cellular broadcast
CB	citizens band (radio)
CBIC	complimentary bipolar integrated circuit
CBR	constant bit rate
CBT	computer based training
CBT	computer-based terminal
CCD	charge-coupled device
CCE	computer controlled equalisation
CCIA	Computer and Communications Industries Association
CCIF	International Telephone Consultative Committee
CCIR	International Radio Consultative Committee
CCITT	International Telegraph and Telephone Consultative Committee
CCTS	Co-ordinating Committee on Satellite Communications
CCTT	cold-cathode trigger tube
CCTV	closed-circuit television
CD	compact disc
CDC	compiler object code
CDI	collector-diffusion isolation
CD-I	compact disc interactive

CDMA	code-division multiple access
CDM	charged-device model
CDU	central display unit
CD-R	compact disc recordable
CD-ROM	compact disc read-only memory
CD-RW	compact disc rewritable
CEEFAX	BBC teletext service
CEMA	Canadian Electrical Manufacture Association
CEN	European Committee on the Co-ordination of Standards
CENEL	European Electrical Standard Co-ordinating Committee
CERDIP	ceramic dual-in-line package
CERN	European Particle Physics Laboratory, Geneva
CESA	Canadian Engineering Standards Association
CFC	chlorofluorocarbon
CFD	computational fluid dynamics
CFR	constant failure rate
CGA	Colour/Graphics Adapter
CGI	common-gateway interface
CH	chain home (radar)
CHL	chain home low (radar)
CIRM	International Marine Radio Association
CISC	complex-instruction-set computer
CISPR	International Special Committee on Radio Interference
CIRC	cross-interleaved Reed-Soloman code (for CDs)
CLCC	ceramic leaded chip carrier
CLOS	common LISP object system
CLS	clear screen display
CML	current-mode logic
CMM	compatibility maturity model
CMO	common-mode operation
CMOS	complementary metal-oxide-semiconductor
CMP	chemical mechanical polishing
CMR	common-mode rejection
CMRR	common-mode rejection ratio
CMV	common-mode voltage
CMVR	common-mode voltage ratio
CNRS	Centre Nationale de la Recherche Scientifique (the French National Science Foundation)
COB	chip on board
COBOL	common business-oriented language
codec	coder-decoder
COFDM	common orthogonal frequency-division multiplexing
COMP	compiler
COMSAT	communications satellite corporation

CORBA	common object-request-broker architecture
COTS	commercial off-the-shelf
CPC	crucially permutable code
CPI	characters per inch
CPM	continuous phase modulation
CPM	control program for microcomputers
CPP	Consultative Panel on Public Information (United Nations)
CPS	characters per second
CPU	central processing unit
CPW	co-planar waveguide
CQ	call to all stations
CR	carriage return
CRC	cyclic redundancy check
CRM	cyclic redundancy model
CROM	control read-only memory
CRO	cathode-ray oscilloscope
CRT	cathode-ray tube
CSA	carrier service area
CSMA/CD	carrier-sense multiple access collision detector
CSN	common services network
CSP	chip scale packaging
CT	cordless telecommunications
CTD	charge transfer device
CTE	coefficient of thermal expansion
CTL	complementary transistor logic
CTS	clear to send
CU	control unit
CVBS	composite video blanking and sync
CVD	chemical vapour deposition
CW	continuous wave
D-A	digital to analogue
DAB	digital audio broadcasting
DAC	digital-to-analogue converter
DACS	Data and Analysis Centre Software
DAL	data access line
DAP	distribute array processor
DAQ	data acquisition
DARPA	Defense Advanced Research Project Agency
DART	dual asynchronous receiver/transmitter
DAS	data acquisition system
DASM	direct access storage medium
DAT	digital audio tape
DAVIC	Digital Audio-Visual Council

DAVID	digital audio-video interactive decoder
dB	decibel
dBi	decibels related to an isotropic antenna
dBm	decibels related to 1 mW
dBw	decibels related to 1V into 600 Ohm
DBMS	database management system
DBPSK	differential-binary phase-shift keying
DBS	direct-broadcasting satellite
DC	direct current
DC	don't care
DCA	Defense Communications Agency
DCA	direct chip attach
DCE	data communications equipment
DCS	digital cellular service
DCS	digital communication system
DCTE	data circuit termination equipment
DCTL	direct-coupled transistor logic
DD	disk drive
DDD	direct distance dialling
DDL	data description language
DECT	digital enhanced cordless communications
DECT	digital European cordless communications
DES	data encryption standard
DEW	distant early warning
DF	direction finding
DFA	digital fault analysis
DFM	design for manufacturability
DFT	design for test
DFT	discrete Fourier transform
DGAC	data generation and acquisition circuit
DI	dielectric isolation
DIANE	Direct Information Access Network for Europe
DIN	Deutsches Industrie Normale (German Standards Institute)
DIL	dual-in-line
DIP	dual-in-line package
DIP	dual-in-line plastic
DMA	direct memory access
DMAC	direct memory access control
DMD	depletion-mode device
DMM	digital multimeter
D-MOS	double-diffused metal-oxide semiconductor
DMS	dynamic mapping system
DMUX	demultiplexer
DNA	digital network architecture

DNS	domain name service
DOA	direction of arrival (signal)
DOS	disc operating system
DP	data processing
DPCM	differential pulse-code modulation
DPDT	double-pole double-throw (switch)
DPE	data processing equipment
DPEK	differential phase-exchange keying
DPM	digital panel meter
DPSK	differential phase-shift keying
DPU	display processing unit
DQA	design quality assurance
DQPSK	differential quadrature phase-shift keying
DR	data rate
DRAM	dynamic random access memory
DS	direct sequence (in spread spectrum systems)
DSB	double sideband (transmission)
DSBSC	double-sideband suppressed carrier
DSBTC	double sideband transmitted carrier
DSDD	double-sided double-density
DSC	direct satellite communication
DSCC	double silk-covered copper
DSIR	Department of Scientific and Industrial Research (UK)
DSL	digital subscriber line
$\Delta\Sigma M$	delta sigma modulation
DSN	deep space network
DSP	digital signal processing
DSP	data signal processor
DSSC	double-sideband suppressed carrier
DSSD	double-sided single-density
DSTC	double-sideband transmitted carrier
DTE	data terminal equipment
DTF	dynamic track following
DTL	diode/transistor logic
DTMF	dual-tone multiple-frequency
DTP	desktop publishing
DTT	digital terrestrial television
DTV	digital television
DUT	device under test
DVB	digital video broadcasting
DVD	digital versatile disc
DVD	digital video disc
DVD-RAM	digital versatile disc random-access memory
DVM	digital voltmeter

DWDM	dense wave division multiplexing
DX	duplex

EAN	European Article Number
EAROM	electrically alterable read-only memory
EBCDIC	extended binary code decimal interchange code
EBIM	electron beam induced current
EBL	electron beam lithography
ECCM	electronic counter-counter measures
ECL	emitter-coupled logic
ECM	electronic counter measures
ECMA	European Computer Manufacturers Association
ECO	electron-coupled oscillator
EDA	electronic design automation
EDC	error detecting code
EDCC	error detection/correction code
EDP	electronic data processing (or processor)
EEPROM	electrically erasable and programmable read-only memory
EEROM	electrically erasable read-only memory
EFL	emitter follower logic
EFM	eight-to-fourteen modulation
EFT	electrical fast transient
EGA	enhanced graphics adaptor
EHF	extremely high frequency
EIA	Electronic Industries Association
EIAJ	Electronic Industries Association of Japan
EIDE	enhanced integrated drive electronics
EIRP	effective isotropic radiated power
ELF	extremely low frequency (< 3 kHz)
ELN	extra-low noise
ELS	electrostatic loudspeaker
EMC	electromagnetic capability
EMD	enhancement-mode device
EMF	electromagnetic field (or force)
EMI	electromagnetic interference
EMP	electromagnetic pulse
EMR	electromagnetic radiation
EMS	energy management system
EMT	electrical module test
EMU	electromagnetic unit
ENG	electronic news gathering
EO	erasable optical
EOC	end of conversion
EOD	end of data

EOF	end of file
EOM	end of message
EOS	electrical overstress
EOT	end of text
EOT	end of transmission
EPLD	erasable programmable logic device
EPOS	electronic point of sale
EPS	encapsulated PostScript
EPG	electronic programme guide
EPM	equipment performance monitoring
EPROM	erasable programmable read-only memory
EPSRC	Engineering and Physical Sciences Research Council
ERCIM	European Research Consortium for Information and Mathematics
ERDA	Elastic Recoil Detection Analysis
EROM	erasable read-only memory
ESA	European Space Agency
Esc	escape
ESD	electrostatic discharge
ESDA	electronic system design automation
ESL	essential service line
ESPRIT	European Strategic Programme for Research and Development in Information Technology
ESR	equivalent series resistance
ESREF	European Symposium on Reliability of Electronic Devices, Failure Physics and Analysis
ESS	electronic switching system
EST	echo suppressor terminal
ESTEC	European Space Research and Technology Centre
ETA	estimated time of arrival
ETD	estimated time of departure
ETS	European Telecommunications Standards
ETS	equivalent time sampling
ETSI	European Telecommunications Standards Institute
EUREKA	European Research and Co-ordination Agency
EUT	equipment under test
eV	electron Volt
EXTND	extended data transfer
F	Farad
FA	failure analysis
FAA	Federal Aviation Authority
FAB	fabrication automation bay
FAMOS	floating-gate avalanche metal-oxide semiconductor

FAMOX	floating-gate avalanche-rejection semiconductor
FAT	file allocation table
fax	facsimile
FC-AL	fibre channel-automated loop
FCD	file control block
FCC	Federal Communication Commission
FCFS	first come first served
FD	fully depleted (Silicon)
FDC	floppy disc controller
FDD	floppy disc drive
FDDI	fibre-distributed data interface
FDM	frequency-division multiplex
FDMA	frequency-division multiple access
FDTD	finite difference time domain
FEC	forward error correction
FED	field emission display
FEK	frequency exchange keying
FEN	far end noise
FET	field effect transistor
FEXT	far end cross talk
FFT	fast Fourier transform
FG	frame ground
FH	frequency hopping (in spread-spectrum systems)
FIB	focussed ion beam
FIFO	first in, first out
FIMBE	focussed ion beam molecular beam epitaxy
fit	failure rate, one in 10^9 hours
FIT	failure in time
FKE	Institute for Solid-State Electronics (Austria)
FM	frequency modulation
FMEA	failure mode effects analysis
FO	fibre optic
FOC	fibre optic communications
FORTRAN	formula translator (programming language)
FOT	fibre optic transmission
FP	flat pack
FP	floating point
FPGA	field-programmable gate array
FPLA	field-programmable logic array
FPMCM	field-programmable multi-chip module
F-PROM	field-programmable read-only memory
FRAM	high-density, sub-micrometer ferroelectric memory
FSK	frequency shift keying
FSM	frequency shift modulation

FTMIS	floating-gate tunnel-metal-insulator semiconductor
FTP	file transfer protocol (Internet)
FTTH	fibres to the home
FX	effects
FX	foreign exchange
GaAs	gallium arsenide
GaAlAs	gallium aluminium arsenide
GATT	General Agreement on Tariffs and Trade
GB	grid bias
Gbyte	gigabyte (= 1024 Mbytes)
GCC	Global Climate Coalition (Washington, D.C.)
GCI	ground-controlled interception
GDI	graphics device interface
GDP	graphic display processor
GDS	graphic data system
GEE	grid navigation system (radar)
GF	ground fixed
GIGO	garbage in, garbage out
GIF	graphic image file
GIF	graphics interchange format
GL	gun laying (radar)
GM	ground mobile
GMPCS	global mobile positioning communications by satellite
GOP	group of pictures (motion comprehension)
GP	general purpose
GPIB	general-purpose interface bus
GHz	gigaHertz
GP	general purpose
GPIS	general-purpose interface bus
GPS	global positioning system (satellite)
GSM	global system for mobile communications
GUI	graphical user interface
HASS	highly accelerated stress testing
HAST	highly accelerated stress test
HALT	highly accelerated life testing
HBT	heterojunction bipolar transistor
HCMOS	highly scaled high-speed MOS
HCP	hard copy printer
HD	hard disc
HDI	high-density interconnect
HDL	hardware description language
HDLC	high-level data link control

HDMOS	high-density metal oxide semiconductor
HDSL	high-bit-rate digital-subscriber line
HDSS	holographic data storage system
HDT	high-density interconnect
HDTV	high-definition television
HEMT	high-electron-mobility transistor
hex	hexadecimal
HF	high frequency
HFC	high-frequency choke
HIC	hybrid integrated circuit
HiNIL	high noise-immunity logic
HIPPA	high-performance parallel interface
HITFET	highly-integrated temperature-protected FET
HLL	high-level language
HLL	high-level logic
H-MOS	high MOS
HSM	hermetically-sealed microcircuit
HTL	high-threshold logic
HTML	hypertext markup language
HTTP	hypertext transfer protocol
HUD	head-up display
HV	high voltage
Hz	Hertz
IAEA	International Atomic Energy Agency
IANA	Internet Assigned Numbers Agency
IARU	International Amateur Radio Union
IATA	International Air Transport Association
IBA	International Broadcasting Authority
IBE	ion beam etching
IBM	International Business Machines
IC	integrated circuit
ICAO	International Civil Aviation Organisation
ICE	in-circuit emulator
ICL	International Computers Limited
ICPO	International Criminal Police Organisation (Interpol)
IDE	integrated drive electronics (interface)
IDL	interface definition language
IDPM	Institute of Data Processing Management
IDS	input-data strobe
IDX	integrated digital exchange
IEC	infused emitter coupling
IEC	International Electrotechnical Commission
IEC	Interzonal Electrotechnical Commission

IEE	Institution of Electrical Engineers (UK)
IEEE	Institute of Electrical and Electronics Engineers (USA)
IEMP	International Electronics Manufacturing Technology (Symposium)
IERE	Institution of Electronic and Radio Engineers
IES	Institute of Environmental Sciences (USA)
IF	intermediate frequency
IFAC	International Federation of Automatic Control
IFF	identification, friend or foe
IFH	Institüt für Hochfrequenztechnik (Darmstadt)
IGBT	insulated-gate bipolar transistor
IGFET	insulated-gate field-effect transistor
ILB	Inner Lead Bonding
I^2L	integrated-injection logic
ILS	instrument landing system
IMAPS	International Microelectronics Assembly and Packaging Society
IMF	infant mortality factor
IMP	interconnected mesh power system
IMPATT	impact avalanche transit time (diode)
InP	indium phosphide
I/O	input/output
IOP	input/output processor
IP	Internet protocol
IPC	Institute for Interconnecting and Packaging Electronics Circuits (USA)
IPs	intermodulation products
IPL	initial program loader
IR	infra-red radiation
IRD	integrated receiver/decoder
IRE	Institute of Radio Engineers (USA) (now IEEE)
IRQ	interrupt request
ISA	industry-standard architecture (PC Bus)
ISA	Instrument Society of America
ISAM	indexed sequential access method
ISB	independent sideband (transmission)
ISCC	international service co-ordination standard
ISD	international subscriber dialling
ISDN	integrated services digital network
ISDT	integrated services digital terminal
ISDX	integrated services digital exchange
ISFET	ion-sensitive field-effect transistor
ISHM	International Society for Hybrid Microelectronics
ISL	integrated Schottky logic

ISMC	International Society for Measurement and Control
ISO	International Standards Organisation
ISOC	International Service Co-ordination Centre
ISP	Internet service provider
ISPBX	Integrated Services Private Branch Exchange
ISV	independent software vendor
IT	information technology
ITE	international telephone exchange
ITU	International Telecommunications Union
JAN	joint army/navy
JCL	job control language
JEDEC	Joint Electron Device Engineering Council
JEMIMA	Japan Electric Measuring Instruments Manufacturers' Association
JESSI	Joint European Submicron Silicon Project
JFET	junction field-effect transistor
JI	junction isolation
JPCA	Japanese Printed Circuit Board Association
JPCD	just-perceptible colour difference
JPEG	Joint Photographic Experts Group
JUGFET	junction-gate field-effect transistor
JVM	Java Virtual Machine
K	kilOhm
KBD	keyboard
kbit	kilobit
kbyte	kilobyte
KGD	known-good die
kHz	kiloHertz
kΩ	kilOhm
kR	kilOhm
KSR	keyboard send/receive
kV	kiloVolt
LAN	local area network
LAP	link access procedure
LAP	line access point
LAPUT	light-activated programmable unijunction transistor
LARAM	line-addressable random-access memory
LASCR	light-activated silicon controlled rectifier
LASER	light amplification by stimulated emission of radiation
LCA	life cycle analysis
LCC	leaded chip carrier

LCD	liquid crystal display
LDM	limited-distance modem
LEC	light energy converter
LED	light-emitting diode
LEED	low-energy electron diffraction
LF	low frequency (30 kHz to 300 kHz)
LFSR	linear-feedback shift register
LGA	lead-grid array
LIC	linear integrated circuit
LIFO	last in, first out
LLCC	leadless chip carrier
LLD	lightly doped drain
LMDS	local multipoint distribution service
LMM	Laboratoire de Microstructures et de Microélectronique
LMWS	licensed millimetre wave service
LNA	low-noise amplifier
LNB	low-noise block
LNR	low-noise receiver
LO	local oscillator
LOCMOS	locally oxidised CMOS
LORAN	long-range air navigation
LOS	line-of-sight
LP	line printer
LP	long play (gramophone record)
LPCVD	low-pressure chemical vapour deposition
LPE	liquid-phase epitaxy
LPM	library of parameterised modules
LPTTL	low-power transistor/transistor logic
LPTV	low-power television
LRL	logical record length
LRU	least replacement unit
LRU	line replacement unit
LS	lower sideband
LSB	least significant bit
LSD	least significant digit
LSI	large-scale integration
LT	low tension
LUF	lowest usable frequency
LUT	look-up table
LZW	Lempel-Ziv-Welsh (compression)
M	megOhm
mA	milliAmp
µA	microAmp

Mac	Macintosh
MAC	multiply-accumulate
MADT	micro-alloy diffused transistor
MAP	manufacturing automatic protocol
MAP	minimum acceptable performance
MAR	memory address register
MAR	mercury arc rectifier
MASER	microwave amplification by stimulated emission of radiation
MASTER	multimedia-aids-teleconferencing
MATV	master antenna TV
MBA	multiple-beam antenna
MBE	molecular beam epitaxy
Mbit	megabit
MBR	master boot record
MBR	memory buffer register
Mbyte	megabyte
MCM	multi-chip module
MCM-C	multi-chip module - ceramic dielectric
MCM-D	multi-chip module - organic dielectric
MCM-L	multi-chip module - laminate
MCM-Si	multi-chip module - silicon substrate
MCR	minimum cell rate
MCT	mobile communications terminal
MCU	microcontroller (unit)
MDS	microprocessor development system
MEDIA	Microelectronics Development for European Applications
MEMS	microelectromechanical systems
MESFET	metallised-semiconductor field-effect transistor
MF	medium frequency (300 kHz to 3 MHz)
μF	microFarad
MFD	microFarad
MFLOPS	millions of floating-point instructions per second
MFP	mini flat pack
MFSK	multi-frequency shift keying
MHL	microprocessor host loader
MHz	megahertz
MIC	microwave integrated circuit
MIDI	musical instrument digital interface
MIMD	multiple instruction, multiple data
MIME	multipurpose Internet mail extensions
MIN	multistage interconnection networks
MIPS	millions of instructions per second
MIS	metal oxide/silicon

MISD	multiple instruction, single data
MLA	microprocessor language assembler
MLB	multilayer board
MLE	microprocessor language editor
MLE	maximum likelihood estimation
MLS	microwave landing system
MMDS	multichannel, multipoint distribution system
MMDS	multipoint microwave distribution system
MMF	magnetomotive force
MMI	man/machine interface
MMIC	monolithic microwave integrated circuit
MMU	memory management unit
MMX	multimedia extensions
MNCS	multipoint network control system
MNOS	metal nitride oxide semiconductor
MΩ	megOhm
modem	modulator/demodulator
MODHIC	modular hybrid integrated circuit
MOS	metal oxide semiconductor
MOSFET	metal oxide semiconductor field-effect transistor
MOST	metal oxide semiconductor transistor
µP	microprocessor
MPEG	Motion Picture Experts Group
MPU	microprocessor unit
MPX	multiplex
MR	magnetorestrictive
MR	megOhm
MSB	most significant bit
MSC	mobile switching centre
MSD	most signicant digit
MS DOS	Microsoft Disk Operating System
MSI	medium-scale integration
MSIN	multistage interconnection network
MSK	medium shift keying
MST	multiservice terminal
MTBF	mean time between failures
MTD	mass tape duplicator/verifier
MTR	magnetic tape recorder
MTTF	mean time to failure
MTTFF	mean time to first failure
MTTR	mean time to repair
MUMPS	Massachussetts General Hospital Utility Multi-Programming System
MUSA	multiple-unit steerable antenna (or array)

MUX	multiplexer
mV	milliVolt
MV	megaVolt
μV	microVolt
nA	nanoAmp
NAND	NOT AND gate
NASA	National Aeronautics and Space Administration
NBS	National Bureau of Standards (USA)
NC	network computer
NCC	National Computing Centre
NDP	numeric data processor
NDRO	non-destructive readout
NEMA	National Electrical Manufacturers Association
NEN	near-end noise
NEP	noise equivalent power
NEXT	near-end cross-talk
nF	nanoFarad
NF	noise figure
NFB	negative feedback
NIC	network information centre
NIC	negative impedance converter
NiCad	nickel/cadmium cell
NICAM	near-instantaneous compression and audio multiplex
NIST	National Institute of Standards and Technology
NIT	network information table
n-MOS	n-channel metal-oxide semiconductor
NMRR	normal mode rejection ratio
NOP	no operation
NOR	inverted OR gate
NORAD	North American Air Defence Command
NPL	National Physical Laboratory (UK)
NRL	Naval Research Laboratory (USA)
NRZ	non-return zero
NRZI	non-return to zero inverted
NSPE	National Society of Professional Engineers (USA)
NTC	negative temperature coefficient
NTIA	National Telecommunications Information Administration
NTSC	National Television Systems Committee (USA)
NTU	network termination unit
nV	nanoVolt
Ω	Ohm
OAM	operation and maintenance

OB	optical buffer
OB	outside broadcast
OBL	one block lookahead
OBW	occupied bandwidth
OCR	optical character reader
OCR	optical character recognition
ODA	open-document architecture
ODI	open datalink interface
ODS	output datalink strobe
OEE	overall equipment effectiveness
OEM	original equipment manufacturer
OFDM	orthogonal frequency division multiplexing
OHS	occupational health system
OLB	outer lead bonding (in tape automated bonding)
OLE	object linking and embedding
OLR	off-line reader
OMR	optical mark reader
OOBE	out-of-box experience
OODMS	object-oriented database management system
OOK	on/off keying
OOP	object-oriented planning
OOP	object-oriented programming
OOPS	object-oriented programming system
OPAL	operational performance analysis language
OP AMP	operational amplifier
ORACLE	ITV teletext service
OS	operating system
OSCAR	Orbiting Satellite Carrying Amateur Radio
OSD	on-screen display
OSI	open systems interconnection
OSPSK	offset phase shift keying
OTA	operational transconductance amplifier
OTDM	optical time-division multiplexing
OTDMA	optical time-division multiple access
OTDR	optical time-domain reflectometer
OWF	optical working frequency
OXC	optical cross connects
PA	power amplifier
PA	public address
PABX	private automatic branch exchange
PACE	plasma-assisted chemical etching
PAD	packet assembly/disassembly
PADT	post-alloy diffused transistor

PAL	phase alternation, line (colour TV system)
PAM	pulse-amplitude modulation
PAR	partial amplitude response
PAR	precision approach radar
PAR	program aid routine
PASCAL	computer language
PAT	portable appliance tester
PAT	program association table (SI)
PAVD	plasma-assisted vapour deposition
PBGA	plastic ball-grid array
PBX	private branch exchange
PC	printed circuit
PC	personal computer
PCA	principal components analysis
PCB	printed circuit board
PCC	plastic chip carrier
PC-DOS	Personal Computer Disk Operating System
PCI	peripheral component interconnect
PCL	printer control language
PCM	pulse code modulation
PCMCIA	Personal Computer Memory Card International Association
PCN	personal communication network
PCS	personal communication system
PD	partially depleted (silicon)
PD	phase-change disc
PD	photodetector
PD	potential difference
PD	public domain
PDC	personal digital cellular
PDM	product data management
PDM	pulse duration modulation
PDN	provate data network
PDP	plasma display panel
PEC	photo-electric cell
PEM	plastic encapsulated microcircuit
PEP	peak envelope power
PERT	program evaluation and review technique
PES	packetised elementary stream
pF	picoFarad
PF	power factor
PFM	pulse frequency modulation
PFR	power failure restart
PGA	pin grid array (package)
PHS	personal headphone system

PI	peripheral interface
PIA	peripheral interface adaptor
PID	packet identity
PIL	precision in-line (colour picture tube)
PIM	pulse interval modulation
PIND	particle impact noise detection
PIPO	parallel in, parallel out
PISO	parallel in, series out
PIV	peak inverse voltage
PLA	programmable logic array
PLC	programmable logic controller
PLCC	plastic leadless chip carrier
PLD	programmable logic device
PLD	programmable logic design
PLL	phase-locked loop
PLMN	public land mobile network
PM	phase modulation
PMG	permanent magnet generator
p-MOS	p-channel metal-oxide semiconductor
PMR	private mobile radio
PMT	program map table
POTS	plain old telephone service
p-p	peak-to-peak
PPI	plan position indicator
PPI	programmable peripheral interface
PPGA	plastic pin gate array
ppm	parts per million
PPM	peak programme meter
PPM	periodic permanent magnet
PPM	pulse position modulation
PPTC	polymer positive temperature co-efficient
PPV	pay-per-view
PQFP	plastic quad flat package
PRACL	page-replacement algorithm and control logic
PRB	pseudo-random binary (sequence)
PRESTEL	commercial teletext service
PRF	pulse repetition frequency
PROM	programmable read-only memory
PRTP	pseudo-random test pattern
PRTPG	pseudo-random test pattern generator
PSD	power spectral density
PSI	program status indicator
PSK	phase-shift key(ing)
PSN	packet switched network

PSP	programme service provider
PSPDN	packet-switched public data network
PSTN	public switched telephone network
PSW	processor status word
PTB	plated-through board
PTFE	polytetrafluoroethylene
PTH	plated-through-holes
PTT	postal, telegraph and telephone (authority)
PUT	programmable unijunction transistor
PVQ	pyramid vector quantiser
PWB	printed wiring board
PWM	pulse-width modulation
PWS	programmable workstation
QA	quality assurance
QAGC	quiet automatic gain control
QAM	quadrature amplitude modulation
QC	quality control
QFD	quality, function, deployment
QFP	quad flat pack
QISAM	queued indexed sequential access method
QMB	quick make and break (switch)
QML	qualified manufacturers list
QoS-A	quality of service (architecture)
QPL	qualified parts list
QPP	quiescent push-pull
QPSK	quadrature phase-shift keying
QUIP	quad in-line package
R	Ohm
RAC	Reliability Analysis Centre
RACE	Research and Technology Development in Advanced Communication Technologies in Europe
RADAR	radio detection and ranging
RALU	register and arithmetic/logic unit
RAM	random-access memory
RAS	row address select
RASSP	rapid programming of application-specific signal processors
RC	resistance/capacitance
RCTL	resistor-coupled transistor logic
RDB	radio data broadcast
RDF	radio direction finding
RDP	reliable data protocol

RDS	radio data services
RDSR	relative data service request
rec	receiver
RETMA	Radio, Electronic and Television Manufacturers Association
RF	radio frequency
RFC	radio frequency choke
RFI	radio-frequency interference
RFC	request for comments
RGB	red/green/blue
RIGFET	resistive insulated-gate field-effect transistor
RIM	read-in mode
RISC	reduced-instruction-set computer
RJE	remote job entry
RLC	run-length coding
RLL	run-length limited
R&M	reliability and maintenance
RMS	root mean square
ROM	read-only memory
RPROM	reprogrammable read-only memory
RRD	reversed resistance design
RMM	read-mostly mode
RS	Reed-Solomon coding
RSA	real sector access
RSGB	Radio Society of Great Britain
RST	running status table (part of SI)
RTB	return to base
RTB	reverse translation buffer
RTC	real-time clock
RTFL	real-time fault location
RTL	resistor/transistor logic
RTO	regenerative thermal oxidiser
RTS	request to send
R/W	read/write
RV	random vibration
Rx	receiver
RZ	return to zero
SA	spectrum analyser
SA	signature analysis
SAGE	semi-automatic ground environment (radar)
SAM	scanning auger microscopy
SAR	synthetic aperture radar
SAW	surface acoustic wave

SAZ	signature analyser
SBS	silicon bilateral switch
SC	semiconductor
SC	short circuit
SCA	subchannel adapter
SCAT	scanning acoustic tomography
SCM	subcarrier multiplexing
SCR	silicon controlled rectifier
SCSI	small computer system interface
SCTE	Society of Cable Telecommunications Engineers (USA)
SDH	synchronous digital hierarchy
SDLC	synchronous data-link control
SDMA	space-division multiple access
SDT	service description table (part of SI)
SECAM	système en couleurs à memoire (colour TV system)
SEM	scanning electron microscope
seq	sequential
S/H	sample and hold
SIA	Semiconductor Industry Association (USA)
Si	silicon
SI	service information (or housekeeping details added on to the video, audio or multimedia data stream)
SIC	specific inductive capacity
SID	Society for Information Display (USA)
SIL	single in-line
SIMD	single instruction, multiple data
SIMM	single in-line memory module
SIO	serial input/output
SiO_2	silicon dioxide
SIP	single in-line package
SIPO	serial in, parallel out
SISD	single instruction, single data
SISO	serial in, serial out
SLIC	subscriber line interface card
SLR	sideways looking radar
SMA	surface mount assembly
SMA	type of co-axial connector
SMART	stress marginality and accelerated reliability test
SMART	surface modification by atomic resolution terraces
SMATV	satellite master antenna television
SMB	type of co-axial connector
SMC	surface mounting component
SMC	type of co-axial connector
SMD	surface mount device

SME	Society of Manufacturing Engineers
SMP	symmetrical multiprocessor
SMPS	switched-mode power supply
SMS	subscriber management system
SMT	surface-mount technology
SMT	systems management terminal
SMTA	Surface Mount Technology Association (USA)
SMTP	simple message-transfer protocol
SMPTE	Society of Motion Picture and Television Engineers
SN	signal-to-noise
SNOBOL	string-oriented symbolic language
SNR	signal-to-noise ratio
SOHO	small office/home office
SOI	silicon on insulator
SOIC	small-outline integrated-circuit package
SOM	self-organising map
SOP	small outline package
SOP	Swiss outline package
SOS	silicon-on-sapphire
SP	stack printer
SPC	statistical process control
SPCS	satellite personal communication service
SPICE	simulated program with integrated-circuit emphasis
SPURS	spurious responses
SQL	structured query language
SQUID	superconducting quantum interference device
SRAM	stand-alone random-access memory
SRAM	static random-access memory
SRBP	synthetic resin bonded paper
SRC	Science Research Council (UK)
SSB	single sideband (broadcast)
SSBSC	single sideband, suppressed carrier
SSDD	single-sided, double-density
SSI	small-scale integration
SSIN	single-stage interconnection network
SSSD	single-sided single-density
SSM	scanning squid microscope
SSTC	single-sideband transmitted-carrier
SSTV	slow-scan television
ST	stuffing table (part of SI)
STB	set-top box
STD	subscriber trunk dialling
SUS	silicon universal switch
SWR	standing wave ratio

Sx	simplex
SYSOP	system operator
TAB	tape automated bonding
TAT	Transatlantic Telephone Transmission (cable)
TBMT	transmitter buffer empty
TC	temperature coefficient
TCP	transmission control protocol
TCP/IP	transmission control protocol/Internet protocol
TDD	time division duplex
TDM	time division multiplexing
TDMA	time division multiple-access
TDR	time domain reflectometry
TDSR	transmitter data service request
TDT	time domain transmission
TELEX	teletypewriter exchange (service)
TEM	transmission electron microscope
TEM	transversal electromagnetic
TFT	thin-film transistor
THD	total harmonic distortion
TIA	Telecommunications Industry Association (USA)
TIFF	tagged image file format
TIQ	total integrated quality
T^2L	transistor/transistor logic
TLB	translation lookaside buffer
TLM	transmission-line matrix
TMOS	T-metal oxide/semiconductor
TMR	triple modular redundancy
TODL	tunable optical delay line
TOP	technical and office protocol
TOS	top of stack
TOT	time offset table (part of SI)
TP	test point
TPA	transient program area
TQFP	thin quad flat pack
TQM	total quality management
TRAPPATT	trapped-plasma avalanche transit time (diode)
TRF	tuned radio frequency (receiver)
TRL	transistor/resistor logic
TSL	tri-state logic
TSOP	thin small-outline package
TSS	time shared system
TTL	transistor/transistor logic
TTTN	tandem tile trunk network

TTY	teletypewriter
TV	television
TVS	transient voltage suppressor
TWAIN	toolkit without an important name
TWT	travelling wave tube
Tx	transmitter
UART	universal asynchronous receiver/transmitter
UDP	user datagram protocol
UHF	ultra high frequency (300–3000 MHz)
UHP	ultra high purity
UJT	unijunction transistor
ULA	uncommitted logic array
ULSI	ultra large-scale integration
UMTS	universal mobile telecommunication system
UNCOL	universal computer-oriented language
UNESCO	United Nations Educational, Scientific and Cultural Organisation
UNI	user network interface
UPC	universal peripheral controller
URCLK	universal receiver clock
URL	universal resource locator
USART	universal synchronous/asynchronous receiver/transmitter
USRT	universal synchronous receiver/transmitter
UTC	co-ordinated universal time
UTCLK	universal transmitter clock
UTP	ultra-thin package
UUT	unit under test
VAB	voice answer back
VAN	value added network
VBR	variable bit rate
VCO	voltage-controlled oscillator
VCR	video cassette recorder
VCSEL	vertical-channel surface-emitting laser
VCT	voltage-to-current transactor
VCXO	voltage-controlled crystal oscillator
VDG	video display generator
VDI	visual display interface
VDP	vapour deposition polymerisation
VDR	voltage-dependent resistor
VDSL	very-high-speed digital subscriber line
VDU	visual display unit
VE	value engineering

VESA	Video Electronics Standards Association
VFO	variable-frequency oscillator
VGA	video graphics array
VGA	video graphics adaptor
VHDL	VHSIC hardware description language
VHF	very high frequency (30–300 MHz)
VHSIC	very-high-speed integrated circuit
VIC	video interface chip
VIDEOTEX	interactive computer-controlled medium using TV and telephone
VIEWDATA	now PRESTEL
VIL	vertical injection logic
VIP	versatile information processor
VIP	virtual information projection
VLC	variable-length coding
VLF	very low frequency (3–30 kHz)
VLSI	very-large-scale integration
VM	virtual memory
VMA	valid memory address
VMM	virtual machine monitor
VMS	virtual machine system
VMOS	vertical groove MOS
VOM	Volt/Ohm meter
VPA	valid peripheral address
VPN	virtual private network
VPT	vapour-phase technology
VR	virtual reality
VRAM	video random-access memory
VRC	vertical redundancy check
VRML	virtual-reality modelling language
VS	virtual storage
VSAT	very-small-aperture terminal
VSB	vestigial sideband transmission
VSOP	very small outline package
VTR	video tape recorder
WAIS	wide-area information servers
WAN	wide-area network
WARC	World Advisory Radio Conference
WATS	wide-area telephone service
WDM	wavelength-division multiplexing
WIMPS	windows, icons, menus and pointer
WORM	write once, read many times
WP	word processor

WPC	write precompensation
WPM	words per minute
WS	working set
WSTS	world semiconductor trade statistics
WT	wireless telegraphy
WWW	worldwide web
WYSIWYG	what you see is what you get
XML	extensible markup language
XMS	extended memory specification
XNOR	exclusive-NOR gate
XOR	exclusive-OR gate
xtal	crystal
YIG	yttrium/iron garnet
ZF	zero frequency
ZIF	zero-insertion force